Linear Programming

Linear Programming

Ronald I. Rothenberg

North Holland
New York • Oxford

Elsevier North Holland, Inc.
52 Vanderbilt Avenue, New York, New York 10017

Distributors outside the United States and Canada:

Thomond Books
(A Division of Elsevier/North-Holland Scientific Publishers, Ltd.)
P.O. Box 85
Limerick, Ireland

Library of Congress Cataloging in Publication Data

Rothenberg, Ronald I.   1936–
    Linear programming.

    Bibliography: p.
    Includes index.
    1. Linear programming. I. Title.
T57.74.R67   519.7'2   79-25896
ISBN 0-444-00325-8

Desk Editor   Philip Schafer
Copy Editor   Barry Levine
Design   Edmée Froment
Cover design   Paul Agule Design
Production Manager   Joanne Jay
Compositor   Photo Graphics, Inc.
Printer   Haddon Craftsmen

Manufactured in the United States of America

This book is dedicated to
the memory of my father, Archie H. Rothenberg

Contents

Preface

The field of *Linear Programming*, largely developed by G. Dantzig and his associates in the 1940's, is concerned with problems of optimization. The most common method of approach to solving problems in this field—the simplex algorithm—was formulated by Dantzig. Certain refinements in technique and notation, and contributions to the theory of duality were later made by A. W. Tucker.

Linear programming problems deal with determining the best possible allocation of available resources to meet certain specifications. In particular, they may deal with situations where a number of resources, such as materials, machines, people, and land, are available and are to be combined to yield several products.

In linear programming, the goal is to determine a permissible allocation of resources that will maximize or minimize some numerical quantity, such as profit or cost.

Strictly speaking, linear programming is a special case of the broader field of *Mathematical Programming*. Linear programming deals with that class of mathematical programming problems in which the relations among the variables are linear. Both the constraint relations (i.e., the *restrictions*) and the equation containing the function to be "optimized" must be linear in form.

In a typical maximum problem, a manufacturer may wish to use available resources to produce several products. The manufacturer, knowing how much profit is made for each unit of product produced, would wish to produce that particular combination of products that would maximize the total profit.

An example of a minimum problem is as follows: A company owning several mines with varying grades of ore is given an order to supply certain quantities of each grade; how can the company satisfy the requirements in such a way that the cost of operating the various mines is minimized?

Transportation problems comprise a special class of linear programming problems. In a typical problem of this type one may be interested in finding the least expensive (minimum cost) way of transporting large quantities of a product from a number of origins to a number of destinations.

Assignment problems are related to transportation problems. An example of this type of problem is determining the best way to assign n applicants to n jobs, given ratings of the applicants with respect to the different jobs.

This book will deal with the study of the types of problems described above. The emphasis will be on formulating the problem, mathematically analyzing and finally solving it, and then interpreting the solution.

The main computational technique in linear programming is the simplex algorithm. It is presented in this book by means of tableaus (i.e., tables) arranged in "compact tableau" format. The form of analysis used requires little more than a good knowledge of basic algebra. This approach is due in large part to A. W. Tucker, Professor Emeritus of Princeton University, and the compact tableau is often referred to as the "Tucker tableau." Also included in the book is a discussion (in Chapter 7) of techniques pertaining to the older "extended tableau" format.

In Chapter 8, a section dealing with using a "packaged" computer program ("canned program") for solving linear programming problems has been included. The packaged program, entitled EASYLP and developed by Professor J. D. Herniter of Boston University, involves the conversational programming concept, in which the user communicates with the computer through the medium of a time-sharing terminal. Printouts of the solutions of several model linear programming problems are presented. It should be noted that using the EASYLP package requires very little knowledge of computer programming or operation; in short, it is remarkably easy to use.

The major part of Chapter 8 deals with real-world applications of the linear programming model. Several real-world applications from economics, business, operations research, and engineering appear in this chapter. In addition, real-world models from the above fields appear frequently in the chapters on sensitivity analysis (Chapter 6) and transportation and assignment problems (Chapter 9).

It is anticipated that this book will appeal to students taking linear programming courses on many different levels, i.e., from the sophomore to graduate school level. The book should be useful to students and practitioners in the fields of Mathematics, Economics, Business Administration, Accounting, Operations Research and Systems Analysis (Quantitative Methods for Management), Probability and Statistics, Computer Science, Mathematical Methods in the Social Sciences, and Engineering.

The book contains over 150 problems which are solved in detail within the text. The subject matter is developed in progressive fashion; theorems, definitions, and methods (computational algorithms) are clearly stated at the appropriate points. Each chapter concludes with a set of supplementary problems (labeled S. P.), together with answers. These provide a review of the material covered in each chapter.

Difficult concepts are treated gingerly, and from more than one point of view when that is deemed helpful. The author's intent is to clarify, not to confuse. The book will be accessible to those with a fairly good mathematical background in conventional algebra and analytic geometry. A knowledge of matrix algebra, particularly that part concerned with notation and manipulation, will be helpful. In order to make the book attractive to people not familiar with matrix algebra, material on this subject has been included (Chapters 2 and 3). This material may be regarded as review material for those with more advanced backgrounds.

Bibliographical references are listed at the end of each chapter; these are indicated within the text by brackets, i.e., [ ].

The topics covered in this book are keyed to many of the topics covered in modern courses and textbooks dealing with Linear Programming, Finite Mathematics, and Operations Research. This book can serve either as a supplementary text in a course bearing one of the above titles or as a textbook in a course dealing with linear programming. Material from this book (most of Chapters 1, 3, 4, 5, and 9 and parts of Chapters 7 and 8) has comprised the major portion of a one-semester course in linear programming given at Queens College in New York City. It is estimated that there is enough material in this book for between 1 and 1.5 full-semester, undergraduate-level courses.

FORMAT

Each chapter is divided into titled sections, which are labeled 1, 2, 3, etc. (e.g., Chapter 4 contains Sections 1 through 6, inclusive). Theorems, Definitions, Methods (Computational Algorithms), and Figures (Illustrations) are labeled consecutively within each chapter according to the number of the chapter. Thus, in Chapter 9 we have Theorems 9.1, 9.2, 9.3, etc.; Definitions 9.1, 9.2, 9.3, etc.; Methods 9.1, 9.2, etc.; and Figures 9.1 and 9.2.

The worked-out problems within each chapter, which are both computational and theoretical, are labeled consecutively according the the number of the chapter. Thus, in Chapter 4 we have Problems 4.1, 4.2, 4.3, etc. Supplementary problems are given at the end of each chapter. Answers are provided for all such problems, while partial solutions and hints are provided for many of them. The supplementary problems are also labeled consecutively according to the number of the chapter in which they are located. In addition, they bear the prefix label S. P. Thus, in Chapter 4 they are labeled S. P. 4.1, S. P. 4.2, S. P. 4.3, etc.

In many places throughout the book I make reference to so-called standard linear programming problems. I often use the same labeling for such problems, regardless of which chapter the problem is located in. Thus, the objective function is labeled (1), the m main constraints are labeled (2.1), (2.2), . . . , (2.m), and the n nonnegativity constraints are labeled (3.1), (3.2), . . . , (3.n). Examples of this labeling scheme can be found in Problems 1.5, 1.7, and 1.11, after Theorem 3.2, and at the beginning of Section 5 in Chapter 5.

Variations of this labeling scheme also occur throughout the book. In Chapter 4, the standard linear programming problem in so-called "standard slack" form ("equality form") has the main constraints—which are now *equations*—labeled (2.1), (2.2), . . . , (2.m) as before, while the nonnegativity constraints—which now number $(n + m)$—are labeled (3.1), (3.2), . . . , (3.n), (3.$n + 1$), . . . , (3.$n + m$). This labeling scheme, with its variations, should not offer any difficulty to the reader, since it has not done so in my experience as a teacher of linear programming over the past few years. It is employed mainly to provide a simple referencing system for standard linear programming problems.

For the worked-out problems, I occasionally labeled items (e.g., Equations, Inequalities, Constraints) during the course of the solutions. My labeling proceeded consecutively, usually starting with (1) or a similar starting label, such as (i) or (a). For example, in Problem 1.1 of Chapter 1, the labeling is (1), (2), etc., while in Problem 2.19, the labeling is (1), (2), and (3). Likewise, I employ similar labeling schemes in Proofs of Theorems (e.g., in the Proof of Theorem 2.10, I label items (1), (2), (3), etc.) and also for the material within the body of the text (e.g., see the material immediately preceding the statement of Problem 2.19). I found the individual labeling within Problems, Proofs of Theorems, and elsewhere to be a great help in cross-referencing, and in highlighting important items.

Ronald I. Rothenberg

Acknowledgments

I gratefully acknowledge the comments and suggestions of the following reviewers:

Dr. Gary D. Peterson, Pacific Lutheran University;
Prof. Loren D. Meeker, University of New Hampshire;
Dr. Abraham M. Glicksman.

I am also grateful to Dr. Jae K. Shim of the Queens College Accounting and Information Systems Department for his general comments and help with respect to computer usage, and to Dr. Jerome D. Herniter of the Boston University School of Management for his permission to include print-outs of his packaged computer program, EASYLP. The constructive help of Nick Detsis and Ken Pospisil of the Queens College Computer Center, and of my former student, Robert Sterlacci, are greatly appreciated.

The author is grateful to A. Sreckovitch, J. Snyder, P. Rubin, and M. Green for their typing skills. Thanks are extended to the staff of Elsevier North Holland, Inc., in particular to Editors Kenneth J. Bowman and Phil Schafer. The author is also grateful to the many students at Queens College, who took his course in Linear Programming and Game Theory over the years. Their participation in the course provided the main inspiration for the writing of this book. Finally, I would like to indicate my appreciation for the patience and understanding of my wife, Olga.

Linear Programming

Chapter One

Scanning Extreme Points

1. INTRODUCTORY REMARKS. THE SETUP FOR A TYPICAL PROBLEM

In this chapter, we shall explore some of the key aspects of linear programming situations. Two main theorems will be stated and then used in solving problems. Proofs and a more elaborate discussion of the major theory will be postponed until a later chapter.

Linear programming problems occur throughout the fields of economics, business, and engineering. Such problems deal with the best possible allocation of available resources to meet certain specifications. In particular, they relate to situations where a number of resources, such as materials, machines, people, and land, are available and are to be combined to yield several products. The goal of linear programming is to determine the permissible allocation of resources that will maximize or minimize some numerical quantity, such as profit or cost.

Linear programming is really just a part of the broader field of mathematical programming. Linear programming deals with that class of mathematical programming problems in which the relations among the variables are linear. Both the constraint relations (i.e., the *restrictions*) and the equation containing the function to be "optimized" must be linear in form. The term *optimize* can mean either to maximize or to minimize.

We shall now consider a typical maximum problem where the situation and the numbers involved are rather simple. This problem, which deals with the manufacture of furniture, is typical of maximum problems in that it deals with an optimal allocation of resources. Variations of this problem appear on page 19 of Reference [1] and page 17 of [2].

Problem 1.1: A Maximum Problem

A furniture firm manufactures chairs and tables. Both items must be processed by two machines, A and B. Production of one chair requires 2 hours from machine A and 1 hour from machine B. Each table requires

1 hour from machine A and 2 hours from machine B. The profit realized by selling one chair is \$3, while for a table the figure is \$4. Machines A and B are each available for at most 12 hours per day. How many chairs and tables should be made per day so as to maximize profits? Develop a mathematical formulation.

Solution (partial)

We are not yet in a position to solve this problem completely, but we can formulate it mathematically without too much trouble.

Let x equal the number of chairs and y the number of tables that are to be produced per day. The daily profit u, in dollars, is thus given by

$$u = 3x + 4y \text{ (to be maximized).} \tag{1}$$

Here $3x$ is the profit realized by selling x chairs, and $4y$ is the corresponding profit from selling y tables.

Now let us consider the distribution of resources, namely, the time available per day from the machines. The production of x chairs requires $2x$ hours from machine A and $1 \cdot x$ or x hours from B. Production of y chairs requires $1 \cdot y$ or y hours from A and $2y$ hours from B. Thus, our two *main constraints* are

$$2x + y \leq 12 \quad \text{(machine } A) \tag{2}$$

and

$$x + 2y \leq 12 \quad \text{(machine } B). \tag{3}$$

In addition, we have two nonnegativity constraints:

$$x \geq 0 \tag{4}$$

and

$$y \geq 0. \tag{5}$$

Thus, the mathematical formulation indicates that we should try to maximize u, expressed by Equation (1), subject to the constraints (2), (3), (4), and (5).

The symbol $\leq$, which will occur frequently in this book, is perhaps new to some readers. The statement $a \leq b$ means a is "less than or equal to" b (b is "greater than or equal to" a). An equivalent way of writing this is $b \geq a$. The statement $c < d$ ($d > c$) means c is "less than" d (d is "greater than" c).

In Inequality (2), $2x$ is the time required from machine A to produce x chairs: 2 hours per one chair times x chairs equals $2x$ hours. The quantity y in Inequality (2) is the time required from machine A to produce y tables: 1 hour per one table times y tables equals $1 \cdot y$ or y hours. Thus, the total time required from machine A to produce x chairs and y tables is $2x + y$, and this has to be less than or equal to the total

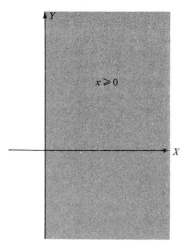

FIGURE 1.1 The set S_4.

time available per day, namely, 12 hours. This is the meaning of the
mathematical statement $2x + y \leq 12$.

2. GRAPHICAL CONSIDERATIONS

Before we pursue the complete solution of the above problem, let us
digress for a moment. First, we shall try to represent in the plane all
those points (x,y) that simultaneously satisfy the four constraint inequalities. We can do this by stages. The points that satisfy (4) and (5),
respectively, are indicated by the shading in Figures 1.1 and 1.2. Thus,
$x \geq 0$ refers to all those points that lie to the right of and on the Y axis,
while $y \geq 0$ refers to the set of points which lie above and on the X axis.

FIGURE 1.2 The set S_5.

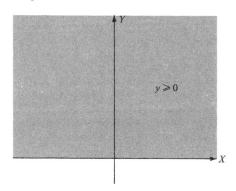

We denote these sets as S_4 and S_5. Now let us consider Inequalities (2) and (3). In order to graphically represent (2), first we plot the equation $2x + y = 12$, which has x intercept 6 and y intercept 12. This is, of course, the equation of a straight line. Next, we determine the set of points which satisfy the strict inequality $2x + y < 12$. To do this, we choose an arbitrary point that does not satisfy the equation part, say $(0,0)$, and test it to see if it satisfies the strict inequality. Here we obtain

$$2 \cdot 0 + 0 = 0 < 12,$$

and thus the strict inequality is satisfied. It turns out that all points satisfying $2x + y < 12$ will lie on the same side of $2x + y = 12$ as the point $(0,0)$. Thus, we obtain Figure 1.3. We proceed in a similar fashion (again testing, say, the point $(0,0)$) and develop Figure 1.4.

Note: The point $(0,0)$ is useful as a test point only if it does not lie on the boundary line. If $(0,0)$ lies on the boundary line, i.e., if $(0,0)$ satisfies the equation part, then we have to use another test point.

We call the sets that satisfy Constraints (2) and (3), S_2 and S_3, respectively. These are indicated in Figures 1.3 and 1.4 by the shading.

The set of points which simultaneously satisfy Inequalities (2), (3), (4), and (5) is called the *constraint set*, S_c. In set notation we have

$$S_c = S_2 \cap S_3 \cap S_4 \cap S_5,$$

where the symbol $\cap$ indicates intersection of sets. The set S_c for Problem 1.1 is given in Figure 1.5 by the shaded quadrilateral $OABC$.

Points contained within the constraint set are called *feasible points*.

FIGURE 1.3 The set S_2.

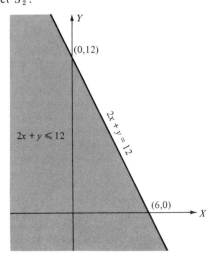

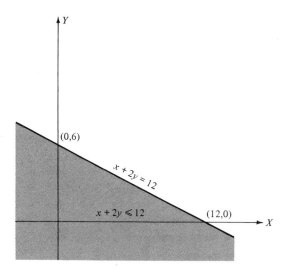

FIGURE 1.4 The set S_3.

The terms constraint set and feasible point are defined in more detail later in this chapter (Definitions 1.2 and 1.3).

We can graphically generate the constraint set easily by referring to the arrows in Figure 1.5. The arrows clearly indicate that S_c is the set of those points that simultaneously lie within the sets S_2, S_3, S_4, and S_5.

The points O, A, B, and C are called extreme points (corner points or vertices). The coordinates of point B, namely, $(4,4)$, are obtained by solving $2x + y = 12$ and $x + 2y = 12$ simultaneously.

FIGURE 1.5 The constraint set S_c for Problem 1.1.

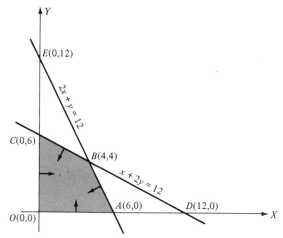

We can now complete the solution of Problem 1.1 if we make use of the very important Theorem 1.1, which is given below. After the theorem, some of the terms that appear in its statement are discussed by means of notes (more precise definitions of these terms will be given at the appropriate time later in the development).

Theorem 1.1. *Let u be a linear function defined over a bounded, polyhedral, convex set* S_c*. Then u takes on its maximum (and minimum) value at an extreme point of* S_c*.*

Notes:
a. A general definition of the term ''extreme point'' is given in Definition 3.2. A discussion of extreme points pertaining to linear programming problems is given in the pages that follow the definition.
b. For the case of two independent variables (here, x and y), the general equation for a linear function is

$$u = c_1 x + c_2 y,$$

where c_1 and c_2 are constants.
c. A bounded, two-dimensional set is a set that can be totally contained within a circle of finite radius whose center is at the origin.
d. A convex set S is a set such that for each pair of points P_1 and P_2 belonging to S, the entire line segment from P_1 to P_2 is contained in S.
e. The word *polyhedral* refers to the fact that the boundaries of the set S_c are straight lines in the two-dimensional case. In the three-dimensional case the boundaries would be planes.

Problem 1.2
Complete the solution of Problem 1.1.

Solution

First, we note that the constraint set S_c is a bounded, polyhedral, convex set. Thus, Theorem 1.1 is applicable. All we have to do is evaluate our linear function u (also called the objective function) at the extreme points of the constraint set:

Extreme Point	Value of $u = 3x + 4y$
$O(0,0)$	0
$A(6,0)$	18
$B(4,4)$	28
$C(0,6)$	24

Thus, the maximum value of u is 28, achieved at the extreme point

$B(4,4)$. In words, the maximum profit is $28 per day, and this occurs when we produce four chairs and four tables per day.

Problems 1.1 and 1.2 dealt with a typical maximizing situation. In Problem 1.3 we shall consider a typical minimum problem, but before we tackle that we need another major theorem to guide us. Theorem 1.2, which follows, is different from Theorem 1.1 in that it deals with the case where the constraint set is unbounded.

Theorem 1.2. *Given that the polyhedral, convex set S_c is unbounded. Suppose that the linear function u is bounded from below (above) with respect to the set S_c. Then the minimum (maximum) value of u will occur at an extreme point of S_c.*

Notes:
a. An unbounded set is one that is not bounded (refer to Note (c) after Theorem 1.1). Put simply, an unbounded set extends indefinitely in certain directions. In the problem that follows, the set extends indefinitely along the X axis, along the Y axis, and along innumerable other lines (e.g., the 45 degree line $y = x$).
b. When we say that the function u is bounded from *below* with respect to a set, we mean that u is greater than or equal to some real number K (in symbols, $u \geq K$) for *all* points in the set. In the example that follows, $u \geq 0$ for all points in the set; thus, it is bounded from *below* by zero and, in fact, by any number less than zero.

Problem 1.3: A Minimum Problem with Unbounded Constraint Set

A company owns mines A and B. Mine A is capable of producing 1 ton of high-grade ore, 4 tons of medium-grade ore, and 6 tons of low-grade ore per day. Mine B can produce 2 tons of each of the three grades of ore per day. The company requires at least 60 tons of high-grade ore, 120 tons of medium-grade ore, and 150 tons of low-grade ore. If it costs $200 per day to work mine A and $300 per day to work mine B, how many days should each mine be operated if the company wishes to minimize costs?

Solution

Let x denote the number of days mine A should be operated and y the corresponding number of days for mine B. Then the cost function u is given by

$$u = 200x + 300y \quad \text{(to be minimized)}. \tag{1}$$

The "at least" requirements of the different ore types are stated mathematically as the following three main constraints:

$$1 \cdot x + 2y \geq 60 \quad \text{(high)}, \tag{2}$$

$$4x + 2y \geq 120 \quad \text{(medium)}, \tag{3}$$

and

$$6x + 2y \geq 150 \quad \text{(low)}. \tag{4}$$

In addition, we have the two nonnegativity constraints

$$x \geq 0 \tag{5}$$

and

$$y \geq 0 \tag{6}$$

We can now plot the constraint set in the xy plane as in Problem 1.1. For example, the set S_2, which consists of the points (x, y) that satisfy (2), is given in Figure 1.6 (shaded region). First, we plot $x + 2y = 60$, which has x intercept 60 and y intercept 30. We observe that $(0,0)$ does not satisfy the strict inequality part $(>)$ of (2) (the symbol $\not>$ means "not greater than"):

$$0 + 0 = 0 \not> 60.$$

Thus, S_2 consists of the points on the opposite side of $x + 2y = 60$ from the point $(0,0)$. The points on the line $x + 2y = 60$ are, of course, also in S_2.

In the same way, we can plot Inequalities (3), (4), (5), and (6). Thus, we arrive at Figure 1.7—a graphical representation of the constraint set S_c (shaded region). Note that S_c is an unbounded set.

It is important to locate the extreme points. As an example, extreme point $E(15,30)$ is obtained by solving the equation parts of Inequalities

FIGURE 1.6 The set S_2 corresponding to Inequality (2).

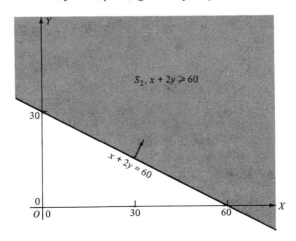

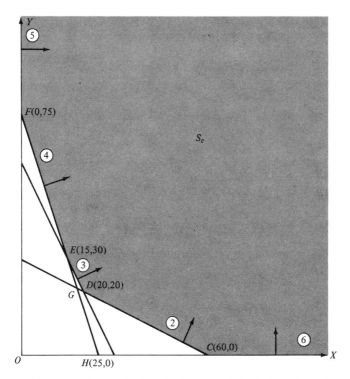

FIGURE 1.7 The set (shown shaded) for Problem 1.3. Encircled numbers denote the equation parts of the constraints, i.e., ② represents $x + 2y = 60$, ③ represents $4x + 2y = 120$, ④ represents $6x + 2y = 150$, ⑤ represents $x = 0$, and ⑥ represents $y = 0$.

(3) and (4) simultaneously:

$$4x + 2y = 120$$
$$6x + 2y = 150.$$

Subtracting the first equation from the second yields $2x = 30$; thus, $x = 15$ and then $y = 30$ from either of the above equations.

Theorem 1.2 applies to this problem, since the constraint set S_c is unbounded, but the linear cost function u has a lower bound on S_c. (For a function u to have the number K as a lower bound with respect to a set S, it is necessary that $u \geq K$ for all points in the set S.) To see this, observe that (2), (3), (4), (5), and (6) must each be satisfied for all points (x, y) in the constraint set. In particular, (5) and (6) must be satisfied. Thus, referring to Equation (1), we have $200x \geq 0$ and $300y \geq 0$; hence,

$$u = 200x + 300y \geq 0,$$

i.e., $u \geq 0$, which means that u is bounded from below by the real number 0 for all points (x, y) in the constraint set.

The following table gives the extreme points in S_c and the corresponding values of u:

Extreme Point	Value of $u = 200x + 300y$
$C(60,0)$	12,000
$D(20,20)$	$4,000 + 6,000 = 10,000$
$E(15,30)$	$3,000 + 9,000 = 12,000$
$F(0,75)$	22,500

Thus, the minimum value of the cost function u is \$10,000, and this occurs when both mines are operated for 20 days. We note that this causes (2) and (3) to be satisfied as equations, but (4) is satisfied as a strict inequality:

$$6x + 2y = 120 + 40 = 160 > 150.$$

Problems 1.1 and 1.3 had graphical representations in the plane because only two main variables were involved (namely, x and y). In the case of three main variables, we can represent S_c on a three-dimensional diagram. The next problem, which we shall refer to as the "nut problem," involves such a representation. It is a maximum problem for which the constraint set is bounded.

Problem 1.4: The "Nut Problem"

A confectionery has 900 pounds of peanuts and 600 pounds of walnuts in stock. It sells three different mixtures of the nuts. The first mixture, which consists of all peanuts, sells for 25¢ per pound. The second mixture is two-thirds peanuts and one-third walnuts by weight, and it sells for 40¢ per pound. The third mixture is one-quarter peanuts and three-quarters walnuts by weight, and it sells for 50¢ per pound. How many pounds of each mixture should the confectionery produce to maximize sales revenue?

Solution

Let x denote the amount of the first mixture in pounds, and likewise let y and z denote the amounts of the second and third mixtures, respectively. The sales revenue function, in dollars, is given by

$$u = \tfrac{1}{4}x + \tfrac{2}{5}y + \tfrac{1}{2}z \quad \text{(to be maximized).} \tag{1}$$

The selling prices were converted to reduced ratios (in dollars) because it is easier to work with ratios than with decimals if one is doing hand calculations. Of course, we could just as well have expressed the sales revenue function in cents: $u = 25x + 40y + 50z$. Here we have a case of a linear function in three variables. The general form for the three-variable case is $u = c_1 x + c_2 y + c_3 z$, where the c_i's are constants. This is just a logical extension of what we had before for two variables.

The main constraints are arrived at by considering the total resources available:

$$1 \cdot x + \tfrac{2}{3} y + \tfrac{1}{4} z \leq 900 \quad \text{(peanuts)} \tag{2}$$

and

$$0 \cdot x + \tfrac{1}{3} y + \tfrac{3}{4} z \leq 600 \quad \text{(walnuts).} \tag{3}$$

Finally, we have the three nonnegativity constraints

$$x \geq 0, \tag{4}$$

$$y \geq 0, \tag{5}$$

and

$$z \geq 0. \tag{6}$$

We shall now determine the extreme points of the constraint set in a new and systematic way. We take three constraints at a time and solve them as equations. Geometrically, this is equivalent to finding the intersection point of three planes.

For example, with Constraints (2), (3), and (4) we obtain the following system of three equations:

$$x + \tfrac{2}{3} y + \tfrac{1}{4} z = 900$$

$$\tfrac{1}{3} y + \tfrac{3}{4} z = 600$$

$$x = 0.$$

This system has the solution $(0,1260,240)$, as can be seen by solving the three equations simultaneously. We then check to see if the remaining constraints are satisfied as inequalities. They are, as we see by substitution:

$$y = 1260 > 0 \quad \text{and} \quad z = 240 > 0.$$

Thus, the point $(0,1260,240)$ is indeed an extreme point, and we thus evaluate the function u at this point:

$$u = \tfrac{1}{4} \cdot 0 + \tfrac{2}{5} \cdot 1260 + \tfrac{1}{2} \cdot 240 = 624.$$

We continue in this way with each *combination* of three equations from the five constraints and develop the following table:

Constraints	Point	Check	u
(2), (3), (4)	$A(0,1260,240)$	Yes	624 (max)
(2), (3), (5)	$B(700,0,800)$	Yes	575
(2), (3), (6)	$(-300,1800,0)$	No	
(2), (5), (6)	$E(900,0,0)$	Yes	225
(3), (4), (5)	$C(0,0,800)$	Yes	400
(3), (4), (6)	$F(0,1800,0)$	No	
(2), (4), (5)	$(0,0,3600)$	No	
(2), (4), (6)	$D(0,1350,0)$	Yes	540
(3), (5), (6)	No solution	No	
(4), (5), (6)	$O(0,0,0)$	Yes	0

Thus, we observe that the maximum value of u is \$624, and this is achieved at point A, where $x = 0$, $y = 1260$, and $z = 240$. In other words, the maximum sales revenue occurs when the confectionery decides to produce none of the first mixture (just peanuts), 1260 pounds of the second mixture, and 240 pounds of the third mixture. The diagram of the bounded, three-dimensional constraint set for this problem is given in Figure 1.8.

FIGURE 1.8 The three-dimensional constraint set for Problem 1.4. Coordinates of labeled points: $O(0,0,0)$, $A(0,1260,240)$, $B(700,0,800)$, $C(0,0,800)$, $D(0,1350,0)$, $E(900,0,0)$, and $F(0,1800,0)$.

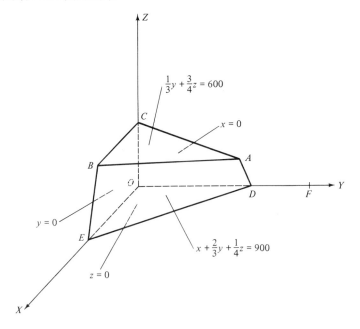

It is instructive to go through some of the calculations needed to develop the above table. Consider the case where Constraints (2), (3), and (6) are solved as equations:

$$x + \tfrac{2}{3}y + \tfrac{1}{4}z = 900$$

$$\tfrac{1}{3}y + \tfrac{3}{4}z = 600$$

$$z = 0.$$

Substituting $z = 0$ into the second equation yields $y = 1800$, and then substitution of these values into the first equation yields $x = -300$. However, this means that Constraint (4) is not satisfied as an inequality; therefore, the point $(-300, 1800, 0)$ is not in the constraint set.

If we solve Constraints (3), (4), and (6) as equations, we obtain

$$\tfrac{1}{3}y + \tfrac{3}{4}z = 600$$

$$x = 0$$

$$z = 0$$

This yields the point $(0, 1800, 0)$, which is plotted as point F in Figure 1.8. Here we note that Constraint (5) is satisfied, since $y = 1800$ is greater than zero. However, Constraint (2) is not satisfied, as the following calculation shows:

$$1 \cdot x + \tfrac{2}{3}y + \tfrac{1}{4}z = 0 + 1200 + 0 = 1200 \nless 900,$$

i.e., 1200 is greater than 900, not less than 900 as required. Thus, point F is not in the constraint set.

As another illustration, consider solving Constraints (3), (5), and (6) as equations. Here we have

$$\tfrac{1}{3}y + \tfrac{3}{4}z = 600$$

$$y = 0$$

$$z = 0.$$

This leads to the inconsistency $0 = 600$, which means that there is no solution for (x, y, z) to the above system of equations.

Notes: a. The previous problem could have been solved merely by listing all the possible extreme point candidates and then testing them; i.e., the sketch of S_c was not needed, except that it did indicate that S_c was a bounded set. This last fact meant that Theorem 1.1 was applicable.

It is interesting to observe that the number of extreme point candidates is no greater than the number of combinations of five things (total number of constraints) taken three at a time. Here three is the number of variables and also the dimensional order of the space. Since the equation for

$C(k,r)$—the number of combinations of k things taken r at a time—is

$$C(k,r) = \frac{k!}{r!(k-r)!} \,,$$

for our case we have

$$C(5,3) = \frac{5!}{3!2!} = 10 \text{ combinations.}$$

The symbol ! means *factorial*. For whole numbers greater than zero, we have $1! = 1$, $2! = 2\cdot1 = 2$, $3! = 3\cdot2\cdot1 = 6$, $4! = 4\cdot3\cdot2\cdot1 = 24$, etc. Also, $0!$ is *defined* to be 1. In Problem 1.4 we listed all of the ten combinations in the first column of the table ((2), (3), (4); (2), (3), (5); (2), (3), (6); etc.) Of these, nine provided extreme point candidates; however, only six of the nine candidates turned out to be actual extreme points. We thus see that $C(5,3)$ is an *upper bound* for the number of extreme point candidates.

b. Solving a system of three equations in three unknowns is geometrically equivalent to finding the intersection point of three planes in three-dimensional space. In two dimensions, the corresponding process would involve solving two equations in two unknowns, which is equivalent to finding the intersection point of two straight lines. This was illustrated in Problems 1.2 and 1.3.

We perform the check part of the calculation (seeing if the remaining inequalities are satisfied) in order to determine if the intersection point is in fact in the constraint set. In Problem 1.4, for example, we see that $A(0,1260,240)$ is in S_c, but that $F(0,1800,0)$ is not.

The method of working out Problem 1.4 indicates a general method for solving linear programming problems. However, this process of listing extreme point candidates and then testing them has several drawbacks. One is that real-world problems often contain a large number of variables and constraints, sometimes in the hundreds. Such problems would involve testing huge numbers of extreme point candidates. For example, if we had 15 constraints and 10 variables, the upper bound for the number of extreme point candidates would be 3003, as the following calculation indicates:

$$C(15,10) = \frac{15!}{10!\cdot5!} = \frac{15\cdot14\cdot13\cdot12\cdot11}{5\cdot4\cdot3\cdot2\cdot1} = 3003.$$

(Note that $15! = 15\cdot14\cdot13\cdot12\cdot11\cdot10!$; we cancel out the 10! with the 10! in the denominator.)

Another drawback is that the process of listing extreme point candidates is usually of little value in determining whether or not the constraint set is unbounded. Moreover, even if we knew that S_c were unbounded, we would still have to find out whether or not the function u was bounded with respect to S_c (see Theorem 1.2).

Fortunately, there is a method—the *simplex algorithm*—capable of systematically finding a solution to a linear programming problem without first analyzing the value of u at all extreme points. In this method, we are guided to move from one extreme point to another in such a way that the value of the function u is improved at each step, until we arrive at the solution of the problem. The number of extreme points involved in this search will usually be a small fraction of the total number. In addition, the simplex algorithm will reveal if the constraint set of a linear programming problem is empty, or if the function u is unbounded. We shall postpone further consideration of the simplex algorithm until Chapter 4.

3. GENERAL NOTATION

In this section, we shall develop some general notation and definitions for two of the standard forms of linear programming problems.

Problem 1.5

Develop the *standard* form for a linear programming problem of maximum type.

Solution

First, we have the equation for maximizing the linear function u, which is expressed in terms of n variables:

$$\text{Maximize } u = c_1 x_1 + c_2 x_2 + \cdots + c_n x_n. \tag{1}$$

Next, we have the m main constraints to which u is subjected:

$$a_{11} x_1 + a_{12} x_2 + \cdots + a_{1n} x_n \le b_1 \tag{2.1}$$

$$a_{21} x_1 + a_{22} x_2 + \cdots + a_{2n} x_n \le b_2 \tag{2.2}$$

$$\cdots\cdots\cdots\cdots\cdots\cdots\cdots\cdots\cdots\cdots$$

$$\cdots\cdots\cdots\cdots\cdots\cdots\cdots\cdots\cdots\cdots$$

$$a_{m1} x_1 + a_{m2} x_2 + \cdots + a_{mn} x_n \le b_m. \tag{2.m}$$

Finally, we have the n nonnegativity constraints placed on x_1 through x_n:

$$x_1 \ge 0 \tag{3.1}$$

$$x_2 \ge 0 \tag{3.2}$$

$$\cdots\cdots\cdots$$

$$\cdots\cdots\cdots$$

$$x_n \ge 0. \tag{3.n}$$

In Section 5 of Chapter 2, we will express the standard maximum problem in matrix-vector form.

Notes:

a. Here n denotes the number of variables. In Problems 1.1 and 1.3, $n = 2$, while in Problem 1.4, $n = 3$. Furthermore, m denotes the number of main constraints. There were $m = 2$ main constraints in both Problems 1.1 and 1.4, but $m = 3$ in Problem 1.3.

b. The a_{ij} terms a_{11}, a_{12}, a_{21}, a_{22}, etc., denote the *constant* coefficients that multiply the variables x_1, x_2, etc. In a typical a_{ij} coefficient, the first subscript (i) refers to the constraint label, while the second subscript (j) indicates that variable x_j is multiplied. For example, a_{21} appears in the second main constraint and it multiplies x_1. It should be mentioned that the notation in (1), (2), and (3) above is consistent with the standard matrix notation used for linear programming problems. Such notation is discussed further in Chapter 2.

Problem 1.6

Express the terms of Problem 1.4 in the general notation of Problem 1.5.

Solution

The variables x, y, and z are replaced by x_1, x_2, and x_3. Thus, we see that $c_1 = 1/4$, $c_2 = 2/5$, and $c_3 = 1/2$.

Examining the $m = 2$ main constraints, we see the following correspondence:

$$a_{11} = 1, \ a_{12} = \tfrac{2}{3}, \ a_{13} = \tfrac{1}{4}, \ b_1 = 900$$
$$a_{21} = 0, \ a_{22} = \tfrac{1}{3}, \ a_{23} = \tfrac{3}{4}, \ b_2 = 600.$$

Note that the total number of original variables $n = 3$.

Problem 1.7

Develop the *standard* form for a linear programming problem of minimum type.

Solution

First, we have the equation for minimizing the linear function u, which is expressed in terms of n variables:

$$\text{Minimize } u = c_1 x_1 + c_2 x_2 + \cdots + c_n x_n. \tag{1}$$

(Sometimes, we shall use w as the symbol for the linear function to be minimized; in particular, we shall do this in Chapter 5.)

The m main constraints are as follows:

$$a_{11}x_1 + a_{12}x_2 + \cdots + a_{1n}x_n \geq b_1 \qquad (2.1)$$

$$a_{21}x_1 + a_{22}x_2 + \cdots + a_{2n}x_n \geq b_2 \qquad (2.2)$$

$$\cdots\cdots\cdots\cdots\cdots\cdots\cdots\cdots\cdots\cdots\cdots\cdots\cdots\cdots$$

$$\cdots\cdots\cdots\cdots\cdots\cdots\cdots\cdots\cdots\cdots\cdots\cdots\cdots\cdots$$

$$a_{m1}x_1 + a_{m2}x_2 + \cdots + a_{mn}x_n \geq b_m. \qquad (2.m)$$

Note that the inequalities are of the "$\geq$" type (left-hand side is greater than or equal to the right-hand side), whereas in Problem 1.5 they were of the "$\leq$" type.

Finally, we have the n nonnegativity constraints placed on x_1 through x_n:

$$x_1 \geq 0 \qquad (3.1)$$

$$x_2 \geq 0 \qquad (3.2)$$

$$\cdots\cdots\cdots$$

$$\cdots\cdots\cdots$$

$$x_n \geq 0. \qquad (3.n)$$

Problem 1.8

Express the terms of Problem 1.3 in the general notation of Problem 1.7.

Solution

The variables x and y are replaced by x_1 and x_2. Thus, we observe that $c_1 = 200$ and $c_2 = 300$.

Examining the $m = 3$ constraints, we see the following correspondence:

$$a_{11} = 1,\ a_{12} = 2,\ b_1 = 60$$

$$a_{21} = 4,\ a_{22} = 2,\ b_2 = 120$$

$$a_{31} = 6,\ a_{32} = 2,\ b_3 = 150.$$

We also note that the total number of original variables $n = 2$.

Problem 1.9

Determine an upper bound for the number of extreme points for a linear programming problem in standard form.

Solution

We shall employ the symbolism of Problems 1.5 and 1.7. Thus, m is the number of main constraints, and n is the number of nonnegativity constraints. Hence, the total number of constraints is $(m + n)$.

We shall generalize the approach used in Problem 1.4; the reader is advised to refer to the table given in that problem.

An upper bound for the number of extreme points, as well as for the number of extreme point candidates, is determined by finding the number of ways of choosing n of the constraint inequalities and attempting to solve them as equations.

In Problem 1.4, where $n = 3$, we chose three constraint inequalities from the total of five ($m + n = 2 + 3 = 5$) and then attempted to solve them simultaneously as equations.

In general, the total number of choices of n equations drawn from a total of ($m + n$) constraint inequalities is identical to the number of combinations of n things when drawing from a total of ($m + n$) things. This number is the upper bound for the number of extreme point candidates, and thus also for the number of extreme points. The symbol for the number of combinations of ($m + n$) things taken n at a time is $C(m + n, n)$. (Refer to Note (a) after Problem 1.4 for a comment on combinations.)

Hence, we have

$$\text{Upper bound} = C(m + n, n) = \frac{(m + n)!}{m!n!}.$$

In Problem 1.4 the total number of inequalities was $m + n = 2 + 3 = 5$. Thus, the number of combinations was

$$C(m + n, n) = C(5,3) = \frac{5!}{3!2!} = \frac{5 \cdot 4 \cdot 3!}{3! \cdot 2 \cdot 1} = \frac{5 \cdot 4}{2 \cdot 1} = 10.$$

The first column of the table given in Problem 1.4 lists the ten combinations of the five constraints taken three at a time. In that problem we saw that the upper bound, 10, was clearly greater than the number of extreme point candidates (9) and the number of extreme points (6).

Problem 1.10

Discuss and solve the following maximum problem, and identify the values of m and n. Indicate the upper bound for the number of extreme points as determined from the equation of Problem 1.9.

$$\text{Maximize } u = x + y \tag{1}$$

$$\text{subject to } 2x + 3y \le 24 \tag{2}$$

$$5x + 3y \le 30 \tag{3}$$

$$x \ge 0, \tag{4}$$

and

$$y \ge 0. \tag{5}$$

Solution

Since we have two main variables, x and y, we choose two constraints at a time and solve them as equations. Observe that $n = 2$ and $m = 2$. The upper bound for the number of extreme points is thus

$$C(4,2) = \frac{4!}{2!2!} = \frac{4 \cdot 3}{2 \cdot 1} = 6.$$

We form a table as in Problem 1.4. Note that the first column lists the combinations of the four constraints (2), (3), (4), and (5) taken two at a time. Some sample calculations pertaining to the table will be carried out later.

Constraints	Point	Check	u
(2), (3)	$B(2,20/3)$	Yes	$8\frac{2}{3}$ (max)
(2), (4)	$A(0,8)$	Yes	8
(2), (5)	$D(12,0)$	No	
(3), (4)	$E(0,10)$	No	
(3), (5)	$C(6,0)$	Yes	6
(4), (5)	$O(0,0)$	Yes	0

A diagram of the bounded, two-dimensional constraint set for this problem is given in Figure 1.9 (shaded region).

The solution to the problem is as follows: u is maximized at point B, where $x = 2$ and $y = 20/3$; the maximum value of u is $8\frac{2}{3}$. Now let us work through two sample calculations. To determine the point $B(2,20/3)$, we take the equations corresponding to Constraints (2) and (3) and solve

FIGURE 1.9 The constraint set for Problem 1.10. Encircled numbers indicate the equality parts of the constraints. Thus,②indicates $2x + 3y = 24$,③indicates $5x + 3y = 30$,④indicates $x = 0$,⑤indicates $y = 0$.

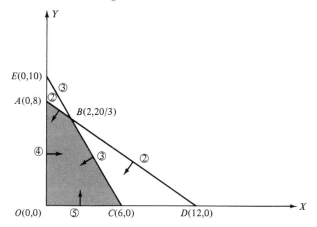

them simultaneously (geometrically, each equation corresponds to a line, and solving the two equations simultaneously is equivalent to finding the intersection point of two lines):

$$2x + 3y = 24$$
$$5x + 3y = 30.$$

We obtain $x = 2$ and $y = 20/3$, which satisfy Constraints (4) and (5), since x and y are both positive. Thus, B is in the constraint set and is an extreme point. In the figure, the equation parts of the constraints are denoted by encircled numbers. Thus, ② represents the line $2x + 3y = 24$, while ③ represents the line $5x + 3y = 30$.

Let us now consider what happens when we attempt to solve Constraints (3) and (4) as equations:

$$5x + 3y = 30 \quad \text{and} \quad x = 0.$$

This yields $x = 0$ and $y = 10$, i.e., the point $(0,10)$. These values satisfy Constraint (5) as an inequality, since 10 is a positive number, but Constraint (2) is not satisfied, as the following calculation shows:

$$2x + 3y = 0 + 30 = 30 > 24.$$

Thus, the point $(0,10)$ (which is plotted as point E in Figure 1.9) is not in the constraint set. In other words, the two lines ③ and ④ have an intersection point that is outside the constraint set. Thus, the point E is an extreme point candidate that turns out not to be an extreme point.

In this problem the upper bound, 6, is exactly equal to the number of extreme point candidates. The number of extreme points is 4.

It is appropriate at this time to give some general definitions which pertain to standard linear programming problems. The reader is advised to refer to Problems 1.5 (maximum case) and 1.7 (minimum case).

Definition 1.1. The linear function $u = c_1 x_1 + c_2 x_2 + \cdots + c_n x_n$ is called the *objective function*.

Definition 1.2. The set of points $(x_1, x_2, \ldots, x_n)$ which satisfy Constraints (2.1) through $(2.m)$ and (3.1) through $(3.n)$ of Problem 1.5 (or Problem 1.7) is called the *constraint set, S_c*.

In general, the set of *all* such points with n components is known as an n-dimensional space. (A brief treatment of this concept appears in Chapter 2.) The constraint set S_c is a subset of such an n-dimensional space.

Definition 1.3. Those points that lie in the constraint set are called *feasible points*.

Notes:
   a. Clearly, it is difficult to visualize the constraint set if n is greater than 3. We may think of a feasible point $(x_1, x_2, \ldots, x_n)$ as a collection of x_i values which cause all the constraints to be satisfied.
   b. In Problem 1.10, the constraint set was indicated by the shaded region of Figure 1.9. The points on the boundary-line segments AB, BC, CO, and OA were also in the constraint set. In that case visualization was easy, since $n = 2$ (two dimensions).
   c. In Figure 1.8, we saw that we could visualize the three-dimensional constraint set ($n = 3$) of Problem 1.4 without too much trouble.
   d. In Chapter 2 we shall see that the constraint sets for the standard maximum and minimum problems are also *convex*. It is appropriate to delay the precise definition of convex set until then (however, see Note (d) after Theorem 1.1).

4. THE DIET PROBLEM

One of the classical linear programming problems of minimum type is the diet problem. Here we are given information on the nutritional contents of several foods. For example, we might know how many grams of carbohydrate, fat, and protein are contained in one unit of each food under consideration. The unit may be a mass unit (gram, kilogram, or pound) or might refer to a container (e.g., a box of cookies or a quart container of milk). We are also given the minimum requirement for each nutrient and the cost per unit of food. The problem is to determine the distribution of foods that will result in minimum cost and satisfy the nutrient requirements.

In the following problems, we shall first pose the diet problem in general terms and then consider a specific example involving specific nutrients and foods.

It should be noted that Stigler [3] considered a diet problem involving 77 foods and 9 nutrients. The diet was for one person over a 1-year period. The linear programming solution yielded a diet with a minimum cost of $39.67 (for the year 1939).

Problem 1.11: General Diet Problem

Formulate the equation and inequalities for determining the minimum-cost diet that satisfies certain minimum requirements for various nutrients. Suppose the symbolism is as follows:

m = the number of nutrients
n = the number of foods
a_{ij} = the number of grams of the ith nutrient in one unit of the jth food
b_i = the minimum number of grams of the ith nutrient required
c_j = the cost per unit of the jth food
x_j = the number of units of the jth food to be purchased

Solution

Let u be the total cost of purchasing the various foods. Thus, we wish to minimize u, where

$$u = c_1x_1 + c_2x_2 + \cdots + c_nx_n \tag{1}$$

(here, e.g., c_1x_1 is the cost of purchasing x_1 units of food 1: c_1 dollars per unit of food 1 times x_1 units equals c_1x_1 dollars).

The statements involving minimum nutrient requirements are given as the following inequalities:

$$a_{11}x_1 + a_{12}x_2 + \cdots + a_{1n}x_n \geq b_1 \text{ (nutrient 1)} \tag{2.1}$$

$$a_{21}x_1 + a_{22}x_2 + \cdots + a_{2n}x_n \geq b_2 \text{ (nutrient 2)} \tag{2.2}$$

. . . .

. . . .

$$a_{m1}x_1 + a_{m2}x_2 + \cdots + a_{mn}x_n \geq b_m \text{ (nutrient } m). \tag{2.m}$$

We also have the common-sense statements that $x_1, x_2, \ldots, x_n$ must not be negative:

$$x_1 \geq 0 \tag{3.1}$$

$$x_2 \geq 0 \tag{3.2}$$

. . . .

. . . .

$$x_n \geq 0. \tag{3.n}$$

It should be noted that this problem has exactly the same form as the standard linear programming problem of minimum type (see Problem 1.7). In addition, the b_i's are all positive.

Problem 1.12: A Specific Diet Problem.

A dietitian is preparing a mixture of two foods, A and B. Each unit of food A contains 10 grams of carbohydrate, 20 grams of fat, and 15 grams of protein, and costs 25¢. Each unit of food B contains 25 grams of carbohydrate, 10 grams of fat, and 20 grams of protein, and costs 40¢. Determine the minimum-cost mixture of foods A and B, subject to the constraints that the diet must contain at least 400 grams of carbohydrate, 500 grams of fat, and 300 grams of protein.

Solution (Set up, but not solved completely.)

Let x be the number of units of food A and y the number of units of food B.

The cost function u, in dollars, is thus given by

$$u = \tfrac{1}{4}x + \tfrac{2}{5}y \tag{1}$$

(here, e.g., the cost of x units of food A is calculated as follows: $\frac{1}{4}$ dollars per unit times x units equals $\frac{1}{4} \cdot x$ dollars).

We, of course, wish to minimize u. The carbohydrate, fat, and protein requirements are as follows:

$$10x + 25y \geq 400 \text{ (carbohydrate)} \tag{2.1}$$

$$20x + 10y \geq 500 \text{ (fat)} \tag{2.2}$$

and

$$15x + 20y \geq 300 \text{ (protein)}. \tag{2.3}$$

In addition, x and y are nonnegative:

$$x \geq 0 \tag{3.1}$$

and

$$y \geq 0. \tag{3.2}$$

It is instructive to analyze (2.1). From food A, 10 grams of carbohydrate are supplied per unit; thus, in x units, 10 times x or $10x$ grams of carbohydrate are supplied. From food B, 25 grams of carbohydrate are supplied per unit; thus, in y units, 25 times y or $25y$ grams of carbohydrate are supplied. The total amount of carbohydrate supplied from foods A and B is thus $10x + 25y$, and this is required to be at least as great ($\geq$) as the minimum amount of carbohydrate required, namely, 400 grams.

The complete solution of the above problem involves the same type of setup as in Problem 1.3, i.e., we would be led to scanning the extreme points of a two-dimensional, unbounded constraint set. The solution would be found at one of the extreme points, since the function u is bounded from below (Theorem 1.2).

REFERENCES FOR CHAPTER 1

1. Gass, S. *Linear Programming*. Third edition. New York: McGraw-Hill, 1969.
2. Kwak, N. K. *Mathematical Programming with Business Applications*. New York: McGraw-Hill, 1973.
3. Stigler, G. J. "The Cost of Subsistence." *J. Farm Econ.* vol. 27, 1945.

SUPPLEMENTARY PROBLEMS

S.P. 1.1:   Solve the following problem. Sketch the constraint set.

Maximize $u = 4x + 3y$

subject to $x + 3y \leq 12$

$2x + y \leq 19$

and

$$x, y \geq 0.$$

S.P. 1.2:   Solve the following problem. Sketch the constraint set.

Minimize $u = 4x + 5y$

subject to $x + 3y \geq 12$

$2x + y \geq 19$

and

$$x, y \geq 0.$$

S.P. 1.3:   Solve the following problem. Sketch the constraint set.

Maximize $u = 2x + 7y$

subject to $x + 3y \leq 12$
$2x + y \leq 19$

and

$$x \leq 9, \; y \leq 3, \; x \geq 0, \; y \geq 0.$$

S.P. 1.4:   Solve the following problem. Sketch the constraint set.

Minimize $u = 7x + 4y$

subject to $8x + 2y \geq 12$

$3x + 3y \geq 15$

$x + 5y \geq 15$

and

$$x, y \geq 0.$$

S.P. 1.5:   Solve the following problem using the method of Problem 1.4.

Maximize $u = 2x + 3y + 5z$

subject to $12x + 9y + 3z \leq 1200$

$5y + 10z \leq 1000$

and

$$x, y, z \geq 0.$$

S.P. 1.6:   Given the same constraint set as in S.P. 1.5, find the solution if the goal is to maximize $v = 2x + 6y + 3z$.

S.P. 1.7:   Given the same constraint set as in S.P. 1.5, find the solution if the goal is to minimize $w = 4x - 3y - 2z$.

S.P. 1.8:   Find the solution to Problem 1.3 if it costs $500 per day to work mine A and $200 per day to work mine B.

S.P. 1.9:   Find the solution to Problem 1.4 if the sales revenue function is $v = 0.8x + 0.6y + 0.4z$ (everything else is unchanged).

S.P. 1.10:   Find the solution to Problem 1.12.

S.P. 1.11: A tailor has available 20 sq yd of cotton, 15 sq yd of wool, and 10 sq yd of silk. A suit requires 2 sq yd of cotton, 2 sq yd of wool, and 0.5 sq yd of silk. A dress requires 2 sq yd of cotton, 1 sq yd of wool, and 2 sq yd of silk. If a suit sells for $80 and a dress for $50, how many of each garment should the tailor make in order to maximize income?

S.P. 1.12: An oil company requires 12,000, 20,000, and 15,000 barrels of high-, medium-, and low-grade oil, respectively. Refinery A produces 100, 300, and 200 barrels per day of high-, medium-, and low-grade oil, respectively, while refinery B produces 200, 400, and 100 barrels per day of high-, medium-, and low-grade oil, respectively. (a) If refinery A costs $400 per day, and refinery B costs $300 per day to operate, how many days should each be run to minimize costs while satisfying requirements? (b) Rework the problem if both refineries cost $350 per day to operate.

S.P. 1.13: A dietitian is to prepare a mixture of two foods, A and B, that provides at least 600 g of protein and 800 g of carbohydrate. Each unit of food A contains 20 g of protein and 30 g of carbohydrate, and costs 40¢. Each unit of food B contains 25 g of protein and 20 g of carbohydrate, and costs 60¢. Determine the minimum-cost mixture of foods A and B that provides the specified nutrient requirements.

S.P. 1.14: A furniture company manufactures chairs, beds, and tables with the aid of workers A, B, and C. The time (in hours) that each worker spends on each of these items and the corresponding profits are shown in the following table:

	A	B	C	Profit, $
Chair	5	0	3	15
Bed	3	6	0	20
Table	0	3	4	30

Employees A and B can work at most 40 hr per week, while employee C can work at most 30 hr. How many chairs, beds, and tables should be built in order to maximize profits?

S.P. 1.15: A box manufacturer makes small and large boxes from large pieces of cardboard. The large boxes require 4 sq ft per box, while the small boxes require 3 sq ft per box. The manufacturer is required to make at least three large boxes and at least twice as many small boxes as large boxes. (a) If 60 sq ft of cardboard is in stock, and if the profits on the small and large boxes are $2 and $3 per box, respectively, how many of each should be made in order to maximize the total profit? (b) Rework the problem if the unit profits are $1.50 and $2, respectively, for small and large boxes.

S.P. 1.16: An automobile manufacturer makes standard and compact cars. The factory is divided into an assembly shop (shop 1) and a finishing shop (shop 2). The work requirements in shop 1 are 5 man-hours per standard car and 3 man-hours per compact car. In shop 2, 3 man-hours are required for both standard and compact cars. In shop 1,

150 man-hours per week are available, while 120 man-hours per week are available in shop 2. If the profit is $400 on each standard car and $300 on each compact car, how many cars of each type should be produced so as to maximize profits? (Assume that all cars produced can be sold.)

S.P. 1.17: Assume that a pound of a type of meat contains 1 unit of carbohydrate, 4 units of vitamins, and 10 units of protein, while a pound of a certain vegetable contains 2 units of carbohydrate, 4 units of vitamins, and 2 units of protein. The respective costs for the meat and the vegetable are $1.98 and $0.54 per pound. The minimum daily requirements for a certain diet are 8 units of carbohydrate, 20 units of vitamins, and 15 units of protein. (a) If there is a palatability requirement that at least 0.6 lb of meat must be eaten, what is the minimum-cost diet? (b) Determine the solution if the palatability requirement is at least 0.8 lb of meat.

S.P. 1.18: An investor has $50,000 available to invest in three kinds of securities: low-risk bonds yielding 5% annual interest; medium-risk growth stock yielding 8% annual interest; and high-risk speculative stock yielding an anticipated annual interest of 16%. To account for the risk aspect, the investor requires that at most $25,000 be invested in speculative stock and that the amount invested in both growth and speculative stock should be at most $30,000. How much should the investor invest in each of the three securities to maximize the annual interest return? (*Suggestion*: Use a three-dimensional graphical approach.)

S.P. 1.19: A manufacturer produces two types of bolts, I and II, using lathes, drill presses, and polishers. The machinery requirements for each type of bolt (in hours) and daily machine capacities are given in the following table:

Bolt	Lathes	Drill Presses	Polishers
I	.02	.02	.025
II	.01	.04	.050
Daily machine capacity (hr)	80	98	150

The profits for type I and type II bolts are 12¢ and 15¢ per bolt, respectively. What should the daily production of each bolt be in order to maximize profits? (Assume that all bolts made can be sold.)

ANSWERS TO SUPPLEMENTARY PROBLEMS

S.P. 1.1: Max $u = 39$ at $x = 9$, $y = 1$.
S.P. 1.2: Min $u = 41$ at $x = 9$, $y = 1$.
S.P. 1.3: Max $u = 27$ at $x = 3$, $y = 3$.
S.P. 1.4: Min $u = 21$ at $x = \frac{1}{3}$, $y = \frac{14}{3}$.
S.P. 1.5: Max $u = 650$ at $x = 75$, $y = 0$, $z = 100$.
S.P. 1.6: Max $v = 840$ at $x = 0$, $y = 120$, $z = 40$.
S.P. 1.7: Min $w = -440$ at $x = 0$, $y = 120$, $z = 40$.

S.P. 1.8: Minimum costs equals \$13,500; operate mine A for 15 days and mine B for 30 days.

S.P. 1.9: Maximum sales revenue equals \$880; $x = 700$ lb, $y = 0$, $z = 800$ lb.

S.P. 1.10: Minimum cost equals \$8.3125 when $x = 21.25$ units and $y = 7.5$ units.

S.P. 1.11: Max $u = \$635.71$, $x = 5.71$, $y = 3.57$. ("Rounded-off" *integer* solution: $u = \$630$, $x = 6$, $y = 3$. For this solution, all constraints must be satisfied!)

S.P. 1.12: (a) Min $u = \$33,000$, $x = 60$, $y = 30$.
(b) Min $u = \$31,500$, $x = 60$, $y = 30$.

S.P. 1.13: Min $u = \$12.00$, $x = 30$, $y = 0$.

S.P. 1.14: Max $u = \$283.33$, $x = 5.10$, $y = 4.83$, $z = 3.67$. (Rounded-off integer solution: $u = \$265$, $x = y = 5$, $z = 3$.) *Alternate solution:* Max $u = \$283.33$, $x = 0$, $y = 2.92$, $z = 7.5$. (Rounded-off integer solution: $u = \$270$, $x = 0$, $y = 3$, $z = 7$.)

S.P. 1.15: (a) Maximum profit equals \$42.00; $x_1 = 12$, $x_2 = 6$. (b) Maximum profit equals \$30.00; attained at both $x_1 = 12$, $x_2 = 6$ and $x_1 = 16$, $x_2 = 3$.

S.P. 1.16: Maximum profit equals \$13,500 per week; $x_1 = 15$ standard cars per week, $x_2 = 25$ compact cars per week.

S.P. 1.17: (a) Minimum cost equals \$3.60; $x_1 = \frac{5}{8}$ lb, $x_2 = \frac{35}{8}$ lb. (b) Minimum cost equals \$3.852; $x_1 = 0.8$ lb, $x_2 = 4.2$ lb.

S.P. 1.18: *Hint:* don't forget the overall requirement $x + y + z \leq 50,000$. Maximum annual interest equals \$5,400; $x = \$20,000$, $y = \$5,000$, $z = \$25,000$.

S.P. 1.19: Maximum daily profit equals \$534.00; $x_1 = 3700$ per day, $x_2 = 600$ per day.

Chapter Two

A Review of Matrices and Vectors

1. INTRODUCTORY REMARKS

In this chapter we shall study the matrix and vector notation used in linear programming problems. Examples that employ new symbolism will be provided whenever possible.

There are two main purposes to this chapter: (1) to familiarize the student with manipulations involving matrices and vectors and (2) to lay the groundwork for Chapter 3—"Convex Geometry"—where we shall develop the theory pertaining to Theorem 1.1 and, in addition, attempt to develop a geometric intuition concerning linear programming problems.

The reader who is familiar with the basics of matrix and vector notation should feel confident to skip ahead to Section 4 of this chapter, where the discussion of convex geometry begins.

In Chapter 4, a fairly independent algebraic approach to linear programming is presented. In that chapter we shall encounter the all important simplex algorithm, which will form one of the most important computational methods of this book. It is possible to virtually ignore the material on convex geometry and still gain a sound appreciation of linear programming. Thus, readers who wish to take this route (and who have been exposed to matrix and vector manipulations) are urged to jump ahead to Chapter 4, while retaining the option of referring back to Chapters 2 and 3 if unfamiliar symbols and concepts (from convex geometry) are encountered in that and later chapters.

2. MATRICES AND THEIR PROPERTIES

Definition 2.1. The general form of a matrix with m rows and n columns is

$$A = \begin{pmatrix} a_{11} & a_{12} & \cdots & a_{1n} \\ a_{21} & a_{22} & \cdots & a_{2n} \\ \vdots & \vdots & & \vdots \\ a_{m1} & a_{m2} & \cdots & a_{mn} \end{pmatrix}.$$

We shall usually denote matrices by uppercase letters (A, B, etc.). Lowercase letters, such as a, b, c, etc., denote real numbers. Another name for a real number (frequently used during discussions on matrices) is *scalar*.

The numbers in the array, such as a_{11}, a_{12}, ..., a_{mn}, are called the *elements* of the matrix. The double subscript is used to denote the location of an element. The first subscript gives the row and the second the column in which the element is located. Thus, a_{12} refers to the element in row 1 and column 2. The symbol a_{ij} refers to a typical element of the matrix located in row i and column j.

Matrix A is called an m by n (or $m \times n$) matrix.

Two matrices A and B are said to have the same *shape* if they have the same respective number of rows and columns.

Two matrices A and B are *equal* if they have the same shape and if the corresponding elements are equal.

Problem 2.1

Interpret the statement that matrices A and B are equal if A and B are given as follows:

$$A = \begin{pmatrix} a_{11} & a_{12} & a_{13} \\ a_{21} & a_{22} & a_{23} \end{pmatrix} \quad \text{and} \quad B = \begin{pmatrix} 3 & 2 & 1 \\ 4 & 6 & 5 \end{pmatrix}.$$

Solution

The equality of matrices A and B means that they have the same shape (in the present case, both are 2 by 3) and that the corresponding elements are equal. Thus,

$$a_{11} = 3; \quad a_{12} = 2; \quad a_{13} = 1;$$

$$a_{21} = 4; \quad a_{22} = 6; \quad a_{23} = 5.$$

Definition 2.2. A matrix with one row is called a *row vector*. A matrix with one column is called a *column vector*.

Thus, in general, we may refer to a row vector as a 1 by n matrix and a column vector as an m by 1 matrix. We shall usually denote vectors by lowercase letters with a bar above, e.g., $\bar{a}$. Other symbols used in the literature include a letter with an arrow above, e.g., $\vec{a}$; a letter with a line below, e.g., $\underline{a}$; or a letter in boldface type, e.g., $\mathbf{a}$.

For a k-component row vector $\bar{a}$, we have the following symbolism:

$$\bar{a} = (a_1, a_2, \ldots, a_k).$$

Here the elements are referred to as components and only a single subscript is used. Thus, the components of $\bar{a}$ are a_1, a_2, a_3, etc.

For a k-component column vector $\bar{b}$, we have the symbolism

$$\bar{b} = \begin{pmatrix} b_1 \\ b_2 \\ \vdots \\ b_k \end{pmatrix}.$$

The components of $\bar{b}$ are b_1, b_2, . . . , b_k. As in the case of row vectors, here we also have a single subscript notation.

In this book we shall often use both row and column vectors, although the column notation will be used more frequently than the row notation. Since it is rather awkward to write out columns, *a row with square brackets enclosing the components will be used to represent a column vector.* Thus, the k-component column vector $\bar{b}$ previously considered would have the following representation:

$$\bar{b} = [b_1, b_2, \ldots, b_k].$$

It is very important to remember this notational convention.

Definition 2.3: Matrix Addition. First of all, matrix addition of matrices A and B is defined only if A and B have the same shape. Thus, for A and B of the same shape, the *sum* of A and B, written $A + B$, is the matrix obtained by adding the corresponding elements of A and B.

Problem 2.2

Find $A + B$ if

$$A = \begin{pmatrix} 2 & 6 & 1 \\ 4 & 3 & 8 \end{pmatrix} \quad \text{and} \quad B = \begin{pmatrix} 6 & -3 & 4 \\ 2 & 5 & 1 \end{pmatrix}.$$

Solution

$$A + B = \begin{pmatrix} 2+6 & 6+(-3) & 1+4 \\ 4+2 & 3+5 & 8+1 \end{pmatrix} = \begin{pmatrix} 8 & 3 & 5 \\ 6 & 8 & 9 \end{pmatrix}.$$

Here matrices A, B, and $A + B$ are all 2 by 3.

Theorem 2.1. *For matrices A, B, and C (all of the same shape) we have the following two properties:*

a. $(A + B) + C = A + (B + C) = A + B + C$,

i.e., addition of matrices is associative;

b. $A + B = B + A$,

i.e., addition of matrices is commutative.

The matrix for which all the elements are zero is called the zero matrix and is denoted by **0**. *We thus have a third property:*

c. $A + 0 = 0 + A = A$.

Definition 2.4: Scalar Multiplication. The product of a scalar (real number) k and a matrix A, written as kA, is the matrix obtained by multiplying each element of A by k. Thus, if A is a typical 2 by 3 matrix, we have

$$kA = k \begin{pmatrix} a_{11} & a_{12} & a_{13} \\ a_{21} & a_{22} & a_{23} \end{pmatrix} = \begin{pmatrix} ka_{11} & ka_{12} & ka_{13} \\ ka_{21} & ka_{22} & ka_{23} \end{pmatrix} .$$

Note that A and kA have the same shape.

Definition 2.5. a. Negative of a Matrix. The *negative* of a matrix is obtained as follows:

$$-A = (-1)A,$$

i.e., $-A$ is equal to the scalar -1 times A.

b. Matrix Subtraction. For matrix subtraction we have that

$$A - B = A + (-B),$$

i.e., $A - B$ is equivalent to adding A and $(-B)$, where $(-B)$, or $-B$, is as defined in Definition 2.5(a).

Problem 2.3

Given that $A = \begin{pmatrix} 6 & 5 & 4 \\ 1 & -2 & 3 \end{pmatrix}$ and $B = \begin{pmatrix} 3 & 8 & 5 \\ 2 & 1 & 1 \end{pmatrix}$, express $3A$, $-A$, $-B$, and $A - B$.

Solution

$$3A = 3 \begin{pmatrix} 6 & 5 & 4 \\ 1 & -2 & 3 \end{pmatrix} = \begin{pmatrix} 3 \cdot 6 & 3 \cdot 5 & 3 \cdot 4 \\ 3 \cdot 1 & 3 \cdot (-2) & 3 \cdot 3 \end{pmatrix}$$

$$= \begin{pmatrix} 18 & 15 & 12 \\ 3 & -6 & 9 \end{pmatrix} .$$

$$-A = (-1) \begin{pmatrix} 6 & 5 & 4 \\ 1 & -2 & 3 \end{pmatrix} = \begin{pmatrix} -6 & -5 & -4 \\ -1 & +2 & -3 \end{pmatrix} .$$

$$-B = (-1) \begin{pmatrix} 3 & 8 & 5 \\ 2 & 1 & 1 \end{pmatrix} = \begin{pmatrix} -3 & -8 & -5 \\ -2 & -1 & -1 \end{pmatrix} .$$

$$A - B = A + (-B) = \begin{pmatrix} 3 & -3 & -1 \\ -1 & -3 & 2 \end{pmatrix} .$$

Definition 2.6: Matrix Multiplication. Let A and B be matrices such that the number of columns of A is equal to the number of rows of B. Then AB (read as A times B), which is the symbol for the product of A and B, is a matrix that has the same number of rows as A and the same number of columns as B. To be specific, suppose A is m by p and B is p by n; then the matrix AB is m by n.

The rule for computing a typical element of AB will now be considered. Let us rename AB as C. The definition for the computation of c_{ij}—the element in the ith row and jth column of C—is as follows:

$$c_{ij} = a_{i1}b_{1j} + a_{i2}b_{2j} + \cdots + a_{ip}b_{pj}.$$

There are p terms in this sum.

Figure 2.1 illustrates the rule for computing c_{ij} by "multiplying" the ith row of A by the jth column of B.

Note: If the number of columns of A is not equal to the number of rows of B (e.g., if A is m by p and B is r by n, where $p \neq r$), then the product AB is not defined.

It is helpful to illustrate matrix multiplication with some worked-out problems.

Problem 2.4

Compute $AB = C$ if A and B are given by

$$A = \begin{pmatrix} 1 & 3 \\ 2 & 5 \\ 4 & 6 \end{pmatrix} \quad \text{and} \quad B = \begin{pmatrix} 4 & 3 \\ 1 & 2 \end{pmatrix}.$$

FIGURE 2.1 Matrix multiplication: $AB = C$.

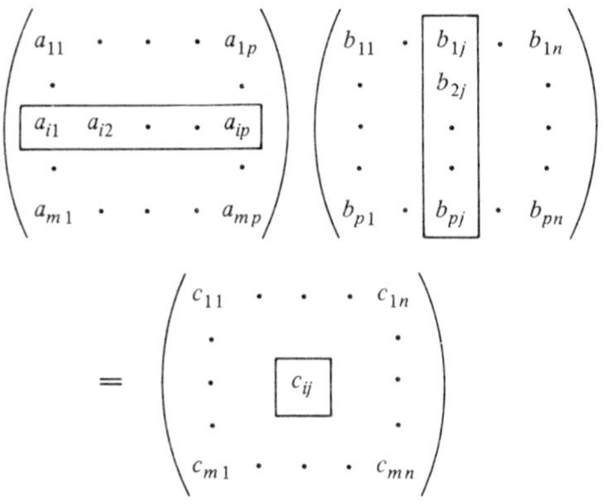

$$\begin{pmatrix} a_{11} & \cdots & a_{1p} \\ & & \\ a_{i1} & a_{i2} & \cdots & a_{ip} \\ & & \\ a_{m1} & \cdots & a_{mp} \end{pmatrix} \begin{pmatrix} b_{11} & \cdot & b_{1j} & \cdot & b_{1n} \\ & & b_{2j} & & \\ & & & & \\ b_{p1} & \cdot & b_{pj} & \cdot & b_{pn} \end{pmatrix}$$

$$= \begin{pmatrix} c_{11} & \cdots & c_{1n} \\ & & \\ & c_{ij} & \\ & & \\ c_{m1} & \cdots & c_{mn} \end{pmatrix}$$

Solution

$$c_{11} = 1 \cdot 4 + 3 \cdot 1 = 7 \quad \text{(row 1 of } A \text{ ''times'' column 1 of } B).$$

$$c_{12} = 1 \cdot 3 + 3 \cdot 2 = 9 \quad \text{(row 1 of } A \text{ ''times'' column 2 of } B).$$

$$c_{21} = 2 \cdot 4 + 5 \cdot 1 = 13 \quad \text{(row 2 of } A \text{ ''times'' column 1 of } B).$$

For calculation of c_{22}, refer to the boxed-in row 2 of A and the boxed-in column 2 of B in the above representation of A and B:

$$c_{22} = 2 \cdot 3 + 5 \cdot 2 = 16.$$

$$c_{31} = 4 \cdot 4 + 6 \cdot 1 = 22 \quad \text{(row 3 of } A \text{ ''times'' column 1 of } B).$$

$$c_{32} = 4 \cdot 3 + 6 \cdot 2 = 24 \quad \text{(row 3 of } A \text{ ''times'' column 2 of } B).$$

Thus, $AB = C$ is the 3 by 2 matrix given by

$$AB = \begin{pmatrix} 7 & 9 \\ 13 & 16 \\ 22 & 24 \end{pmatrix}.$$

Note that BA is not defined.

Problem 2.5

Compute AB if A and B are given by

$$A = \begin{pmatrix} 1 & 2 \\ 0 & 3 \end{pmatrix} \quad \text{and} \quad B = \begin{pmatrix} 1 & 3 \\ 2 & 4 \end{pmatrix}.$$

Solution

$$AB = \begin{pmatrix} 1 \cdot 1 + 2 \cdot 2 & 1 \cdot 3 + 2 \cdot 4 \\ 0 \cdot 1 + 3 \cdot 2 & 0 \cdot 3 + 3 \cdot 4 \end{pmatrix} = \begin{pmatrix} 5 & 11 \\ 6 & 12 \end{pmatrix}.$$

Problem 2.6

Compute $A\bar{x}$ if A and $\bar{x}$ are given by

$$A = \begin{pmatrix} 4 & 2 & 3 \\ 1 & 5 & 6 \end{pmatrix} \quad \text{and} \quad \bar{x} = \begin{pmatrix} 4 \\ 3 \\ 1 \end{pmatrix}.$$

Solution

Notice that $\bar{x}$ is a three-component column vector, which is equivalent to a 3 by 1 matrix. We have

$$A\bar{x} = \begin{pmatrix} 4 \cdot 4 + 2 \cdot 3 + 3 \cdot 1 \\ 1 \cdot 4 + 5 \cdot 3 + 6 \cdot 1 \end{pmatrix} = \begin{pmatrix} 25 \\ 25 \end{pmatrix}.$$

Following the notational convention mentioned earlier, we could equivalently write $\bar{x}$ and $A\bar{x}$ as $\bar{x} = [4, 3, 1]$ and $A\bar{x} = [25, 25]$, respectively.

Problem 2.7

Compute $A\bar{x}$ if A and $\bar{x}$ are given by

$$A = \begin{pmatrix} a_{11} & a_{12} & a_{13} \\ a_{21} & a_{22} & a_{23} \end{pmatrix} \quad \text{and} \quad \bar{x} = \begin{pmatrix} x_1 \\ x_2 \\ x_3 \end{pmatrix}.$$

Solution

This problem is similar to Problem 2.6, except that values have not been assigned to the various terms. We have

$$A\bar{x} = \begin{pmatrix} a_{11}x_1 + a_{12}x_2 + a_{13}x_3 \\ a_{21}x_1 + a_{22}x_2 + a_{23}x_3 \end{pmatrix}.$$

Thus, $A\bar{x}$ is a two-component column vector whose first and second components are $a_{11}x_1 + a_{12}x_2 + a_{13}x_3$ and $a_{21}x_1 + a_{22}x_2 + a_{23}x_3$, respectively.

Theorem 2.2. *Matrix multiplication satisfies the following properties:*

    a. $(AB)C = A(BC) = ABC$ (associative law)
    b. $A(B + C) = AB + AC$ (distributive law)
    c. $(B + C)A = BA + CA$ (distributive law)
    d. $k(AB) = (kA)B = A(kB)$ (k is a scalar).

It is assumed that the products and sums in Theorem 2.2 are defined.

Definition 2.7: Square Matrix. A matrix with the same number of rows and columns is called a *square matrix*. If a square matrix has n rows and n columns, it is said to be of order n.

The square matrix having ones along the main diagonal (the diagonal running from the upper left to the lower right) and zeros elsewhere is called the *identity matrix* (unit matrix) and is denoted by I or I_n. For example, the third-order identity matrix is

$$I = \begin{pmatrix} 1 & 0 & 0 \\ 0 & 1 & 0 \\ 0 & 0 & 1 \end{pmatrix} \quad \text{(order 3)}.$$

Definition 2.8: Transpose. The transpose of a matrix A is the matrix formed from A by interchanging rows and columns in such a way that row i of A becomes column i of the transposed matrix. The transpose is denoted by A^t and is n by m if A is m by n.

If the element in row i and column j of A^t is denoted by a_{ij}^t or $(a^t)_{ij}$, then

$$a_{ij}^t = a_{ji}.$$

Problem 2.8

Give the transpose of a general 2 by 3 matrix.

Solution

Let

$$A = \begin{pmatrix} a_{11} & a_{12} & a_{13} \\ a_{21} & a_{22} & a_{23} \end{pmatrix}.$$

Then

$$A^t = \begin{pmatrix} a_{11} & a_{21} \\ a_{12} & a_{22} \\ a_{13} & a_{23} \end{pmatrix}.$$

According to the symbolism discussed above, we may write

$$A^t = \begin{pmatrix} a^t_{11} & a^t_{12} \\ a^t_{21} & a^t_{22} \\ a^t_{31} & a^t_{32} \end{pmatrix}.$$

Thus, $a^t_{12} = a_{21}$, $a^t_{31} = a_{13}$, etc.

Problem 2.9

Determine A^t if A is given by

$$A = \begin{pmatrix} 10 & 8 & 7 & 6 \\ 5 & 9 & 2 & 3 \\ 4 & 5 & 8 & 1 \end{pmatrix}.$$

Solution

$$A^t = \begin{pmatrix} 10 & 5 & 4 \\ 8 & 9 & 5 \\ 7 & 2 & 8 \\ 6 & 3 & 1 \end{pmatrix}.$$

We observe that row 1 of A is column 1 of A^t, row 2 of A is column 2 of A^t, etc. Also, $a_{24} = 3 = a^t_{42}$, etc.

Theorem 2.3. *The transpose operation on matrices obeys the following rules:*

a. $(A + B)^t = A^t + B^t$
b. $(A^t)^t\ \ \ \ \ = A$
c. $(kA)^t\ \ \ \ = kA^t$, where k is a scalar
d. $(AB)^t\ \ \ \ = B^t A^t$.

Note: Property (d), which is fairly important, states that the transpose of a product equals the product of the transposes, *but in the opposite order*.

Problem 2.10

Verify property (d) of Theorem 2.3 for the matrices

$$A = \begin{pmatrix} 2 & 3 \\ 1 & 4 \end{pmatrix} \quad \text{and} \quad B = \begin{pmatrix} 1 & 5 \\ 3 & 0 \end{pmatrix} .$$

Solution

$$AB = \begin{pmatrix} 2 \cdot 1 + 3 \cdot 3 & 2 \cdot 5 + 3 \cdot 0 \\ 1 \cdot 1 + 4 \cdot 3 & 1 \cdot 5 + 4 \cdot 0 \end{pmatrix} = \begin{pmatrix} 11 & 10 \\ 13 & 5 \end{pmatrix} . \tag{1}$$

Thus,

$$(AB)^t = \begin{pmatrix} 11 & 13 \\ 10 & 5 \end{pmatrix} . \tag{2}$$

Now,

$$A^t = \begin{pmatrix} 2 & 1 \\ 3 & 4 \end{pmatrix} \quad \text{and} \quad B^t = \begin{pmatrix} 1 & 3 \\ 5 & 0 \end{pmatrix} .$$

Thus,

$$B^t A^t = \begin{pmatrix} 1 \cdot 2 + 3 \cdot 3 & 1 \cdot 1 + 3 \cdot 4 \\ 5 \cdot 2 + 0 \cdot 3 & 5 \cdot 1 + 0 \cdot 4 \end{pmatrix} = \begin{pmatrix} 11 & 13 \\ 10 & 5 \end{pmatrix} . \tag{3}$$

Comparing Equations (2) and (3), we see that property (d) of Theorem 2.3 is verified.

3. VECTORS AND THEIR PROPERTIES

As already indicated in Definition 2.2, a matrix with a single row is called a row vector, while a matrix with a single column is called a column vector. The latter will usually be represented by a *row* enclosed in square brackets.

It is convenient to give vectors a geometrical interpretation. This is particularly simple if we are dealing with two- or three-dimensional spaces. In Figure 2.2, we represent the two-component column vector $\bar{b}$ = $[b_1, b_2]$ as a point in two-dimensional space whose x_1 coordinate is b_1 and whose x_2 coordinate is b_2. Here X_1 and X_2 refer to the mutually perpendicular axes. The X_1 and X_2 axes are also traditionally labeled as the X and Y axes. We will often refer to two-dimensional space as "the plane."

Thus, we associate the point (b_1, b_2) in two-dimensional space with the vector $\bar{b} = [b_1, b_2]$. It is also useful to represent the vector $\bar{b}$ by *a directed line segment from the origin (i.e., O) to the point that characterizes the vector.* Therefore, the vector $\bar{b}$ also determines a direction.

The directed line segment labeled $\bar{b}_{tr}$ in Figure 2.2 is called a *"translated vector"* corresponding to $\bar{b}$. It is *not* a vector, since it does not

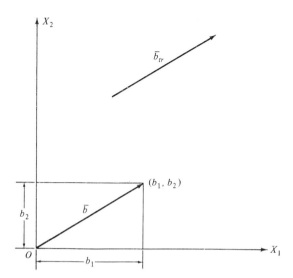

FIGURE 2.2 Geometrical representation of the two-component column vector b $= [b_1, b_2]$.

emanate from the origin—its initial point (starting point) can be anywhere in the plane. It is parallel to the line segment representing $\bar{b}$, and its direction and length are the same as the direction and length of the line segment representing $\bar{b}$.

Thus, for a vector, the initial point is at the origin, while for a translated vector, the initial point can be anywhere in the plane.

The situation is similar in three-dimensional space. For example, the three-component column vector $\bar{b} = [b_1, b_2, b_3]$ can be represented as a point in three-dimensional space whose x_1, x_2, and x_3 coordinates are b_1, b_2, and b_3, respectively. The X_1, X_2, and X_3 axes are mutually perpendicular. Here, too, we can again think of the vector $\bar{b}$ as a directed line segment from the origin to the point that characterizes the vector. The translated vector device can also be used here.

It should be noted that the above discussion pertained only to column vectors. Actually, there is no geometric distinction between column and row vectors, i.e., they are geometrically equivalent. In this book we shall use column vectors more often than row vectors, and our geometric representation of vectors will usually be for column vectors.

Another point worthy of discussion is the notation for the geometrical representation of the vector. In Figure 2.2, the point that characterizes the column vector $\bar{b}$ is labeled (b_1, b_2), and not $[b_1, b_2]$. The only reason for this is the traditional use of parentheses to represent points in analytic geometry.

Just as the vector $\bar{b} = [b_1, b_2]$ can be considered a point in two-dimensional space, the vector $\bar{b} = [b_1, b_2, \ldots, b_n]$ can be considered a point in n-dimensional space. We will discuss the abstract concept known as n-dimensional space in more detail later in this section.

Three operations pertaining to vectors are of importance. Two of them—addition (subtraction), and scalar multiplication—are defined in the same way as for general matrices. The third operation is forming the *dot product* (scalar product or inner product).

Definition 2.9. The dot product $\bar{a} \cdot \bar{b}$ of the two n-component vectors $\bar{a}$ and $\bar{b}$ is defined by the following equation:

$$\bar{a} \cdot \bar{b} = a_1 b_1 + a_2 b_2 + \cdots + a_n b_n. \tag{1}$$

Equation (1) holds regardless of whether $\bar{a}$ and $\bar{b}$ are row or column vectors. The dot product is a scalar (real number). There are equivalent expressions for the dot product depending upon the nature of the vectors $\bar{a}$ and $\bar{b}$. If $\bar{a}$ and $\bar{b}$ are both column vectors, the dot product is equivalent to $\bar{a}^t \bar{b}$, where the term $\bar{a}^t \bar{b}$ denotes the *matrix product* of the n-component row vector $\bar{a}^t$ times the n-component column vector $\bar{b}$. On the other hand, if $\bar{a}$ and $\bar{b}$ are both row vectors, the dot product is equivalent to $\bar{a} \bar{b}^t$. Finally, if $\bar{a}$ is a row vector and $\bar{b}$ is a column vector, the dot product is equivalent to $\bar{a} \bar{b}$. We shall illustrate the case that most frequently occurs in this book, namely, when $\bar{a}$ and $\bar{b}$ are both column vectors. Following the rule for matrix multiplication, and noting that $\bar{a}^t$ is 1 by n while $\bar{b}$ is n by 1, we see that $\bar{a}^t \bar{b}$ is a 1 by 1 matrix, i.e., a real number (1 by 1 matrices are equivalent to real numbers). Following the rule for matrix multiplication, we obtain

$$\bar{a}^t \bar{b} = (a_1, a_2, \ldots, a_n) \begin{pmatrix} b_1 \\ b_2 \\ \vdots \\ b_n \end{pmatrix} = a_1 b_1 + a_2 b_2 + \cdots + a_n b_n. \tag{2}$$

Thus, for column vectors, $\bar{a} \cdot \bar{b}$ and $\bar{a}^t \bar{b}$ yield the same numerical result, and we can use either expression to denote the dot product.

It is useful to illustrate geometrically the process of vector addition. We shall do this for the two-dimensional case, but it should be noted that a similar process applies to three dimensions.

To find the vector sum $\bar{a} + \bar{b}$, which we shall denote as $\bar{c}$, we first construct the translated vector $\bar{b}_{tr}$ such that its initial point coincides with the terminal point (arrowhead) of $\bar{a}$. The directed line segment from the origin to the terminal point of $\bar{b}_{tr}$ is then the representation of $\bar{a} + \bar{b} = \bar{c}$. The coordinates of the terminal point are $(a_1 + b_1, a_2 + b_2)$, as should be expected. This process is known as the *triangle law* for adding vectors.

Another method of determining the sum $\bar{a} + \bar{b}$ is the *parallelogram law* for adding vectors. Here we construct a parallelogram with sides parallel to the vectors $\bar{a}$ and $\bar{b}$ (see Figure 2.3; one side of the parallelogram is $\bar{b}_{tr}$, while the other side is the dashed line parallel to $\bar{a}$). The diagonal of the parallelogram which starts at the origin is then the representation of $\bar{a} + \bar{b}$.

If we perform vector addition according to the rules for matrix addition, we are directly led to the correct result. For example, in the two-dimensional case we have

$$\bar{a} + \bar{b} = [a_1, a_2] + [b_1, b_2] = [a_1 + b_1, a_2 + b_2].$$

At this point, instead of continuing to work entirely with symbols, let us consider a numerical example.

Problem 2.11

Compute $\bar{a} + \bar{b}$ if $\bar{a} = [5, 2]$ and $\bar{b} = [1, 6]$.

Solution

Proceeding as discussed, we have

$$\bar{a} + \bar{b} = [5, 2] + [1, 6] = [6, 8].$$

Thus, $\bar{a} + \bar{b}$ is the vector whose x_1 and x_2 components are 6 and 8, respectively. The numerical aspects of this problem are consistent with Figure 2.3.

The geometrical process of vector subtraction can also be demon-

FIGURE 2.3 Vector addition.

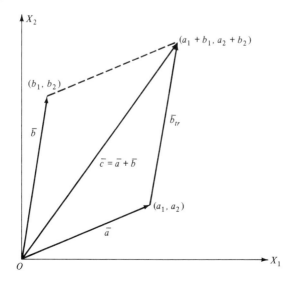

strated by means of the triangle law (refer to Figure 2.4). Suppose that we are given vectors $\bar{a}$ and $\bar{b}$ and that we wish to find $\bar{b} - \bar{a}$. This is the vector that when added to $\bar{a}$ will yield $\bar{b}$, i.e.,

$$\bar{a} + (\bar{b} - \bar{a}) = \bar{b}.$$

Thus, we construct the translated vector $(\bar{b} - \bar{a})_{tr}$ that extends from the terminal point of $\bar{a}$ to the terminal point of $\bar{b}$. The representation for the vector $(\bar{b} - \bar{a})$ is obtained by constructing a directed line segment from the origin which is parallel to and of the same length and direction as $(\bar{b} - \bar{a})_{tr}$.

The process of scalar multiplication is illustrated geometrically in Figure 2.5. For the scalar multiple 3, the vector $3\bar{a}$ is three times as long as vector $\bar{a}$. The directions associated with both vectors are the same. If the scalar multiple is -2, then the vector $-2\bar{a}$ has a direction totally opposite to that of $\bar{a}$, and its length is twice that of $\bar{a}$ (the "length" is the distance from the origin to the terminal point).

If k is a positive real number (scalar), then $\bar{a}$ and $k\bar{a}$ have the same directions; however, if k is a negative scalar, then $\bar{a}$ and $k\bar{a}$ have opposite directions.

Problem 2.12

Given that $\bar{a} = [4,3]$ and $\bar{b} = [2,6]$, compute $\bar{b} - \bar{a}$, $3\bar{a}$, and $-2\bar{a}$.

Solution

$$\bar{b} - \bar{a} = [2,6] - [4,3] = [2 - 4, 6 - 3] = [-2,3].$$

FIGURE 2.4 Vector subtraction.

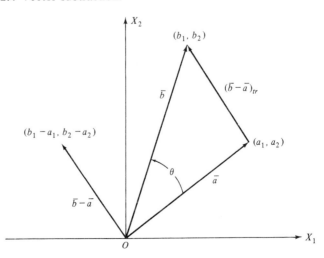

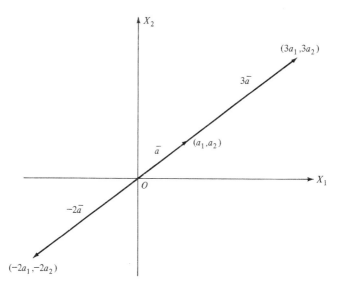

FIGURE 2.5 Scalar multiplication.

This calculation is consistent with Figure 2.4.

$$3\bar{a} = 3[4,3] = [3\cdot4, 3\cdot3] = [12,9],$$
$$-2\bar{a} = -2[4,3] = [-2\cdot4, -2\cdot3] = [-8,-6].$$

The results for $3\bar{a}$ and $-2\bar{a}$ are consistent with Figure 2.5.

Calculations such as those just considered lend themselves to a ready geometrical interpretation in the case of two and three dimensions. In higher dimensions, we lose the geometric interpretation, but the processes of vector addition, vector subtraction, and scalar multiplication are easily performed by means of algebraic equations. For example, the sum $\bar{c}$ of two four-component vectors $\bar{a}$ and $\bar{b}$ is calculated as follows:

$$\bar{c} = \bar{a} + \bar{b} = [a_1,a_2,a_3,a_4] + [b_1,b_2,b_3,b_4]$$
$$= [a_1 + b_1, a_2 + b_2, a_3 + b_3, a_4 + b_4].$$

Such manipulations involving vectors with many components lead to the abstract concept known as n-dimensional Euclidean space. The structure that we shall establish for n-dimensional Euclidean space is algebraic, and the rules that apply are the same as those for two- and three-dimensional space. We shall use two- and (occasionally) three-dimensional diagrams to illustrate various processes in n-dimensional Euclidean space.

Definition 2.10: n-Dimensional Euclidean Space. Consider two typical n-component vectors $\bar{a}$ and $\bar{b}$:

$$\bar{a} = [a_1, a_2, \ldots, a_n] \quad \text{and} \quad \bar{b} = [b_1, b_2, \ldots, b_n].$$

As discussed above, we have the operations of vector addition, scalar multiplication, and forming the dot product:

$$\bar{a} + \bar{b} = [a_1 + b_1, a_2 + b_2, \ldots, a_n + b_n].$$

$$k\bar{a} = [ka_1, ka_2, \ldots, ka_n].$$

$$\bar{a} \cdot \bar{b} = a_1 b_1 + a_2 b_2 + \cdots + a_n b_n.$$

The set of all such vectors (with the above operations) is called n-dimensional Euclidean space and will be denoted by R^n. Equivalent expressions include Euclidean n-space, n-dimensional space, or just n-space.

At this point, it is appropriate to develop some general equations for computing the lengths of vectors and the distances between them. We will allow our knowledge of the properties of two and three dimensions to guide us to general definitions for n dimensions. In particular, let us refer to the two-dimensional diagram of Figure 2.4. When we refer to the "length" or "magnitude" of vector $\bar{a}$, we mean the distance from the origin to the terminal point (denoted by the arrowhead) of vector $\bar{a}$. Let us denote the length of vector $\bar{a}$ by the symbol $\|\bar{a}\|$. In typical mathematical jargon, $\|\bar{a}\|$ is referred to as the *norm* of vector $\bar{a}$. The Pythagorean theorem of geometry leads us to the following equations:

$$\|\bar{a}\| = \sqrt{a_1{}^2 + a_2{}^2} \qquad \text{(two dimensions)}.$$

$$\|\bar{a}\| = \sqrt{a_1{}^2 + a_2{}^2 + a_3{}^2} \quad \text{(three dimensions)}.$$

When we refer to the "distance" between vectors $\bar{a}$ and $\bar{b}$, denoted by $d(\bar{a},\bar{b})$, we mean the distance between the points with coordinates (a_1, a_2) and (b_1, b_2), respectively, if we are working in two dimensions; i.e., $d(\bar{a},\bar{b})$ is the distance between the terminal points of $\bar{a}$ and $\bar{b}$. Again following the Pythagorean theorem, we have

$$d(\bar{a},\bar{b}) = \sqrt{(b_1 - a_1)^2 + (b_2 - a_2)^2} \qquad \text{(two dimensions)}$$

and

$$d(\bar{a},\bar{b}) = \sqrt{(b_1 - a_1)^2 + (b_2 - a_2)^2 + (b_3 - a_3)^2} \quad \text{(three dimensions)}.$$

We note that $\bar{b} - \bar{a} = [b_1 - a_1, b_2 - a_2]$ in two dimensions and thus

$$\|\bar{b} - \bar{a}\| = \sqrt{(b_1 - a_1)^2 + (b_2 - a_2)^2}.$$

In other words, the length of the vector $\bar{b} - \bar{a}$ is equal to the distance between vectors $\bar{a}$ and $\bar{b}$, i.e.,

$$d(\bar{a},\bar{b}) = \|\bar{b} - \bar{a}\|.$$

Let us now extend these ideas to general n-dimensional Euclidean space.

Definition 2.11. The length (norm or magnitude) of the vector $\bar{a}$, written $\|\bar{a}\|$, is defined by

$$\|\bar{a}\| = \sqrt{a_1^2 + a_2^2 + \cdots + a_n^2}.$$

The distance between two vectors $\bar{a}$ and $\bar{b}$, which we write as $d(\bar{a},\bar{b})$, is defined by

$$d(\bar{a},\bar{b}) = \sqrt{(b_1 - a_1)^2 + (b_2 - a_2)^2 + \cdots + (b_n - a_n)^2}.$$

Notes:
a. We see that the length of $\bar{a}$ can also be expressed in terms of a dot product:

$$\|\bar{a}\| = \sqrt{\bar{a} \cdot \bar{a}}.$$

b. The distance $d(\bar{a},\bar{b})$ can also be written in the form of the equation $d(\bar{a},\bar{b}) = \|\bar{b} - \bar{a}\|$.

Problem 2.13

Consider the following two vectors in R^4 (four-dimensional Euclidean space): $\bar{a} = [3, 5, -2, 0]$ and $\bar{b} = [6, -3, 5, 1]$. Compute $\|\bar{a}\|$, $\|\bar{b}\|$, $\bar{a} \cdot \bar{b}$, and $d(\bar{a},\bar{b})$.

Solution

$$\|\bar{a}\| = \sqrt{3^2 + 5^2 + (-2)^2 + 0^2} = \sqrt{9 + 25 + 4 + 0} = \sqrt{38}.$$

$$\|\bar{b}\| = \sqrt{6^2 + (-3)^2 + 5^2 + 1^2} = \sqrt{36 + 9 + 25 + 1} = \sqrt{71}.$$

$$\bar{a} \cdot \bar{b} = 3 \cdot 6 + 5 \cdot (-3) + (-2) \cdot 5 + 0 \cdot 1$$

$$= 18 + (-15) + (-10) + 0 = -7.$$

$$\bar{b} - \bar{a} = [6 - 3, -3 - 5, 5 - (-2), 1 - 0] = [3, -8, 7, 1].$$

Thus,

$$d(\bar{a},\bar{b}) = \|\bar{b} - \bar{a}\| = \sqrt{3^2 + (-8)^2 + 7^2 + 1^2}$$

$$= \sqrt{9 + 64 + 49 + 1} = \sqrt{123}.$$

It is useful to discuss the concept of the angle between two vectors. For example, for the two vectors $\bar{a}$ and $\bar{b}$ shown in Figure 2.4, the angle between them is labeled as θ (theta). The limiting values for θ are 0 degrees and 180 degrees (π radians). An angle of 0 degrees occurs when the two vectors point in the same direction (e.g., the vectors $\bar{a}$ and $3\bar{a}$ in Figure 2.5). An angle of 180 degrees occurs when the vectors point in completely opposite directions (e.g., the vectors $\bar{a}$ and $-2\bar{a}$ in Figure 2.5).

Theorem 2.4. *An important equation relating the lengths of two vectors, their dot product, and the angle between them is as follows:*

$$\bar{a}\cdot\bar{b} = \|\bar{a}\|\,\|\bar{b}\|\cos\theta.$$

Note: On the right-hand side of this equation, the ordinary product of $\|\bar{a}\|$, $\|\bar{b}\|$, and cos θ is implied.

The above relationship, which holds for both two- and three-dimensional spaces, is proved on pages 22–24 of [1]. The term cos θ means the cosine of the angle θ between the vectors.

Although it is difficult to visualize vectors and angles if the dimension of the space is greater than three, the above equation is used for the *definition* of the angle between two vectors in general n-dimensional Euclidean space.

Note: In fact, it can be shown that the cos θ term has the general property $-1 \le \cos\theta \le 1$ regardless of the dimension n (Cauchy–Schwarz inequality). Thus, we can always consider θ as an angle lying between 0 and 180 degrees (recall that cos $0° = +1$, cos $90° = 0$, and cos $180° = -1$).

An important case of the relative directions of two vectors occurs when they are mutually perpendicular, i.e., the angle between them (actually, between the directed line segments corresponding to them) is 90 degrees ($\pi/2$ radians). Since the cosine of 90 degrees is zero, we see from Theorem 2.4 that we can characterize the perpendicularity of two vectors by means of the following theorem.

Theorem 2.5. *If two vectors $\bar{a}$ and $\bar{b}$ are perpendicular, then their dot product is zero. Moreover, the converse is also true, i.e., if the dot product of vectors $\bar{a}$ and $\bar{b}$ is zero, then they are perpendicular.*

Notes:
a. The word *orthogonal* has the same meaning as perpendicular.
b. A special case occurs when one of the vectors is the so-called *zero vector*. This is the vector that has all components equal to zero; it is denoted by $\bar{0}$. Thus,

$$\bar{0} = [0, 0, \ldots, 0].$$

Clearly, the dot product of $\bar{0}$ with any other vector is the number zero. Even though it is difficult to conceive of perpendicularity in this case, we shall *define* the $\bar{0}$ vector to be perpendicular to every other vector.

Theorem 2.5 is actually two theorems: one of the form "if P then Q" and the other of the form "if Q then P." Each is said to be the converse of the other. A shorthand way of indicating this is to say "*P* if and only if *Q*" or "*P* is a necessary and sufficient condition for *Q*." To illustrate this, let us restate the preceding theorem.

Theorem 2.5'. *Two vectors $\bar{a}$ and $\bar{b}$ are perpendicular if and only if their dot product is zero.*

Problem 2.14

Determine the angle θ between the vectors $\bar{a}$ and $\bar{b}$ of Problem 2.12.

Solution

In Problem 2.12 we had $\bar{a} = [4,3]$ and $\bar{b} = [2,6]$. (The relative positions of $\bar{a}$ and $\bar{b}$, and also the angle θ, are illustrated in Figure 2.4.) Thus:

$$\|\bar{a}\| = \sqrt{4^2 + 3^2} = \sqrt{25} = 5.$$

$$\|\bar{b}\| = \sqrt{2^2 + 6^2} = \sqrt{40} = 6.324.$$

$$\bar{a}\cdot\bar{b} = 4\cdot2 + 3\cdot6 = 26.$$

$$\therefore \cos\theta = \frac{\bar{a}\cdot\bar{b}}{\|\bar{a}\|\,\|\bar{b}\|} = \frac{26}{5\cdot6.324} = 0.822.$$

$$\therefore \theta = 34.7 \text{ degrees (from a trigonometry table)}.$$

Problem 2.15

Determine the angle θ between the two-component vectors $\bar{a} = [2,1]$ and $\bar{b} = [-2,4]$. Draw a sketch.

Solution

Here we have

$$\bar{a}\cdot\bar{b} = 2\cdot(-2) + 1\cdot4 = 0.$$

Thus, $\bar{a}$ and $\bar{b}$ are perpendicular, i.e., $\theta = 90$ degrees. The appropriate sketch is given in Figure 2.6.

4. ELEMENTARY CONVEX GEOMETRY

We now introduce the useful concept of a *direction vector* from one vector to another vector. For example, the direction vector from vector $\bar{a}$ to vector $\bar{b}$ in Figure 2.4 is the vector $\bar{b} - \bar{a}$, the direction of which is from the terminal point of $\bar{a}$ to the terminal point of $\bar{b}$. We also note that the direction vector from $\bar{b}$ to $\bar{a}$ is $\bar{a} - \bar{b}$; this vector, of course, has a direction completely opposite to that of $\bar{b} - \bar{a}$:

$$\bar{a} - \bar{b} = -(\bar{b} - \bar{a}).$$

It is important to remember that we have assigned a dual geometrical

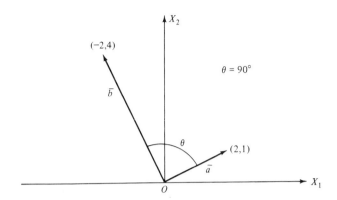

FIGURE 2.6 Diagram of vectors $\bar{a}$ and $\bar{b}$ and angle θ between them for Problem 2.15.

meaning to vectors. Sometimes we interpret a vector as being a point in n-dimensional Euclidean space, where the components of the vector are identical to the corresponding coordinates of the point. The point in question is also called the terminal point. In our other interpretation, we consider the vector as determining a *direction*—from the origin to the terminal point. When we are interested in the direction aspect, we usually put an arrowhead at the terminal point.

Both of the interpretations just given are relevant to our upcoming discussion of hyperplanes and line segments.

Consider the following linear equation in n variables $x_1, x_2, \ldots, x_n$:

$$c_1 x_1 + c_2 x_2 + \cdots + c_n x_n = b \tag{1}$$

If we define the column vectors $\bar{c}$ and $\bar{x}$ as

$$\bar{c} = [c_1, c_2, \ldots, c_n]$$

and

$$\bar{x} = [x_1, x_2, \ldots, x_n],$$

then we can write the following equivalent forms of Equation (1):

$$\bar{c} \cdot \bar{x} = b \tag{2}$$

and

$$\bar{c}^t \bar{x} = b. \tag{3}$$

By the way, we assume that $\bar{c} \neq \bar{0}$ unless otherwise indicated; in other words, at least one of the c_i in Equation (1) is not zero.

Definition 2.12: Hyperplane. The set of all points in R^n which satisfy Equation (1) is called a hyperplane.

In two- and three-dimensional space (also referred to as 2-space and 3-space, respectively), we have the following equations for hyperplanes:

$$c_1x_1 + c_2x_2 = b \quad \text{(2-space)}$$

and

$$c_1x_1 + c_2x_2 + c_3x_3 = b \quad \text{(3-space)}.$$

Thus, in R^2 (i.e., 2-space) the hyperplanes are straight lines, while in R^3 they are planes. We see from these examples that the dimension of the hyperplane is one less than the dimension of the containing space. For example, the line $c_1x_1 + c_2x_2 = b$ in 2-space has dimension one. In general, the dimension of a hyperplane lying in n-dimensional Euclidean space is $(n - 1)$, i.e., one less than the dimension of the containing space. The proof of this statement is difficult relative to the level of this text and thus will not be given.

Theorem 2.6. *The vector $\bar{c}$ is perpendicular to the hyperplane $\bar{c} \cdot \bar{x} = b$.*

Theorem 2.6 means that $\bar{c}$ is perpendicular to the *direction vector* between any two distinct points in the hyperplane. The proof of this theorem will be given after the following two examples (Problems 2.16 and 2.17).

Problem 2.16

Find the equation of the hyperplane in 2-space (R^2) for which $\bar{c} = [2,1]$ and $b = 20$. Draw a sketch. Show that $\bar{c}$ is perpendicular to the hyperplane.

Solution

We know that the vector $\bar{c} = [2,1]$ and that $\bar{x} = [x_1,x_2]$. Thus,

$$\bar{c} \cdot \bar{x} = 2x_1 + x_2,$$

and the equation for the hyperplane is

$$2x_1 + x_2 = 20,$$

which is the equation for a straight line. Let us locate the two vectors (points) corresponding to the intercepts:

When $x_1 = 0, x_2 = 20$ (x_2 intercept).

When $x_2 = 0, x_1 = 10$ (x_1 intercept).

Denoting these two vectors as $\bar{P}$ and $\bar{Q}$, we have $\bar{P} = [0,20]$ and $\bar{Q} = [10,0]$. The hyperplane, the vector $\bar{c}$, and the vectors $\bar{P}$ and $\bar{Q}$ are plotted in Figure 2.7.

Rather than proving in general that $\bar{c}$ is perpendicular to the hyper-

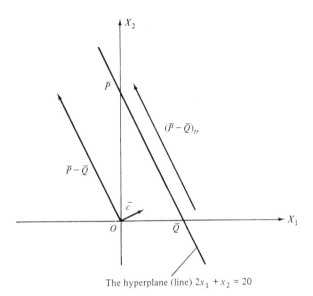

The hyperplane (line) $2x_1 + x_2 = 20$

FIGURE 2.7 A hyperplane in two-dimensional space. $\bar{P}$ has coordinates $(0,20)$. $\bar{Q}$ has coordinates $(10,0)$. Vector $\bar{c} = [2,1]$.

plane, we shall show that $\bar{c}$ is perpendicular to the direction vector from $\bar{Q}$ to $\bar{P}$, namely, $\bar{P} - \bar{Q}$, which is also shown in Figure 2.7. Thus,

$$\bar{P} - \bar{Q} = [0,20] - [10,0] = [-10,20].$$
$$\bar{c} \cdot (\bar{P} - \bar{Q}) = [2,1] \cdot [-10,20] = 2 \cdot (-10) + 1 \cdot 20 = 0$$

Since the dot product of $\bar{c}$ and $\bar{P} - \bar{Q}$ is zero, the two vectors are perpendicular.

Another way of proving that $\bar{c}$ and $\bar{P} - \bar{Q}$ are perpendicular is by showing that the product of their slopes is equal to -1. We shall use the letter m to denote slope. Remembering that *slope* is defined as the change in x_2 divided by the corresponding change in x_1, we have

$$m_{\bar{c}} = \tfrac{1}{2},$$

$$m_{\bar{P}-\bar{Q}} = \frac{20 - 0}{0 - 10} = -2,$$

and

$$m_{\bar{c}} \cdot m_{\bar{P}-\bar{Q}} = \tfrac{1}{2}(-2) = -1.$$

In conclusion, we see that $\bar{c}$ is perpendicular to the hyperplane.

Problem 2.17

Find the equation of the hyperplane in 3-space (R^3) for which $\bar{c} = [2,2,3]$ and $b = 60$. Draw a sketch of the portion of the hyperplane in the first octant.

Solution

Referring to the general equation of the hyperplane in 3-space which follows Definition 2.12,

$$c_1 x_1 + c_2 x_2 + c_3 x_3 = b,$$

we have

$$\text{or} \quad 2x_1 + 2x_2 + 3x_3 = 60. \tag{1}$$

The intercept points (vectors) may now be calculated from Equation (1):

When $x_1 = x_2 = 0$, $x_3 = 20$ (point $\bar{K}$).

When $x_1 = x_3 = 0$, $x_2 = 30$ (point $\bar{L}$).

When $x_2 = x_3 = 0$, $x_1 = 30$ (point $\bar{M}$).

We see that the equation of the hyperplane—Equation (1)—is the equation of an actual plane in three-dimensional space. The sketch for the first octant is shown in Figure 2.8. The vector $\bar{c} = [2,2,3]$ is perpendicular to the hyperplane (vector $\bar{c}$ is not shown in the figure).

FIGURE 2.8 A hyperplane in three-dimensional space.

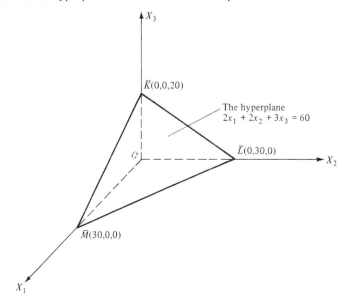

Note: It is useful at this point to state several properties of the dot product:

$$\bar{u}\cdot\bar{v} = \bar{v}\cdot\bar{u} \qquad \text{(commutative law).} \qquad (1)$$

$$\bar{u}\cdot(\bar{v} + \bar{w}) = \bar{u}\cdot\bar{v} + \bar{u}\cdot\bar{w} \qquad \text{(distributive law).} \qquad (2)$$

$$(\bar{v} + \bar{w})\cdot\bar{u} = \bar{v}\cdot\bar{u} + \bar{w}\cdot\bar{u} \qquad \text{(distributive law).} \qquad (3)$$

$$k(\bar{u}\cdot\bar{v}) = (k\bar{u})\cdot\bar{v} = \bar{u}\cdot(k\bar{v}) \quad \text{(k is a scalar).} \qquad (4)$$

We will now present the general proof of Theorem 2.6.

PROOF OF THEOREM 2.6. Let $\bar{P}$ and $\bar{Q}$ denote any two distinct points in the hyperplane $\bar{c}\cdot\bar{x} = b$.

The direction vector from $\bar{Q}$ to $\bar{P}$ is $\bar{P} - \bar{Q}$ (refer to Figure 2.9 for a diagram of the situation in 2-space). We have to show that $\bar{c}$ is perpendicular to $\bar{P} - \bar{Q}$. Thus, we write

$$\bar{c}\cdot(\bar{P} - \bar{Q}) = \bar{c}\cdot\bar{P} - \bar{c}\cdot\bar{Q} \qquad (1)$$

Since $\bar{P}$ and $\bar{Q}$ are both points in the hyperplane, we have

$$\bar{c}\cdot\bar{P} = b \qquad (2)$$

and

$$\bar{c}\cdot\bar{Q} = b. \qquad (3)$$

Thus, Equation (1) becomes

$$\bar{c}\cdot(\bar{P} - \bar{Q}) = b - b = 0, \qquad (4)$$

FIGURE 2.9 Diagram for the proof of Theorem 2.6.

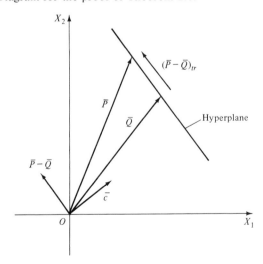

which indicates that $\bar{c}$ is perpendicular to $\bar{P} - \bar{Q}$ and thus to the hyperplane. □

We shall now turn to the mathematical definition of a *line segment* between two points in space. We let the situation in two- and three-dimensional space guide us. Consider the 2-space diagram of Figure 2.10. By the line segment between the points (vectors) $\bar{P}$ and $\bar{Q}$, we mean the set of all points which lie between $\bar{P}$ and $\bar{Q}$ (including the points $\bar{P}$ and $\bar{Q}$ themselves). Let $\bar{x}$ denote any point (vector) between $\bar{P}$ and $\bar{Q}$ inclusive. We would like to express $\bar{x}$ in terms of $\bar{P}$ and $\bar{Q}$. This can be done by means of vector addition, using the triangle law as a guide. Referring to Figure 2.10, we see that

$$\bar{x} = \bar{Q} + t(\bar{P} - \bar{Q}), \tag{1}$$

where t is some number from 0 to 1 inclusive. Thus,

$$\bar{x} = t\bar{P} + (1 - t)\bar{Q}. \tag{2}$$

We note that the variable point $\bar{x}$ reduces to $\bar{Q}$ when $t = 0$ and to $\bar{P}$ when $t = 1$.

The concepts that we have just considered are also valid for 3-space, and we will now use them to *define* the line segment in n-dimensional space.

Definition 2.13. The *line segment* joining two distinct points $\bar{P}$ and $\bar{Q}$ of R^n is the set of all points $\bar{x}$ which satisfy the following equation:

$$\bar{x} = t\bar{P} + (1 - t)\bar{Q}, \quad \text{where} \quad 0 \le t \le 1.$$

We denote the line segment by the symbol $\overline{PQ}$.

The concept of the line segment is basic to the understanding of the concept of *convexity* and can be best illustrated by means of a worked-out problem. However, before considering a specific problem, for the sake of completeness we should present the definition of the *line* in R^n that is determined by two distinct points $\bar{P}$ and $\bar{Q}$. The definition is based on our intuitive concept of what we learned about straight lines when we studied elementary geometry. In the definition that follows, the line extends on indefinitely, and not just between $\bar{P}$ and $\bar{Q}$.

Definition 2.14. The line through two distinct points $\bar{P}$ and $\bar{Q}$ is the set of all points $\bar{x}$ which satisfy the following equation:

$$\bar{x} = t\bar{P} + (1 - t)\bar{Q},$$

where t is any real number.

Thus, the difference between Definitions 2.13 and 2.14 consists in the restrictions placed on t. This difference can be seen geometrically if we refer to Figure 2.10. The line segment $\overline{PQ}$ joining $\bar{P}$ and $\bar{Q}$ consists of those points that lie on the straight line between the points $\bar{P}$ and $\bar{Q}$ (including the points $\bar{P}$ and $\bar{Q}$ themselves). The line through $\bar{P}$ and $\bar{Q}$ includes not only the points on the line segment $\overline{PQ}$, but, *in addition*, points above and to the left of $\bar{P}$ and points below and to the right of $\bar{Q}$. Only a partial sketch of this line can be shown in Figure 2.10, since it extends on indefinitely in both directions.

Problem 2.18

Given the two points $\bar{P} = [1,9]$ and $\bar{Q} = [7,3]$ in 2-space, determine the line segment $\overline{PQ}$ between them, and also the points on the segment for $t = 0, \frac{1}{3}, \frac{1}{2}, \frac{2}{3}$, and 1. Sketch the line segment and the points corresponding to the given values of t.

Solution

A typical point $\bar{x}$ on the segment is given by

$$\bar{x} = t\bar{P} + (1 - t)\bar{Q} = t[1,9] + (1 - t)[7,3].$$

For $t = 0$, $\bar{x} = \bar{Q} = [7,3]$.

For $t = 1$, $\bar{x} = \bar{P} = [1,9]$.

The remaining points are determined from the rules for vector addition as follows:

FIGURE 2.10 A line segment in two-dimensional space.

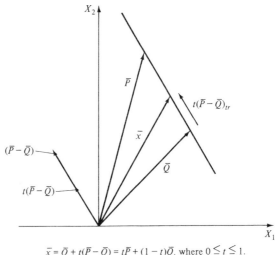

$$\bar{x} = \bar{Q} + t(\bar{P} - \bar{Q}) = t\bar{P} + (1 - t)\bar{Q}, \text{ where } 0 \leq t \leq 1.$$

For $\;t = \frac{1}{3},\; \bar{x} = \frac{1}{3}[1,9] + \frac{2}{3}[7,3] = [\frac{1}{3}\cdot1 + \frac{2}{3}\cdot7, \frac{1}{3}\cdot9 + \frac{2}{3}\cdot3] = [5,5].$

For $\;t = \frac{1}{2},\; \bar{x} = \frac{1}{2}[1,9] + \frac{1}{2}[7,3] = [\frac{1}{2} + \frac{7}{2}, \frac{9}{2} + \frac{3}{2}] = [4,6].$

For $\;t = \frac{2}{3},\; \bar{x} = \frac{2}{3}[1,9] + \frac{1}{3}[7,3] = [\frac{2}{3} + \frac{7}{3}, \frac{18}{3} + \frac{3}{3}] = [3,7].$

The line segment $\overline{PQ}$, highlighting the above five points, is shown in Figure 2.11. It should be particularly noted how the variable point $\bar{x}$ goes uniformly from $\bar{Q}$ to $\bar{P}$ as t goes from 0 to 1.

We now turn to the concept of a *convex set*.

Definition 2.15. A subset (or set) S of R^n is convex if the line segment joining any two points $\bar{P}$ and $\bar{Q}$ of S is contained entirely in S.

In our discussions, we shall use the words set and subset interchangeably.

By convention, we say that any set containing only one point or no points (the null set) is convex. In Figure 2.12, we present diagrams of three typical convex sets in 2-space (shaded regions). The polyhedral, convex sets shown (the first and the third) are typical of the constraint sets of linear programming problems. The three sets are convex because for each of them a line segment joining any two distinct points $\bar{P}$ and $\bar{Q}$ of the set is contained entirely in that set.

In Figure 2.13 several typical nonconvex sets are shown. In each set there are pairs of points for which the connecting line segment is not contained entirely within the set.

The next theorem ties in with the idea that the constraint set S_c for

FIGURE 2.11 Points on a line segment (Problem 2.18).

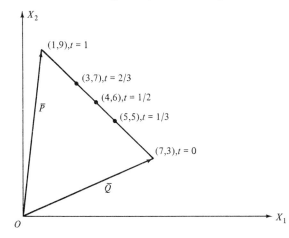

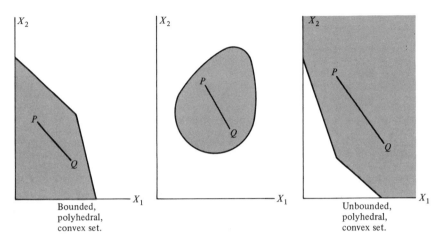

FIGURE 2.12 Typical convex sets in two-dimensional space.

any linear programming problem is a convex set. We shall prove this fact in the following pages.

Theorem 2.7. *The intersection of any number of convex sets is a convex set.*

For those readers who are weak in set theory, we mention that the intersection of three sets A_1, A_2, and A_3 is the set of all those points that simultaneously belong to A_1, A_2, and A_3. This set is denoted by $A_1 \cap A_2 \cap A_3$. In the following, we shall use the symbol $\subset$ to indicate set

FIGURE 2.13 Typical nonconvex sets.

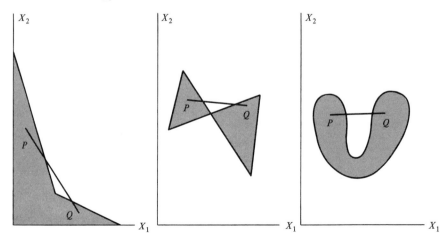

containment. Thus, $A \subset B$ means that the set A *is contained within* the set B, i.e., any point that is in the set A is also in the set B.

Theorem 2.7 is of particular interest to us when the number of convex sets is finite, and so we will prove it for this important special case. Figure 2.14 illustrates the situation for the three convex sets A_1, A_2, and A_3.

PROOF OF THEOREM 2.7 FOR THE CASE OF A FINITE NUMBER OF CONVEX SETS A_1, A_2, A_3, . . . , A_k. Let $\bar{P}$ and $\bar{Q}$ be two distinct points in the intersection set $A_1 \cap A_2 \cap A_3 \cap \cdots \cap A_k$. Thus, the points $\bar{P}$ and $\bar{Q}$ are simultaneously elements of A_1, A_2, A_3, . . . , and A_k. (In Figure 2.14 ($k = 3$), $\bar{P}$ and $\bar{Q}$ are simultaneously elements of A_1, A_2, and A_3.)

The line segment $\overline{PQ}$ is contained in A_1, since $\bar{P}$ and $\bar{Q}$ are both in A_1 and A_1 is convex. (Using symbols, we can write $\overline{PQ} \subset A_1$ to mean that $\overline{PQ}$ is contained in A_1.) Repeating the same argument with respect to all the A_i sets, we have the following set containment statements:

$\overline{PQ} \subset A_1$, since $\bar{P}$ and $\bar{Q}$ are both in A_1 and A_1 is convex. $\qquad$ (1)

$\overline{PQ} \subset A_2$, since $\bar{P}$ and $\bar{Q}$ are both in A_2 and A_2 is convex. $\qquad$ (2)

$\overline{PQ} \subset A_3$, since $\bar{P}$ and $\bar{Q}$ are both in A_3 and A_3 is convex. $\qquad$ (3)

. . . .

$\overline{PQ} \subset A_k$ since $\bar{P}$ and $\bar{Q}$ are both in A_k and A_k is convex. $\qquad$ (k)

Statements (1) through (k) imply that $\overline{PQ}$ is contained in the intersection set, i.e.,

$$\overline{PQ} \subset A_1 \cap A_2 \cap A_3 \cap \cdots \cap A_k.$$

However, this means that the intersection set $A_1 \cap A_2 \cap A_3 \cap \cdots \cap A_k$ is convex, since points $\bar{P}$ and $\bar{Q}$ are both contained in it. $\qquad \square$

FIGURE 2.14 Diagram for Theorem 2.7 ($k = 3$).

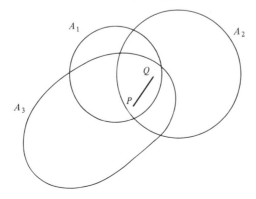

Note: In the proof of Theorem 2.7 we assumed that the intersection set contained more than one point. In the special case where the intersection contains only one point (or no points), we conclude that the intersection set is convex, as was mentioned in the comments following Definition 2.15.

We now turn our attention to a very special type of function—the *linear function*.

Definition 2.16. A function f defined over n-dimensional space is *linear* if it has the following form:

$$f(\bar{x}) = c_1 x_1 + c_2 x_2 + \cdots + c_n x_n. \tag{1}$$

In Equation (1), the terms $c_1, c_2, \ldots, c_n$ are all constants, while $\bar{x}$ is the column vector given by

$$\bar{x} = [x_1, x_2, \ldots, x_n].$$

If we introduce the vector $\bar{c}$ given by $\bar{c} = [c_1, c_2, \ldots, c_n]$, then we obtain the following alternate forms for Equation (1):

$$f(\bar{x}) = \bar{c} \cdot \bar{x} \quad \text{(dot product form)} \tag{2}$$

and

$$f(\bar{x}) = \bar{c}^t \bar{x} \quad \text{(vector times vector form)}. \tag{3}$$

The vector $\bar{c}$ is called a constant vector because all of its components are constants.

By using the dot product form of Equation (2), it is fairly easy to establish the so-called *linearity property*.

Theorem 2.8: Linearity Property. *The linear function f satisfies the following equation:*

$$f(k_1 \bar{x} + k_2 \bar{y}) = k_1 f(\bar{x}) + k_2 f(\bar{y}),$$

where k_1 and k_2 are scalar constants.

PROOF OF THEOREM 2.8. We use the dot product form of the linear function and thus write

$$f(\bar{x}) = \bar{c} \cdot \bar{x}, \tag{a}$$

where $\bar{c}$ is a constant vector. It then follows that

$$f(k_1 \bar{x} + k_2 \bar{y}) = \bar{c} \cdot (k_1 \bar{x} + k_2 \bar{y}). \tag{b}$$

We now make use of the properties of the dot product as stated prior to the proof of Theorem 2.6:

$$\bar{c} \cdot (k_1 \bar{x} + k_2 \bar{y}) = k_1 \bar{c} \cdot \bar{x} + k_2 \bar{c} \cdot \bar{y}. \tag{c}$$

Combining Equations (b) and (c) while at the same time making use of Equation (a) leads us to

$$f(k_1\bar{x} + k_2\bar{y}) = k_1 f(\bar{x}) + k_2 f(\bar{y}). \qquad (d)$$

This is what we wished to prove.   □

Linear functions occur extensively throughout linear programming problems. For example, the objective function u in the standard maximum or minimum problem is a linear function (see Problems 1.5 and 1.7). The left-hand sides of the various inequalities of the standard linear programming problems are also linear functions.

The following theorem also has a bearing on linear programming problems.

Theorem 2.9. *Given a linear function f defined over R^n (i.e., over n-dimensional Euclidean space), suppose that $\overline{PQ}$ is the line segment in R^n connecting distinct points $\bar{P}$ and $\bar{Q}$ and that $f(\bar{P}) < f(\bar{Q})$. Then for any point $\bar{R}$ that lies on the line segment strictly between $\bar{P}$ and $\bar{Q}$ we have $f(\bar{P}) < f(\bar{R}) < f(\bar{Q})$.*

Theorem 2.9′. *Given the same conditions as in Theorem 2.9, except that $f(\bar{P}) = f(\bar{Q})$, we have $f(\bar{R}) = f(\bar{P}) = f(\bar{Q})$.*

PROOFS OF THEOREMS 2.9 AND 2.9′. Since $\bar{R}$ lies on $\overline{PQ}$ strictly between $\bar{P}$ and $\bar{Q}$, we have that

$$\bar{R} = t\bar{P} + (1 - t)\bar{Q}, \qquad (1)$$

where t is a positive number between 0 and 1 (i.e., $0 < t < 1$). Note that $(1 - t)$ is also a positive number between 0 and 1.

Using the linearity property cited in Theorem 2.8, we have

$$f(\bar{R}) = t f(\bar{P}) + (1 - t) f(\bar{Q}). \qquad (2)$$

We can now write $f(\bar{P})$ in the form

$$f(\bar{P}) = f(\bar{P}) \cdot [1] = f(\bar{P}) \cdot [t + (1 - t)] = t f(\bar{P}) + (1 - t) f(\bar{P}). \qquad (3)$$

a. We now assume that the hypothesis of Theorem 2.9 holds, i.e., $f(\bar{P}) < f(\bar{Q})$. If we multiply both sides of this inequality by the positive number $(1 - t)$, we obtain

$$(1 - t) f(\bar{P}) < (1 - t) f(\bar{Q}). \qquad (4)$$

Adding $t f(\bar{P})$ to both sides of (4) gives

$$t f(\bar{P}) + (1 - t) f(\bar{P}) < t f(\bar{P}) + (1 - t) f(\bar{Q}). \qquad (5)$$

Now substituting (2) and (3) into (5), we obtain

$$f(\bar{P}) < f(\bar{R}). \qquad (6)$$

Analogously to Equation (3), we can write $f(\bar{Q})$ as

$$f(\bar{Q}) = tf(\bar{Q}) + (1 - t)f(\bar{Q}). \tag{7}$$

If we multiply both sides of $f(\bar{P}) < f(\bar{Q})$ by the positive number t, we obtain

$$tf(\bar{P}) < tf(\bar{Q}). \tag{8}$$

Adding $(1 - t)f(\bar{Q})$ to both sides of (8) yields

$$tf(\bar{P}) + (1 - t)f(\bar{Q}) < tf(\bar{Q}) + (1 - t)f(\bar{Q}). \tag{9}$$

The substitution of (2) and (7) into (9) results in

$$f(\bar{R}) < f(\bar{Q}). \tag{10}$$

Combining (6) and (10) leads to the conclusion of Theorem 2.9.

b. To prove Theorem 2.9', we merely substitute $f(\bar{Q}) = f(\bar{P})$ into Equation (2). Thus,

$$f(\bar{R}) = tf(\bar{P}) + (1 - t)f(\bar{Q}) = tf(\bar{P}) + (1 - t)f(\bar{P}) \tag{11}$$

$$= f(\bar{P})[t + (1 - t)] = f(\bar{P}) \cdot 1 = f(\bar{P}). \qquad \square$$

We now turn our attention to the geometric interpretation of linear equations and inequalities. We already know that the set of points in R^n which satisfy

$$a_1 x_1 + a_2 x_2 + \cdots + a_n x_n = b \tag{1}$$

is called a hyperplane (see Definition 2.12). Note that the left-hand side of (1) has the form of a linear function. We now consider the following strict inequalities:

$$a_1 x_1 + a_2 x_2 + \cdots + a_n x_n < b \tag{2}$$

and

$$a_1 x_1 + a_2 x_2 + \cdots + a_n x_n > b. \tag{3}$$

The set of points for which (2) or (3) is satisfied is called an *open half-space*. We say that half-spaces (2) and (3) are on opposite sides of hyperplane (1). This will be illustrated in Problem 2.19 for the two-dimensional case. We also say that Equation (1) *partitions* R^n into the three subsets consisting of the points that satisfy (1), (2), and (3), respectively.

The set of points which satisfy

$$a_1 x_1 + a_2 x_2 + \cdots + a_n x_n \leq b \quad (\text{or} \geq b) \tag{4}$$

is called a *closed half-space*; it consists of the points of hyperplane (1) and also those of open half-space (2) or (3), respectively. In this case, the hyperplane is called a *bounding* hyperplane of the closed half-space.

Problem 2.19

Discuss the various open and closed half-spaces determined by the hyperplane $2x_1 + 3x_2 = 24$ in two-dimensional space.

Solution

Here the underlying space is R^2, i.e., the Euclidean plane. The hyperplane is thus the straight line with x_1 intercept 12 and x_2 intercept 8. Referring to Figure 2.15, we note that the hyperplane partitions R^2 into:

$$2x_1 + 3x_2 = 24 \quad \text{(hyperplane)}. \tag{1}$$

$$2x_1 + 3x_2 < 24 \quad \text{(``below'' open half-space)}. \tag{2}$$

$$2x_1 + 3x_2 > 24 \quad \text{(``above'' open half-space)}. \tag{3}$$

To show that (2) lies below and to the left of the hyperplane, we choose an arbitrary point not on the hyperplane, e.g., (0,0), and test inequality (2):

$$2 \cdot 0 + 3 \cdot 0 = 0 < 24.$$

Since (0,0) does satisfy (2), we know that the points on the (0,0) side of (1), i.e., the points below and to the left, comprise open half-space (2).

The closed half-space $2x_1 + 3x_2 \leq 24$ consists of the points of hyperplane (1) and those of open half-space (2), i.e., the points on and below the hyperplane. The closed half-space $2x_1 + 3x_2 \geq 24$ consists of the points on and above the hyperplane.

We note that the hyperplane and various half-spaces of Problem 2.19 are all convex sets. This is true in general, as will be stated by the following theorem.

FIGURE 2.15 A hyperplane and half-spaces in the plane.

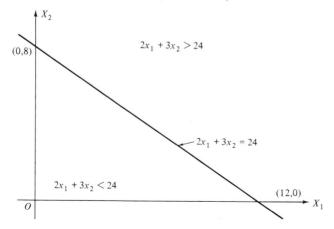

Theorem 2.10. *Hyperplanes, open half-spaces, and closed half-spaces are all convex sets.*

PROOF OF THEOREM 2.10 FOR THE CASES OF (a) A HYPERPLANE, AND (b) A CLOSED HALF-SPACE. a. The general equation of a hyperplane in R^n is

$$a_1x_1 + a_2x_2 + \cdots + a_nx_n = b. \tag{1}$$

The left-hand side of (1) has the form of a linear function of the n-component vector $\bar{x}$; we shall rename it $f(\bar{x})$. Thus, Equation (1) becomes

$$f(\bar{x}) = b. \tag{2}$$

Let $\bar{P}$ and $\bar{Q}$ be two distinct points (vectors) of the hyperplane; thus, they satisfy Equation (2):

$$f(\bar{P}) = b \tag{3a}$$

and

$$f(\bar{Q}) = b. \tag{3b}$$

Now let $\bar{R}$ be any in-between point on the line segment connecting $\bar{P}$ and $\bar{Q}$:

$$\bar{R} = t\bar{P} + (1 - t)\bar{Q} \quad (0 < t < 1). \tag{4}$$

Thus,

$$f(\bar{R}) = tf(\bar{P}) + (1 - t)f(\bar{Q}), \tag{5}$$

since f is a linear function.

Substituting Equations (3a) and (3b) in (5) results in

$$f(\bar{R}) = b[t + (1 - t)] = b. \tag{6}$$

Thus, $\bar{R}$ is also on the hyperplane, and hence the hyperplane is convex. It is important to notice the argument here. Since $\bar{R}$ is *any* arbitrary point on the line segment, it follows that the above demonstration holds for *all* points on the line segment; thus, the whole line segment is contained in the hyperplane, fulfilling the requirement of convexity.

b. Proceeding as in part (a), we may represent the closed half-space by

$$f(\bar{x}) \leq b, \tag{1}$$

where f is a linear function. (The method of proof is unchanged if we start with $f(\bar{x}) \geq b$.)

Let $\bar{P}$ and $\bar{Q}$ be two points in the closed half-space. Thus, we have

$$f(\bar{P}) \leq b \tag{2a}$$

and

$$f(\bar{Q}) \leq b. \tag{2b}$$

Let $\bar{R}$ be any in-between point on the line segment $\overline{PQ}$:

$$\bar{R} = t\bar{P} + (1 - t)\bar{Q} \quad (0 < t < 1). \tag{3}$$

Thus,

$$f(\bar{R}) = t f(\bar{P}) + (1 - t) f(\bar{Q}). \tag{4}$$

Now we will multiply $(2a)$ by the positive number t and $(2b)$ by the positive number $(1 - t)$. The senses of both inequalities are thus maintained, and we have

$$t f(\bar{P}) \le t b \tag{5}$$

and

$$(1 - t) f(\bar{Q}) \le (1 - t) b. \tag{6}$$

Adding (5) and (6) together results in

$$t f(\bar{P}) + (1 - t) f(\bar{Q}) \le b[t + (1 - t)] = b. \tag{7}$$

If we recall Equation (4), we then have

$$f(\bar{R}) \le b. \tag{8}$$

Thus, $\bar{R}$ is also a point in the closed half-space, and hence the closed half-space is convex. □

The proof that an open half-space is convex is similar to the proof of part (b).

We shall now prove that the constraint set for a standard linear programming problem is a convex set. The proof will be for a maximum problem; an almost identical argument holds for a minimum problem.

Theorem 2.11. *The constraint set for a standard linear programming problem of maximum type is convex.*

PROOF OF THEOREM 2.11. The form of the standard maximum problem was given in Problem 1.5. There the constraints were

$$a_{11}x_1 + a_{12}x_2 + \cdots + a_{1n}x_n \le b_1 \tag{2.1}$$

$$a_{21}x_1 + a_{22}x_2 + \cdots + a_{2n}x_n \le b_2 \tag{2.2}$$

$$\cdots\cdots\cdots\cdots\cdots\cdots\cdots\cdots$$

$$\cdots\cdots\cdots\cdots\cdots\cdots\cdots\cdots$$

$$a_{m1}x_1 + a_{m2}x_2 + \cdots + a_{mn}x_n \le b_m \tag{2.m}$$

$$x_1 \ge 0, \tag{3.1}$$

$$x_2 \ge 0 \tag{3.2}$$

$$\cdots\cdots\cdots$$

$$\cdots\cdots\cdots$$

$$x_n \ge 0. \tag{3.n}$$

The sets corresponding to each constraint of (2.1) through (2.m) and (3.1) through (3.n) are *closed half-spaces* in n-dimensional Euclidean space. Let us label these sets as

$$S_{2.1}, S_{2.2}, \ldots, S_{2.m} \quad \text{and} \quad S_{3.1}, S_{3.2}, \ldots, S_{3.n}.$$

Thus, according to Theorem 2.10 all of these sets are *convex*. We recall that the constraint set S_c is the set of points which satisfy each and every one of the $(m + n)$ inequalities above; i.e., S_c is the intersection set of the sets corresponding to the inequalities:

$$S_c = S_{2.1} \cap S_{2.2} \cap \cdots \cap S_{2.m} \cap S_{3.1} \cap S_{3.2} \cap \cdots \cap S_{3.n}.$$

Thus, S_c is the intersection set of convex sets, and by Theorem 2.7 it follows that S_c itself is convex. □

In Chapter 1 we used the phrase *polyhedral, convex set* without precisely defining it (see Theorems 1.1 and 1.2 and Notes (d) and (e) after Theorem 1.1). It is now appropriate to give the definition.

Definition 2.17. The intersection set of a finite number of closed half-spaces in R^n is called a *polyhedral, convex set*.

The use of the word convex is justified because of Theorems 2.10 and 2.7. Also, we see that it is valid to call the constraint set S_c of a standard linear programming problem a polyhedral, convex set because S_c satisfies the condition of Definition 2.17.

Another term which appears in Theorems 1.1 and 1.2 is *bounded set*. A conventional way of defining this is in terms of a set being contained within a sphere of finite radius. However, obviously we must first give a general definition of a sphere in n-dimensional Euclidean space. In the definition which follows, the center of the sphere is at the origin.

Definition 2.18. A closed sphere of finite radius r consists of all those points (vectors) $\bar{x}$ that satisfy the inequality

$$\|\bar{x}\| \le r.$$

We denote this set by the symbol $C.S.\ (r)$.

In the inequality appearing in Definition 2.18, $\|\bar{x}\|$ denotes the distance from the origin to the point $\bar{x}$, i.e.,

$$\|\bar{x}\| = \sqrt{x_1^2 + x_2^2 + \cdots + x_n^2}.$$

In 3-space, the closed sphere is none other than the set of points within and on the surface of a usual sphere of radius r whose center is at the origin. In 2-space, the closed sphere is the set of points within and on the boundary of a *circle* of radius r whose center is at the origin. Here another correct descriptive phrase would be *closed circle,* since the closed sphere in 2-space is, in fact, a closed circle.

In general, a "closed" set is a set consisting of the points that are on the *boundary* as well as in the interior of the set. A closed set thus contains its boundary points.

Definition 2.19. A subset S of R^n is bounded if it is contained within a closed sphere of finite radius (containment means each point within the set S is also a point within the closed sphere that contains S).

An equivalent definition is as follows.

Definition 2.19'. A subset S of R^n is bounded if there is a real positive number r such that $\|\bar{x}\| \leq r$ for every point $\bar{x}$ in the set S.

In Figure 2.16 we present a typical bounded set S in two-dimensional space. It is shown contained within a closed sphere (circle) of radius r.

Now that a bounded set has been defined, it is quite simple to define what *unbounded set* means: An unbounded set is simply a set that is not bounded.

In certain situations in linear programming, it is important to know whether or not a set is bounded. For example, it can be shown that the

FIGURE 2.16 A bounded set in two dimensional space. $\|\bar{x}\| < r$ for all points in the set (S) shown.

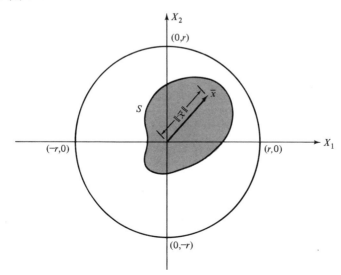

constraint set for a standard maximum problem (see Problem 1.5) is bounded if *all* the a_{ij} terms and the b_i term of one of the main constraints are positive numbers.

5. LINEAR PROGRAMMING PROBLEMS IN MATRIX
AND VECTOR NOTATION

Let us refer to the standard maximum linear programming problem. The vectors will be of the column variety, for which we shall use the horizontal, square bracket notation discussion after Definition 2.2. Thus, referring to Problem 1.5, we have the n-component column vectors $\bar{x}$ and $\bar{c}$ and the m-component column vector $\bar{b}$:

$$\bar{x} = \begin{pmatrix} x_1 \\ x_2 \\ \vdots \\ x_n \end{pmatrix} = [x_1, x_2, \ldots, x_n]$$

$$\bar{c} = [c_1, c_2, \ldots, c_n]$$

and

$$\bar{b} = [b_1, b_2, \ldots, b_m].$$

It is useful at this point to review some vector symbolism that is frequently used. The vector inequality $\bar{d} \geq \bar{e}$ for two n-component vectors means that

$$d_1 \geq e_1, d_2 \geq e_2, \ldots, \text{ and } d_n \geq e_n.$$

If we have $\bar{d} \geq \bar{0}$, where $\bar{0}$ denotes the n-component vector all of whose components are equal to 0 (i.e., the zero vector), that means

$$d_1 \geq 0, d_2 \geq 0, \ldots, \quad \text{and} \quad d_n \geq 0.$$

The symbol A is used to denote the m by n matrix whose elements are the a_{ij} terms of the standard linear programming problem:

$$A = \begin{pmatrix} a_{11} & a_{12} & \ldots\ldots & a_{1n} \\ a_{21} & a_{22} & \ldots\ldots & a_{2n} \\ & \ldots\ldots\ldots\ldots \\ & \ldots\ldots\ldots\ldots \\ a_{m1} & a_{m2} & \ldots\ldots & a_{mn} \end{pmatrix}.$$

The use of the symbol $A_{(i)}$ will also be convenient; it denotes the ith row of matrix A, i.e., $A_{(i)}$ is an n-component row vector:

$$A_{(i)} = (a_{i1}, a_{i2}, \ldots, a_{in}).$$

Another row vector that will appear is the transpose of the vector $\bar{c}$, namely, $\bar{c}^t$:

$$\bar{c}^t = (c_1, c_2, \ldots, c_n).$$

Now referring to the standard maximum linear programming problem (Problem 1.5), we are able to represent it in matrix-vector notation as follows:

$$\text{Maximize} \quad u = \bar{c}^t \bar{x} \tag{1}$$

$$\text{subject to} \quad A\bar{x} \le \bar{b} \tag{2}$$

and

$$\bar{x} \ge \bar{0}. \tag{3}$$

Let us closely examine the separate terms that appear here. The first term could be written as $u = \bar{c} \cdot \bar{x}$ if we preferred dot product notation. Instead, we have vector multiplication of an n-component row vector (or 1 by n matrix) by an n-component column vector (or n by 1 matrix). Inequality (3) is equivalent to Inequalities (3.1) through (3.n) of Problem 1.5.

Finally, in vector inequality (2), the term $A\bar{x}$ indicates that we should multiply the m by n matrix A by the n by 1 matrix $\bar{x}$ to yield the m by 1 matrix $A\bar{x}$. The latter is equivalent to an m-component column vector. To determine the components of $A\bar{x}$, all we have to do is follow the rules of matrix multiplication (e.g., see Problem 2.7). Thus, if we examine the individual components of vector inequality (2), we see that they are identical to Inequalities (2.1) through (2.m) of Problem 1.5. We can also represent the m components of $A\bar{x}$ by means of multiplication of the row vectors $A_{(1)}, A_{(2)}, \ldots, A_{(m)}$, respectively, by the column vector $\bar{x}$. Thus, we end up with the following equivalent way of representing (2.1) through (2.m):

$$A_{(1)}\bar{x} \le b_1 \tag{2.1}$$

$$A_{(2)}\bar{x} \le b_2 \tag{2.2}$$

$$\cdots\cdots\cdots\cdots$$

$$\cdots\cdots\cdots\cdots$$

$$A_{(m)}\bar{x} \le b_m. \tag{2.m}$$

For example, $A_{(1)}\bar{x}$—the first component of $A\bar{x}$—is given by

$$A_{(1)}\bar{x} = (a_{11}, a_{12}, \ldots, a_{1n}) \begin{pmatrix} x_1 \\ x_2 \\ \vdots \\ x_n \end{pmatrix} = a_{11}x_1 + a_{12}x_2 + \cdots + a_{1n}x_n.$$

Another notational convention we shall occasionally use is to write $u(\bar{x})$ in place of u if u happens to be a function of the vector $\bar{x}$. Thus, for example, if u is the symbol for the objective function of a typical linear programming problem (Definition 1.1), we shall sometimes write

$$u(\bar{x}) = c_1 x_1 + c_2 x_2 + \cdots + c_n x_n$$

or

$$u(\bar{x}) = \bar{c}^t \bar{x}.$$

We have already employed this symbolism in our discussion of general linear functions (Definition 2.16).

We are now in a position to define what is meant by the terms *solution* and *value* for a linear programming problem. Tentatively, let us confine our attention to the standard maximum problem.

Definition 2.20. Suppose that there exists a vector $\bar{x}^*$ that satisfies Constraints (2) and (3) and is such that $u(\bar{x}^*) \geq u(\bar{x})$ for all other vectors $\bar{x}$ that satisfy (2) and (3). Then $\bar{x}^*$ is called a *solution* (vector) of the problem, while $u(\bar{x}^*)$ is called the maximum value of $u(\bar{x})$. We sometimes write Max $u(\bar{x}) = u(\bar{x}^*)$.

Let us now develop the matrix-vector form for the standard minimum linear programming problem (see Problem 1.7). We obtain the following:

$$\text{Minimize} \quad u = \bar{c}^t \bar{x} \tag{1}$$

$$\text{subject to} \quad A\bar{x} \geq \bar{b} \tag{2}$$

and

$$\bar{x} \geq \bar{0}. \tag{3}$$

Vector inequality (2) is equivalent to the m main constraints (2.1) through (2.m) of Problem 1.7, while vector inequality (3) is equivalent to the n nonnegativity constraints (3.1) through (3.n) of that problem.

Definition 2.21. Suppose that there exists a vector $\bar{x}^*$ that satisfies Constraints (2) and (3) and is such that $u(\bar{x}^*) \leq u(\bar{x})$ for all other vectors $\bar{x}$ that satisfy (2) and (3). Then $\bar{x}^*$ is called a *solution* (vector) of the problem, while $u(\bar{x}^*)$ is called the *minimum value* of $u(\bar{x})$. We sometimes write Min $u(\bar{x}) = u(\bar{x}^*)$.

In general discussions we sometimes omit the words maximum and minimum and instead use *optimum* or *optimal*. In the same way, we replace maximize and minimize by *optimize*.

Thus, we often refer to $\bar{x}^*$ as an *optimum (optimal) solution* and to $u(\bar{x}^*)$ as the *optimum value* for the objective function.

We close this section by examining a previously worked-out problem in order to illustrate the above terminology and definitions.

Problem 2.20

Illustrate the use of the above symbolism with respect to Problem 1.4.

Solution

This problem is in the form of a standard maximum problem. First let us replace the variables x, y, and z by x_1, x_2, and x_3, respectively. (Note that $m = 2$ and $n = 3$.) For the column vectors we have

$$\bar{x} = [x_1, x_2, x_3]$$

$$\bar{c} = [c_1, c_2, c_3] = [\tfrac{1}{4}, \tfrac{2}{5}, \tfrac{1}{2}]$$

and

$$\bar{b} = [900, 600].$$

The matrix A is given by

$$A = \begin{pmatrix} 1 & \tfrac{2}{3} & \tfrac{1}{4} \\ 0 & \tfrac{1}{3} & \tfrac{3}{4} \end{pmatrix}.$$

The row vectors $A_{(1)}$ and $A_{(2)}$ are

$$A_{(1)} = (1, \tfrac{2}{3}, \tfrac{1}{4})$$

and

$$A_{(2)} = (0, \tfrac{1}{3}, \tfrac{3}{4}).$$

The solution to the problem is achieved at the point $(0,1260,240)$ of Figure 1.8 (labeled point A there). In the new terminology we have

$$\bar{x}^* = [0,1260,240]$$

and

$$u(\bar{x}^*) = \text{Max } u(\bar{x}) = 624.$$

We observe that $\bar{x}^* = [0,1260,240]$ is equivalent to writing

$$x_1^* = 0, \quad x_2^* = 1260, \quad \text{and} \quad x_3^* = 240$$

if we wish to give the solution in terms of components.

In words, we can say that $\bar{x}^* = [0,1260,240]$ is a solution (optimum solution) of the problem. Furthermore, $u(\bar{x}^*)$ is called the maximum value (optimum value) of the objective function $u(\bar{x})$.

REFERENCES

1. Hadley, G. *Linear Algebra*. Reading, Massachusetts: Addison-Wesley, 1961.

SUPPLEMENTARY PROBLEMS

Matrices and Their Properties

For Problems S.P. 2.1 to 2.4, let

$$A = \begin{pmatrix} 3 & -4 & 2 \\ 1 & 0 & 6 \end{pmatrix}, \qquad B = \begin{pmatrix} 5 & 3 & -1 \\ -4 & 2 & 3 \end{pmatrix},$$

$$C = \begin{pmatrix} 4 & 2 & 1 \\ -3 & 2 & 5 \\ 2 & 0 & 3 \end{pmatrix}, \qquad D = \begin{pmatrix} 1 \\ 3 \\ -2 \end{pmatrix} = [1, 3, -2].$$

S.P. 2.1:   Find: (a) $A + B$, (b) $-A$, (c) $A - 2B$, and (d) $A + C$.

S.P. 2.2:   Find: (a) AC, (b) CA, (c) AB, (d) BD, and (e) CD.

S.P. 2.3:   Find: (a) A^t, (b) C^t, (c) $A^t B$, and (d) D^t.

S.P. 2.4:   (a) Verify that $(AC)^t = C^t A^t$. (b) Verify that $CI = C$, where I is the appropriate identity matrix. (c) Verify that $(7A)^t = 7A^t$.

Vectors and Their Properties

S.P. 2.5:   Given that $\bar{a} = [6, 2]$, $\bar{b} = [-3, 2]$, and $\bar{c} = [4, -3]$, compute the following and illustrate them by means of a diagram: (a) $\bar{a} + \bar{b}$; (b) $\bar{a} + \bar{c}$; (c) $\bar{a} - \bar{b}$; (d) $2\bar{a} + 3\bar{b}$; (e) $\bar{c} - \bar{a}$; (f) $\bar{a} - \bar{c}$; (g) $-2\bar{b}$; (h) $\bar{a} \cdot \bar{b}$; (i) $\bar{b} \cdot \bar{a}$; (j) $\bar{a} \cdot (\bar{b} + \bar{c})$.

S.P. 2.6:   Compute the following: (a) $[3, 5, 7, -6] + [-7, 2, 4, 3]$; (b) $[3, 5, 7, -6] + 3[-7, 2, 4, 3]$; (c) $[7, 2, -4, 6] + [8, -4, 5]$; (d) $3[2, 4, -3] + 2[6, -1, -5] - 4[3, 8, 2]$; (e) $[-2, 4, -3] \cdot [1, 5, 7]$.

S.P. 2.7:   Given that $\bar{a} = [4, 3, -2]$ and $\bar{b} = [-2, 1, 4]$, compute the following: (a) $\|a\|$; (b) $\|\bar{b}\|$; (c) $\bar{a} \cdot \bar{b}$; (d) $\cos \theta$; (e) θ.

S.P. 2.8:   Determine which of the following vectors are perpendicular: $\bar{a} = [5, -2, 3]$; $\bar{b} = [2, 1, -4]$; $\bar{c} = [1, 4, 1]$; $\bar{d} = [3, 4.5, -2]$.

Elementary Convex Geometry

S.P. 2.9:   (a) Find the equation of the hyperplane in 2-space for which $\bar{c} = [4, -3]$ and $b = 15$. (b) Find the equation of the hyperplane in 3-space for which $\bar{c} = [-2, 5, 3]$ and $b = -6$.

S.P. 2.10:   Given the two points $\bar{P} = [-2, -3]$ and $\bar{Q} = [5, 1]$ in 2-space, determine the line segment $\overline{PQ}$ between them; determine the points on the segment for $t = 0, \frac{1}{4}, \frac{1}{2}, \frac{3}{4}$, and 1; sketch the line segment and the five points.

S.P. 2.11.   (a) Identify the various open and closed half-spaces determined by the hyperplane $x_1 + 2x_2 = 10$ in 2-space and draw a sketch. (b) Identify the various closed half-spaces determined by the hyperplane $x_1 + 3x_2 + 2x_3 = 12$ in 3-space.

S.P. 2.12:   Prove that an open half-space is a convex set.

Linear Programming Problems in Matrix and Vector Notation

S.P. 2.13: Illustrate the symbolism of Chapter 2, Section 5 with respect to Problem 1.1.

S.P. 2.14: Illustrate the symbolism of Chapter 2, Section 5 with respect to Problem 1.3.

ANSWERS TO SUPPLEMENTARY PROBLEMS

S.P. 2.1: (a) $\begin{pmatrix} 8 & -1 & 1 \\ -3 & 2 & 9 \end{pmatrix}$ (b) $\begin{pmatrix} -3 & 4 & -2 \\ -1 & 0 & -6 \end{pmatrix}$

(c) $\begin{pmatrix} -7 & -10 & 4 \\ 9 & -4 & 0 \end{pmatrix}$ (d) not defined.

S.P. 2.2: (a) $\begin{pmatrix} 28 & -2 & -11 \\ 16 & 2 & 19 \end{pmatrix}$ (b) not defined

(c) Not defined   (d) $\begin{pmatrix} 16 \\ -4 \end{pmatrix} = [16, -4]$

(e) $[8, -7, -4]$.

S.P. 2.3: (a) $\begin{pmatrix} 3 & 1 \\ -4 & 0 \\ 2 & 6 \end{pmatrix}$ (b) $\begin{pmatrix} 4 & -3 & 2 \\ 2 & 2 & 0 \\ 1 & 5 & 3 \end{pmatrix}$

(c) $\begin{pmatrix} 11 & 11 & 0 \\ -20 & -12 & 4 \\ -14 & 18 & 16 \end{pmatrix}$ (d) $(1, 3, -2)$.

S.P. 2.4: (a) $(AC)^t = C^tA^t = \begin{pmatrix} 28 & 16 \\ -2 & 2 \\ -11 & 19 \end{pmatrix}$

(c) $(7A)^t = 7A^t = \begin{pmatrix} 21 & 7 \\ -28 & 0 \\ 14 & 42 \end{pmatrix}$.

S.P. 2.5: (a) $[3, 4]$; (b) $[10, -1]$; (c) $[9, 0]$; (d) $[3, 10]$; (e) $[-2, -5]$; (f) $[2, 5]$; (g) $[6, -4]$; (h) -14; (i) -14; (j) 4.

S.P. 2.6: (a) $[-4, 7, 11, -3]$; (b) $[-18, 11, 19, 3]$; (c) not defined; (d) $[6, -22, -27]$; (e) -3.

S.P. 2.7: (a) $\sqrt{29} = 5.385$; (b) $\sqrt{21} = 4.583$; (c) 3; (d) 0.1216; (e) 83.02 deg = 1.449 rad.

S.P. 2.8: $\bar{a}$ and $\bar{c}$; $\bar{a}$ and $\bar{d}$.

S.P. 2.9: (a) $4x_1 - 3x_2 = 15$; (b) $-2x_1 + 5x_2 + 3x_3 = -6$.

S.P. 2.10: The set of points $\bar{x}$ for which $\bar{x} = t[-2,-3] + (1 - t)[5,1]$, where $0 \le t \le 1$. For $t = 0$, $\bar{x} = [5,1]$; for $t = \frac{1}{4}$, $\bar{x} = [\frac{13}{4},0]$; for $t = \frac{1}{2}$, $\bar{x} = [\frac{3}{2}, -1]$; for $t = \frac{3}{4}$, $\bar{x} = [-\frac{1}{4}, -2]$; for $t = 1$, $\bar{x} = [-2,-3]$.

S.P. 2.11: (a) Open half-spaces: $x_1 + 2x_2 < 10$ ("below"), $x_1 + 2x_2 > 10$

("above"); closed half-spaces: $x_1 + 2x_2 \leq 10$ ("below and on"), $x_1 + 2x_2 \geq 10$ ("above and on"). (b) Closed half-spaces: $x_1 + 3x_2 + 2x_3 \leq 12$ and $x_1 + 3x_2 + 2x_3 \geq 12$.

S.P. 2.12: *Sketch of proof.* Let $f(\bar{x}) < b$ denote the open half-space. For $\bar{R} = t\bar{P} + (1 - t)\bar{Q}$, where $\bar{P}$ and $\bar{Q}$ are in the open half-space, show that $f(\bar{R}) < b$ for $0 < t < 1$.

S.P. 2.13: $\bar{c} = [3,4]$, $\bar{b} = [12,12]$, $A_{(1)} = (2,1)$, $A_{(2)} = (1,2)$, $\bar{x}^* = [4,4]$, and $u(\bar{x}^*) = \text{Max } u(\bar{x}) = 28$.

S.P. 2.14: $\bar{c} = [200,300]$, $\bar{b} = [60,120,150]$, $A_{(1)} = (1,2)$, $A_{(2)} = (4,2)$, $A_{(3)} = (6,2)$, $\bar{x}^* = [20,20]$, and $u(\bar{x}^*) = \text{Min } u(\bar{x}) = 10,000$.

Chapter Three

Convex Geometry

1. INTRODUCTORY REMARKS

In this chapter we shall derive some of the key results of linear programming theory by making use of the techniques and symbolism presented in Chapter 2. The work will be oriented toward geometrical concepts and the manipulation of vectors.

The key principles of linear programming are presented from an algebraic and calculation-oriented point of view beginning in Chapter 4. Also, the comments at the beginning of Chapter 2 are relevant here.

2. CONVEX COMBINATIONS

Suppose that we are working in n-dimensional Euclidean space (denoted as R^n). Let us tentatively think of vectors in R^n as denoting points. We will now define the very useful concept of *convex combination* of vectors.

Definition 3.1: Convex Combination. Let $\bar{a}^1$, $\bar{a}^2$, . . . , $\bar{a}^k$ be k vectors in R^n. If t_1, t_2, . . . , t_k are nonnegative numbers such that

$$t_1 + t_2 + \cdots + t_k = 1,$$

then the linear combination

$$\bar{q} = t_1\bar{a}^1 + t_2\bar{a}^2 + \cdots + t_k\bar{a}^k$$

is called a convex combination of $\bar{a}^1$, $\bar{a}^2$, . . . , $\bar{a}^k$.

Notes:
a. The symbols $\bar{a}^1$, $\bar{a}^2$, . . . , $\bar{a}^k$ denote vectors and should not be confused with the components of a vector. To differentiate from components, note that here the numbers appear as superscripts, not

subscripts. (In some treatises $\bar{a}^2$ is a shorthand symbol for the dot product $\bar{a} \cdot \bar{a}$. This is not the case here.)

b. In practical situations, the number k is at least as great as n—the dimension of the space.

c. From the requirements on the t_i, it follows that each t_i cannot be less than zero or greater than one.

Problem 3.1

Consider the three vectors in two-dimensional space given by

$$\bar{a}^1 = [1,0], \quad \bar{a}^2 = [2,1], \quad \text{and} \quad \bar{a}^3 = [3,0].$$

Determine the convex combination for which $t_1 = t_2 = t_3 = \frac{1}{3}$. Interpret it geometrically.

Solution

Let the relevant convex combination be denoted by $\bar{q}$. We note that the t's add up to one and that each is nonnegative, as required. Thus, $\bar{q}$ is computed as follows:

$$\bar{q} = \tfrac{1}{3}[1,0] + \tfrac{1}{3}[2,1] + \tfrac{1}{3}[3,0] \tag{1}$$

$$\bar{q} = [\tfrac{1}{3} + \tfrac{2}{3} + \tfrac{3}{3}, 0 + \tfrac{1}{3} + 0] = [2,\tfrac{1}{3}]. \tag{2}$$

The vectors $\bar{a}^1$, $\bar{a}^2$, $\bar{a}^3$, and $\bar{q}$ are plotted in Figure 3.1.

It can be shown that any point (including those on the boundary lines) in the triangle drawn in Figure 3.1 can be expressed as a convex combination of the three vertex vectors $\bar{a}^1$, $\bar{a}^2$, and $\bar{a}^3$. For example, any point on the line segment connecting $\bar{a}^1$ and $\bar{a}^3$ can be expressed as a convex combination of just $\bar{a}^1$ and $\bar{a}^3$ (here $t_2 = 0$ and $t_1 + t_3 = 1$).

FIGURE 3.1 Convex combination in two-dimensional space (Problem 3.1).

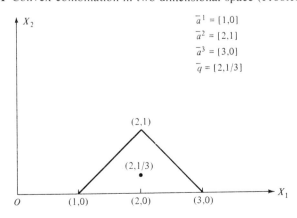

Conversely, it can be shown that any point (vector) expressed in the form of a convex combination of $\bar{a}^1$, $\bar{a}^2$, and $\bar{a}^3$ must lie in the triangle.

Notice also that the vectors $\bar{a}^1$, $\bar{a}^2$, and $\bar{a}^3$ are the extreme points (vertices) of the triangle. Up until now we have had an intuitive notion of what an extreme point is, but we did not give a precise definition of it. In Definitions 3.2 and 3.2' we shall do just that, but first it is instructive to show that any point on a line segment is merely a convex combination of the end points of that segment.

Problem 3.2

Show that any point on the line segment connecting the two points $\bar{P}$ and $\bar{Q}$ of R^n is a convex combination of $\bar{P}$ and $\bar{Q}$. In addition, show that the converse is true.

Solution

a. We see from Definition 2.13 that any point $\bar{x}$ on the line segment is expressible as

$$\bar{x} = t\bar{P} + (1 - t)\bar{Q}, \quad \text{where} \quad 0 \le t \le 1. \tag{1}$$

Letting $t_1 = t$ and $t_2 = 1 - t$, we write

$$\bar{x} = t_1\bar{P} + t_2\bar{Q}.$$

It is clear that

$$0 \le t_1 \le 1 \quad \text{and} \quad 0 \le t_2 \le 1$$

because $0 \le t \le 1$ implies $0 \le 1 - t \le 1$ and hence $0 \le t_2 \le 1$ since $t_2 = 1 - t$. Further,

$$t_1 + t_2 = t + (1 - t) = 1.$$

Thus, we see that t_1 and t_2 possess the required properties and that $\bar{x}$ is indeed expressible as a convex combination of $\bar{P}$ and $\bar{Q}$.

b. To demonstrate the converse, let $\bar{q}$ be a convex combination of $\bar{P}$ and $\bar{Q}$. Thus,

$$\bar{q} = t_1\bar{P} + t_2\bar{Q},$$

where $0 \le t_1$, $0 \le t_2$, and $t_1 + t_2 = 1$. Expressing the latter equation in the form

$$t_2 = 1 - t_1,$$

we can write

$$\bar{q} = t_1\bar{P} + (1 - t_1)\bar{Q}.$$

It only remains to show that $t_1 \le 1$. We begin with the fact that $0 \le t_2$, which can be written as

$$0 \le 1 - t_1.$$

Adding t_1 to both sides, we obtain

$$t_1 \le 1.$$

The demonstration is complete.

Definition 3.2: Extreme Point. A point $\bar{e}$ in a convex set S is called an extreme point of S if $\bar{e}$ cannot be expressed as a convex combination of any other two distinct points of S. (In addition neither t_i in such a convex combination can be zero.)

Equivalently, we can say that $\bar{e}$ is an extreme point if $\bar{e}$ is not an in-between point of any line segment of S. This can be expressed more precisely in the form of the following definition.

Definition 3.2′: Extreme Point. A point $\bar{e}$ in a convex set S is an extreme point of S if there *do not* exist distinct points $\bar{a}$ and $\bar{b}$ in S and a number t (where $0 < t < 1$) such that $\bar{e} = t\bar{a} + (1 - t)\bar{b}$.

We note that the points $\bar{a}^1$, $\bar{a}^2$, and $\bar{a}^3$ of Figure 3.1 do satisfy the stated requirement and are thus extreme points. The reader is advised to refer back to Section 2 of Chapter 1 for other examples of extreme points. We again note that two synonyms for *extreme point* are *corner point* and *vertex*.

As a preface of the important Theorem 3.1, let us recall that a *polyhedral, convex set* (Definition 2.17) is the intersection set of a finite number of closed half-spaces of R^n, while a hyperplane in R^n is the set of points satisfying an equation of the form $c_1 x_1 + c_2 x_2 + \cdots + c_n x_n = b$ (Definition 2.12).

To be specific, a *polyhedral, convex set* is the set of points $\bar{x}$ whose coordinates satisfy all of the following r inequalities, each of which represents a closed half-space:

$$a_{11}x_1 + a_{12}x_2 + \cdots + a_{1n}x_n \le b_1 \tag{1}$$

$$a_{21}x_1 + a_{22}x_2 + \cdots + a_{2n}x_n \le b_2 \tag{2}$$

$$\cdots\cdots\cdots\cdots\cdots\cdots\cdots\cdots\cdots\cdots\cdots\cdots\cdots$$

$$a_{m1}x_1 + a_{m2}x_2 + \cdots + a_{mn}x_n \le b_m \tag{m}$$

$$\cdots\cdots\cdots\cdots\cdots\cdots\cdots\cdots\cdots\cdots\cdots\cdots\cdots$$

$$a_{r1}x_1 + a_{r2}x_2 + \cdots + a_{rn}x_n \le b_r. \tag{r}$$

Here, $r > n$ usually.

The constraint set for the standard maximum problem is of this form (e.g., see Problem 1.5). Inequalities (2.1) through (2.m) in the standard maximum problem have the same form as (1) through (m) here, while the n constraints $x_j \geq 0$ for $j = 1, 2, \ldots, n$ are equivalent to Inequalities $(m + 1)$ through (r) (with $r = m + n$). For example, $x_n \geq 0$ is Inequality (r) with all $a_{rj} = 0$ except for $a_{rn} = -1$. In addition, $b_r = b_{m+n} = 0$. Thus, Inequality (r) can be written in the form $-x_n \leq 0$, which upon multiplication by -1 yields $x_n \geq 0$. (The rule here is as follows: multiplying an inequality by a negative number reverses the sense of the inequality.)

The standard minimum problem also has a constraint set represented by the general list (1) through (r). Again, note that an inequality of $\leq b_i$ form can be converted to one of $\geq b_i'$ form (where $b_i' = -b_i$) if we multiply by -1. We are now in a position to state an important theorem.

Theorem 3.1. *Let S_c denote an appropriate polyhedral, convex set (see the preceding discussion). A point $\bar{e}$ in S_c is an extreme point of S_c if it is the intersection point of n bounding hyperplanes of the set.**

Note: The bounding hyperplanes are determined by the *equation* parts of *some* of the r inequalities in the general list above. (We use the word *some* because a hyperplane of the form $a_{i1}x_1 + a_{i2}x_2 + \cdots + a_{in}x_n = b_i$ may be completely external to S_c.) Thus, one possible bounding hyperplane is given by $a_{11}x_1 + a_{12}x_2 + \cdots + a_{1n}x_n = b_1$. This conforms with the ideas of Chapter 1, where, for example, in Problem 1.1, the four bounding hyperplanes (which in that case are lines) were $2x + y = 12$, $x + 2y = 12$, $x = 0$, and $y = 0$.

Now we are ready for several other key theorems. We shall encounter a new phrase, namely *closed set,* in the discussions that follow. A precise mathematical definition of a closed set is beyond the scope of this book; however, the reader is referred to page 192 of [3] for a definition of this term. For our purposes, a closed set is one that contains all its boundary points (we shall intuitively agree that *boundary points* are points on the boundary of a set). Note that the term boundary points also has a precise definition, for which the interested reader is again referred to page 192 of [3].

Theorem 3.2. *If a closed, bounded, convex set S has a finite number of extreme points, then any point in the set can be written as a convex combination of its extreme points.*

---

* A theorem equivalent to Theorem 3.1 is Theorem 6.3 on P. 199 of [1]. The proof is given there also. Note that the formulation and terminology for linear programming problems are quite different in [1] than in the current book. In fact, understanding the proof requires a background in linear algebra stronger than that which is presumed here.

The proof of Theorem 3.2 is rather advanced and thus will not be undertaken here. The reader is referred to Chapter 6 of [3], which presents many of the definitions, theorems, and examples needed for a more complete understanding of convex geometry.

Shortly, we shall show that Theorem 3.2 applies to a general linear programming problem of maximum type, although we could just as well deal with one of minimum type. The reader should refer to Problem 1.5.

Standard Linear Programming Problem of Maximum Type

$$\text{Maximize} \quad u(\bar{x}) = c_1 x_1 + c_2 x_2 + \cdots + c_n x_n \tag{1}$$

$$\text{subject to} \quad a_{11} x_1 + a_{12} x_2 + \cdots + a_{1n} x_n \leq b_1 \tag{2.1}$$

$$a_{21} x_1 + a_{22} x_2 + \cdots + a_{2n} x_n \leq b_2 \tag{2.2}$$

$$\dots\dots\dots\dots\dots\dots\dots\dots\dots\dots\dots\dots$$

$$\dots\dots\dots\dots\dots\dots\dots\dots\dots\dots\dots\dots$$

$$a_{m1} x_1 + a_{m2} x_2 + \cdots + a_{mn} x_n \leq b_m \tag{2.m}$$

and

$$x_1 \geq 0 \tag{3.1}$$

$$x_2 \geq 0 \tag{3.2}$$

$$\dots\dots\dots$$

$$\dots\dots\dots$$

$$x_n \geq 0. \tag{3.n}$$

We know from Theorem 2.11 that the constraint set S_c for this problem is *convex*. We also know that the number of extreme points is *finite* by virtue of Problem 1.9, where we established a finite upper bound for the number of extreme points. Furthermore, it can be shown that the set S_c is *closed*. Thus, practically all the conditions of Theorem 3.2 are met. If we further hypothesize boundedness, we can then prove the following theorem, which is closely related to Theorem 1.1.

Theorem 3.3. *Suppose that the constraint set S_c of the standard linear programming problem of maximum type is bounded. Then the maximum value of $u(\bar{x})$ is attained at an extreme point of S_c.*

PROOF OF THEOREM 3.3. To make the work more efficient, let us suppose that the finite number of extreme points of S_c is k. Thus, we label the extreme points as follows:

$$\bar{a}^1, \bar{a}^2, \bar{a}^3, \dots, \bar{a}^k.$$

The values of the objective function u at the extreme points are

$$u(\bar{a}^1), u(\bar{a}^2), \ldots, u(\bar{a}^k), \tag{1}$$

respectively. Suppose, without loss of generality, that the maximum value is $u(\bar{a}^1)$, i.e.,

$$u(\bar{a}^1) \geq u(\bar{a}^i) \quad \text{for} \quad i = 2, 3, 4, \ldots, k. \tag{2}$$

Now let $\bar{x}$ be any point (vector) in the constraint set S_c. Thus, by Theorem 3.2,

$$\bar{x} = t_1\bar{a}^1 + t_2\bar{a}^2 + \cdots + t_k\bar{a}^k \tag{3}$$

where

$$t_1 \geq 0, t_2 \geq 0, \ldots, t_k \geq 0 \tag{4}$$

and

$$t_1 + t_2 + \cdots + t_k = 1. \tag{5}$$

Now we evaluate the linear objective function u at $\bar{x}$. We wish to show that $u(\bar{x}) \leq u(\bar{a}^1)$. By virtue of the linearity of u (see Theorem 2.8), we have, after using (3),

$$\begin{aligned} u(\bar{x}) &= u(t_1\bar{a}^1 + t_2\bar{a}^2 + \cdots + t_k\bar{a}^k) \\ &= t_1u(\bar{a}^1) + t_2u(\bar{a}^2) + \cdots + t_ku(\bar{a}^k). \end{aligned} \tag{6}$$

We can write $u(\bar{a}^1)$ in the form

$$u(\bar{a}^1) = 1 \cdot u(\bar{a}^1) = (t_1 + t_2 + \cdots + t_k)u(\bar{a}^1),$$

i.e.,

$$u(\bar{a}^1) = t_1u(\bar{a}^1) + t_2u(\bar{a}^1) + \cdots + t_ku(\bar{a}^1). \tag{7}$$

However, from Inequalities (2) and the fact that all the t_i's are greater than or equal to zero, it follows that

$$t_1u(\bar{a}^1) = t_1u(\bar{a}^1) \tag{8.1}$$

$$t_2u(\bar{a}^1) \geq t_2u(\bar{a}^2) \tag{8.2}$$

$$\cdots \cdots \cdots \cdots \cdots$$

$$\cdots \cdots \cdots \cdots \cdots$$

$$t_ku(\bar{a}^1) \geq t_ku(\bar{a}^k). \tag{8.k}$$

Adding expressions (8.1) through (8.k) together and then employing Equations (6) and (7), we obtain

$$u(\bar{a}^1) \geq u(\bar{x}). \tag{9}$$

Since $\bar{x}$ was an arbitrary point in S_c, the proof is complete. $\square$

Figure 3.2 is a diagram (relevant to Theorem 3.3) that applies in two dimensions; it is drawn for the case $k = 5$.

The same type of proof holds for finding the minimum of some linear function over a bounded constraint set of the same type. Thus, if $u(\bar{a}^2)$ is the minimum of the u values at the extreme points, then $u(\bar{a}^2)$ is also the minimum of u with respect to the entire constraint set. We state this result below as Theorem 3.3'.

Similar versions of the above proof may be found in [2] and [5].

Theorem 3.3'. *Suppose that the constraint set S_c of the standard linear programming problem of minimum type is bounded (see Problem 1.7). Then the minimum value of the linear objective function u is attained at an extreme point of S_c.*

Theorem 1.1 is almost equivalent to Theorem 3.3. To prove the former, what we need to do is establish that a polyhedral, convex set is convex, closed, and has a finite number of extreme points. We know that a polyhedral, convex set is the intersection of a finite number of closed half-spaces. Such a set is convex because of Theorems 2.10 and 2.7. It can also be established that the intersection of a finite number of closed sets is itself a closed set. Thus, a polyhedral, convex set is a closed set because each closed half-space is a closed set. Furthermore, it can be established that the number of extreme points is finite. Hence, we see that all the conditions of Theorem 3.2 are also met for a bounded, polyhedral, convex set. Thus, the proof of Theorem 1.1 is almost identical to that of Theorem 3.3, and we note that the maximum problems of Chapter 1, which involved bounded constraint sets, could have been solved by referring to Theorem 3.3 instead of Theorem 1.1.

FIGURE 3.2 Diagram for Theorem 3.3.

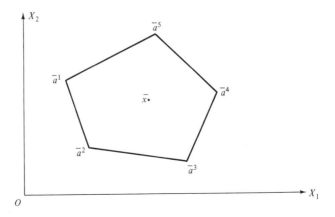

Another point of interest concerning the comparison of Theorems 1.1 and 3.3 is that any polyhedral, convex set can be represented as a constraint set for a *standard* linear programming problem. What is needed is an introduction of some new variables. The next problem illustrates the procedure.

Problem 3.3

Suppose a maximum problem is given as follows:

$$\text{Maximize} \quad u = x_1 + 3x_2 \tag{A}$$

$$\text{subject to} \quad -3x_1 - x_2 \geq -5 \tag{B}$$

and

$$2x_1 + 4x_2 \leq 10. \tag{C}$$

Show that this problem can be represented in the form of a standard linear programming problem of maximum type.

Solution

First of all, we note that the constraint set is determined by Inequalities (B) and (C), each of which represents a closed half-space, as can be seen by referring to the discussion preceding Problem 2.19. The constraint set for the problem contains those points that satisfy both (B) and (C). Thus, the constraint set is the intersection set of the sets (closed half-spaces) corresponding to (B) and (C). In other words, it is the *polyhedral, convex set* (see Definition 2.17) pertaining to (B) and (C).

Next, we refer to Problem 1.5 to observe the inequalities that appear in a standard maximum problem. To convert our problem to standard form, we must do several things. First of all, we must convert Inequality (B) so that its sense is in the other direction. This is no problem; we merely multiply (B) by -1 to obtain

$$3x_1 + x_2 \leq 5 \tag{B'}$$

(Remember that multiplying an inequality by a negative number reverses the sense of the inequality.)

We also must have inequalities expressing the nonnegativity of all the x variables. In our statement of the problem no such restrictions were placed on x_1 and x_2 (variables such as these are called *unrestricted* or *unconstrained*). Thus, we make the following definitions of new variables: Let $x_1 = x_1^+ - x_1^-$ and $x_2 = x_2^+ - x_2^-$, where $x_1^+ \geq 0$, $x_1^- \geq 0$, $x_2^+ \geq 0$, and $x_2^- \geq 0$.

Thus, we express both x_1 and x_2 as the difference between two nonnegative variables. This is allowable, since the *difference* of two nonnegative quantities can be positive, negative, or zero. Thus, we can now say that our linear programming problem has the following equivalent form:

$$\text{Maximize} \quad u = x_1^+ - x_1^- + 3x_2^+ - 3x_2^- \tag{1}$$

$$\text{subject to} \quad 3x_1^+ - 3x_1^- + x_2^+ - x_2^- \le 5 \tag{2}$$

$$2x_1^+ - 2x_1^- + 4x_2^+ - 4x_2^- \le 10 \tag{3}$$

$$x_1^+ \ge 0 \tag{4}$$

$$x_1^- \ge 0 \tag{5}$$

$$x_2^+ \ge 0 \tag{6}$$

$$x_2^- \ge 0. \tag{7}$$

If we compare this problem with Problem 1.5, we see that it is now in standard form, in terms of the four variables x_1^+, x_1^-, x_2^+, and x_2^-.

In the same way as in Problem 3.3, we can establish, in general, that any polyhedral convex set can be represented as a constraint set for a *standard* linear programming problem.

It is of interest now to reconsider the linear function u that is to be maximized or minimized:

$$u = c_1 x_1 + c_2 x_2 + \cdots + c_n x_n.$$

If we let u take on different constant values, we obtain a series of parallel hyperplanes. We shall call these *level* hyperplanes. In 2-space these hyperplanes are, of course, straight lines, while in 3-space they are the usual planes. The vector $\bar{c}$ has the property of being perpendicular to the level hyperplanes. In addition, the vector $\bar{c}$ *points in the direction of greatest increase of* u (actually, greatest increase with respect to distance). One way this can be shown is by using methods of calculus that involve a concept called the *gradient*.

Let us illustrate these ideas for the two-dimensional case (2-space).

Problem 3.4

Consider the linear objective function of Problems 1.1 and 1.2:

$$u = 3x_1 + 4x_2$$

(here we use x_1 and x_2 instead of x and y as our two variables). Plot the linear function for different values of u and interpret the results.

Solution

In Figure 3.3 we plot the objective function for $u = 0, 6, 12, 18, 24,$ and 30. Note that the resulting parallel level hyperplanes are just parallel level lines. The vector $\bar{c}$ is also drawn in the figure. It is seen to be perpendicular to the level hyperplanes and to point in the direction of greatest increase of u.

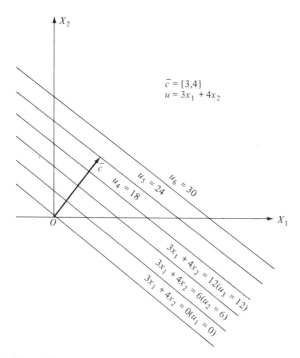

FIGURE 3.3 Plot of level hyperplanes (level lines) for the objective function of Problem 3.4.

The other major theorem of Chapter 1, namely, Theorem 1.2, applied to unbounded constraint sets. The wording there was in terms of polyhedral, convex sets, and while it is true that the constraint sets of standard maximum and minimum linear programming problems are polyhedral, convex sets, it is nonetheless useful to state theorems that apply specifically to the standard problems. These are given now.

Theorem 3.4. *Given a standard linear programming problem of maximum type (Problem 1.5) in which the constraint set S_c is unbounded. If the linear function u is bounded from above, then the maximum value of u will occur at an extreme point of S_c.*

Theorem 3.5. *Given a standard linear programming problem of minimum type (Problem 1.7) in which the constraint set S_c is unbounded. If the linear function u is bounded from below, then the minimum value of u will occur at an extreme point of S_c.*

Note: If all the c_j's in the objective function are nonnegative, then u will be bounded from below by zero, i.e., $u \geq 0$. This occurs as a result of the nonnegativity requirements on all the x_j's. Thus, if we have a standard minimum problem with an unbounded constraint set, we know that the minimum value of u will occur at an extreme point of the constraint set if all the c_j's are nonnegative.

We again remind the reader that Theorems 3.4 and 3.5 are essentially equivalent to Theorem 1.2.

We shall now turn our attention to a theorem that relates to the *simplex algorithm*, which will be considered at length in the next chapter. This algorithm is a process in which one moves from one extreme point of a constraint set to an *adjacent* extreme point where the value of the objective function is improved. (Later we shall use the phrase *basic feasible point* in place of extreme point.) This process is repeated until the solution extreme point is reached or it is determined that the problem does not have a solution.

Theorem 3.6. *Suppose that A is an extreme point of the constraint set for a standard maximum linear programming problem (refer to Problem 1.5). If the objective function u is not maximized at A, then there is an edge of the constraint set, starting at A, along which u increases.*

Notes:
a. A companion theorem holds for the case of minimum problems where u decreases along an edge starting at point A.
b. We have not given a precise definition of edge, although it is intuitively clear as to what is meant in the two- and three-dimensional cases. We shall have more to say about the term edge in the next chapter.

Problem 3.5

Illustrate Theorem 3.6 for the case of a two-dimensional constraint set.

Solution

A typical diagram illustrating the situation for a bounded set is given in Figure 3.4. The constraint set is contained within and on the boundary labeled $ABCDEA$.

Level lines corresponding to different values of the objective function are drawn in the figure. The vector $\bar{c}$ is perpendicular to the level lines and points in the direction of greatest increase of u (it is shown with initial point at the origin). The subscripts on the u's indicate this direction of increase. Thus,

$$u_1 < u_2 < u_3 < \cdots < u_8 < u_9.$$

The arrows marked on the boundary $ABCDEA$ indicate the direction of increase of u along the various edges of the constraint set. We see that extreme points E, A, C, and D correspond to the extreme point

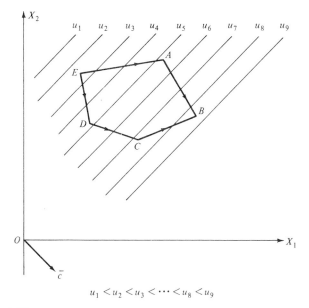

$$u_1 < u_2 < u_3 < \cdots < u_8 < u_9$$

FIGURE 3.4 Diagram pertaining to Theorem 3.6.

mentioned in Theorem 3.6. For example, u is not maximized at A, but u does increase along the edge AB. In the diagram, u is maximized at B and minimized at E.

REFERENCES

1. Cooper, L. and Steinberg, D. *Introduction to Methods of Optimization*. Philadelphia, Pennsylvania: W. B. Saunders Company, 1970.
2. Gass, S. *Linear Programming*. Third edition. New York: McGraw-Hill, 1969.
3. Hadley, G. *Linear Algebra*. Reading, Massachusetts: Addison-Wesley, 1961.
4. Hadley, G. *Linear Programming*. Reading, Massachusetts: Addison-Wesley, 1962.
5. Kemeny, J. G., Mirkil, H., Snell, J. L., and Thompson, G. L. *Finite Mathematical Structures*. Englewood Cliffs, New Jersey: Prentice-Hall, 1959.
6. Marcus, M. *A Survey of Finite Mathematics*. Boston, Massachusetts: Houghton-Mifflin Company, 1969.

SUPPLEMENTARY PROBLEMS

Convex Combinations

S.P. 3.1: Consider the three vectors in two-dimensional space given by $\bar{a}^1 = [2,1]$, $\bar{a}^2 = [4,2]$, and $\bar{a}^3 = [3,5]$. (a) Determine the convex combination for which $t_1 = t_2 = t_3 = \frac{1}{3}$ and interpret it graphically. (b) Rework the problem for $t_1 = \frac{1}{2}$, $t_2 = \frac{3}{8}$, and $t_3 = \frac{1}{8}$.

S.P. 3.2:   Prove Theorem 3.3′.

S.P. 3.3:   Consider the minimum problem

Minimize $\qquad u = 2x + 5y$ $\hfill$ (A)

subject to $\qquad 5x + 4y \geq 7$ $\hfill$ (B)

$\qquad\qquad\qquad 7x - 2y \leq -9$ $\hfill$ (C)

$\qquad\qquad\qquad\qquad x \geq 0.$ $\hfill$ (D)

Represent this problem in the form of a standard linear programming problem of minimum type.

S.P. 3.4:   Consider the linear objective function $u = 200x + 300y$ of Problem 1.3. Plot this linear function for different constant values of u (i.e., plot level lines). Determine the direction of greatest increase of u. Determine the direction of greatest decrease of u.

S.P. 3.5:   Solve Problem 1.1 by drawing level hyperplanes (lines) corresponding to the objective function u.

S.P. 3.6:   Solve Problem 1.3 by drawing level hyperplanes (lines) corresponding to the objective function u.

S.P. 3.7:   Interpret the solution of Problem 1.4 by means of level hyperplanes (planes).

S.P. 3.8:   Given that the three distinct points $\bar{P}_1$, $\bar{P}_2$, and $\bar{P}_3$ are in a convex set S, show that any *convex combination* of the three points is also in S. (One has to show that any linear combination of the form $b_1\bar{P}_1 + b_2\bar{P}_2 + b_3\bar{P}_3$, where each $b_i > 0$ and $b_1 + b_2 + b_3 = 1$, is a point within S.)

S.P. 3.9:   Given k points $\bar{P}_1$, $\bar{P}_2$, . . . , $\bar{P}_k$ in n-dimensional space, we define the *convex polyhedron H* of these points as follows: H is the set of *all* convex combinations of the form $a_1\bar{P}_1 + a_2\bar{P}_2 + \cdots + a_k\bar{P}_k$. (Thus, each $a_i \geq 0$ and $a_1 + a_2 + \cdots + a_k = 1$.) Show that H is a convex set.

S.P. 3.10: Suppose that C is a convex subset of R^n (n-dimensional space). A real-valued function f defined on C is said to be a *convex function* if for any two points $\bar{x}$ and $\bar{y}$ in C and any number h, where $0 \leq h \leq 1$,

$$f[h\bar{x} + (1 - h)\bar{y}] \leq hf(\bar{x}) + (1 - h)f(\bar{y}).$$

(a) Show that the linear function $f(\bar{x}) = \bar{c}^t\bar{x}$ is convex. (b) Suppose that $n = 1$ and that C itself is R^1—the set of real numbers. Prove that $f(x) = x^2$ is a convex function. (Here, x denotes a real number.)

ANSWERS TO SUPPLEMENTARY PROBLEMS

S.P. 3.1:   (a) $[3, \frac{8}{3}]$.   (b) $[\frac{23}{8}, \frac{15}{8}]$.

S.P. 3.2:   *Sketch of Proof.* Suppose, without loss of generality, that $\bar{a}^k$ is the extreme point for which $u(\bar{a}^k) \leq u(\bar{a}^i)$ for $i = 1, 2, \ldots, k - 1$. Follow a procedure similar to that used in the Proof of Theorem 3.3. For any point $\bar{x}$ in S_c show that $u(\bar{x}) \geq u(\bar{a}^k)$.

S.P. 3.3:   Refer to Problem 1.7. Let $y = w - z$, where $w \geq 0$ and $z \geq 0$. Multiply (C) by -1. Obtain the following:

Minimize $u = 2x + 5w - 5z$ (A')

subject to $5x + 4w - 4z \geq 7$ (B')

$-7x + 2w - 2z \geq 9$ (C')

$x \geq 0$ (D')

$w \geq 0$ (E')

$z \geq 0.$ (F')

S.P. 3.4:   The vector $\bar{c} = [200,300]$ points in the direction of greatest increase of u. The vector $-\bar{c} = [-200,-300]$ points in the direction of greatest decrease of u.

S.P. 3.5:   The level line of greatest u value which intersects the constraint set is $3x + 4y = 28$. This occurs at point B, where $x = 4$ and $y = 4$.

S.P. 3.6:   The level line of lowest u value which intersects the constraint set is $200x + 300y = 10,000$. This occurs at point D, where $x = 20$ and $y = 20$.

S.P. 3.7:   Suppose that we plotted level planes of the form $0.25x + 0.4y + 0.5z = k$ in three-dimensional space. The level plane of greatest u value which intersects the constraint set (see Figure 1.8) is $0.25x + 0.4y + 0.5z = 624$; this occurs at the point A, where $x = 0$, $y = 1260$, and $z = 240$.

S.P. 3.8:   *Sketch of Proof.* First show that $\bar{c}$ is in S, where $\bar{c} = [b_1/(b_1 + b_2)] \cdot \bar{P}_1 + [b_2/(b_1 + b_2)] \cdot \bar{P}_2$ and b_1, $b_2 > 0$. Then show that $(b_1 + b_2)\bar{c} + b_3\bar{P}_3$ is in S if each $b_i > 0$ and if $b_1 + b_2 + b_3 = 1$. But this is none other than the desired convex combination $b_1\bar{P}_1 + b_2\bar{P}_2 + b_3\bar{P}_3$. In the same way, one can prove that any finite convex combination of points in a convex set is itself a point in the convex set.

S.P. 3.9:   *Sketch of Proof.* Let $\bar{x}$ and $\bar{y}$ be two points in H. Thus, both $\bar{x}$ and $\bar{y}$ are convex combinations of $\bar{P}_1$ through $\bar{P}_k$. Consider $\bar{Q} = c\bar{x} + (1 - c)\bar{y}$, where $0 < c < 1$. It remains to show that $\bar{Q}$ is in H by showing that $\bar{Q}$ can be expressed in the convex combination form $d_1\bar{P}_1 + d_2\bar{P}_2 + \cdots + d_k\bar{P}_k$.

S.P. 3.10:   *Sketches of Proofs.* (a) Since f is linear, $f[h\bar{x} + (1 - h)\bar{y}] = hf(\bar{x}) + (1 - h)f(\bar{y})$. This satisfies the requirement for a convex function. (b) Let x and y be points in R^1 and suppose that a, $b \geq 0$, where $a + b = 1$. Then work with $af(x) + bf(y) - f(ax + by)$ and show that it equals $ab(x - y)^2$. The latter is greater than or equal to zero, thus showing that f is a convex function.

Chapter Four

The Simplex Algorithm

1. INTRODUCTORY REMARKS

We shall now introduce the simplex algorithm or simplex method, as it is sometimes called. This algorithm provides the main calculational tool of linear programming. The word *algorithm* means a set of rules or a particular routine for attaining the solution to a given problem. The approach we shall use is due, in large part, to Professor Albert W. Tucker [13, 14]; it has become fairly popular in the 1960's and 1970's because of its simplicity. The beauty of the Tucker approach is that it does not require an advanced mathematical background on the part of the student. Little more than a knowledge of basic algebra and the ability to solve sets of simultaneous linear equations is required. In our development, we shall also depend heavily on geometric intuition, and, accordingly, we shall often refer to geometrical diagrams for two- and three-dimensional problems.

The Tucker approach involves the manipulation of a set of linear equations which represent the linear programming problem. Associated with the Tucker approach are a system of tableaus (tables or schemas); each tableau is a shorthand representation of a set of linear equations. We shall refer to such tableaus as *Tucker tableaus*.

The original development of the simplex algorithm goes back to the 1940's. The main member of the original group of linear programming pioneers was Dantzig [2, 3]. His approach was different, although it does lead to equivalent computational processes. In the Dantzig approach, the problem is viewed using a column vector interpretation; a good understanding of this requires a fairly advanced background in linear algebra, which is not presumed with respect to the current book. Associated with the Dantzig approach is a tableau format in which the variables retain a fixed *column* position throughout the manipulations; this tableau is larger than the Tucker tableau, and we shall refer to it as the "extended"

(Dantzig) tableau. The Dantzig approach and tableau are discussed in detail in the textbooks of Hadley [6], Gass [4], and Cooper and Steinberg [1]. (It is interesting to note that some practitioners who use the extended tableau refer to the Tucker tableau as the "condensed" tableau.) Some authors employ a hybrid treatment. For example, Jay E. Strum (*Introduction to Linear Programming*. Holden-Day Incorporated, 1972) uses the Tucker algebraic approach associated with an extended tableau.

At any rate, the development used here (employing the Tucker approach and tableau) has much to recommend it because of its basic simplicity. There does appear to be a trend toward this approach in some recent textbooks. For example, the books of Owen [10, 11], Singleton and Tyndall [12], and Nemhauser and Garfinkel [9] all use variations of it.

It is very significant that Kemeny et al., in the book *Finite Mathematics with Business Applications,* used the Dantzig tableau in their 1962 edition [7], but have switched to the Tucker tableau in their 1972 edition [8].

2. SLACK VARIABLES

Let us refer to the standard maximum linear programming problem (Problem 1.5). We can convert the m inequalities (2.1) through (2.m) into equations by introducing slack variables. For example, in (2.1), we see that b_1 is greater than or equal to the left-hand side of the inequality. Now we let the slack variable t_1 "take up the slack," and we set it equal to b_1 minus the left-hand side. We start with

$$a_{11}x_1 + a_{12}x_2 + \cdots + a_{1n}x_n \leq b_1.$$

Introducing t_1, we obtain

$$a_{11}x_1 + a_{12}x_2 + \cdots + a_{1n}x_n + t_1 = b_1$$

or

$$-(a_{11}x_1 + a_{12}x_2 + \cdots + a_{1n}x_n) + b_1 = t_1.$$

Note that our slack variable $t_1 \geq 0$ if the point $(x_1, x_2, \ldots, x_n)$ is feasible, i.e., is in the constraint set.

It is convenient to rewrite the last equation in terms of $-t_1$. Thus, we simply multiply through by -1 to obtain

$$a_{11}x_1 + a_{12}x_2 + \cdots + a_{1n}x_n - b_1 = -t_1.$$

(This will be the form of a slack equation in the initial tableau, which is discussed in Section 3 and illustrated in Figure 4.4.) Proceeding in the same way with the remaining equalities, we can represent our standard maximum problem in what we shall henceforth call the "standard slack" form:

$$\text{Minimize} \quad u = c_1 x_1 + c_2 x_2 + \cdots + c_n x_n \tag{1}$$

$$\text{subject to} \quad a_{11} x_1 + a_{12} x_2 + \cdots + a_{1n} x_n - b_1 = -t_1 \tag{2.1}$$

$$a_{21} x_1 + a_{22} x_2 + \cdots + a_{2n} x_n - b_2 = -t_2 \tag{2.2}$$

$$\dots\dots\dots\dots\dots\dots\dots\dots\dots\dots\dots\dots\dots\dots\dots$$

$$\dots\dots\dots\dots\dots\dots\dots\dots\dots\dots\dots\dots\dots\dots\dots$$

$$a_{m1} x_1 + a_{m2} x_2 + \cdots + a_{mn} x_n - b_m = -t_m, \tag{2.m}$$

$$x_1 \geq 0 \tag{3.1}$$

$$x_2 \geq 0 \tag{3.2}$$

$$\dots\dots\dots$$

$$\dots\dots\dots$$

$$x_n \geq 0, \tag{3.n}$$

and

$$t_1 \geq 0 \tag{3.n+1}$$

$$t_2 \geq 0 \tag{3.n+2}$$

$$\dots\dots\dots$$

$$\dots\dots\dots$$

$$t_m \geq 0. \tag{3.n+m}$$

Notice that we now have an additional m nonnegativity constraints involving the slack variables t_1 through t_m.

Problem 4.1

Represent Problem 1.1 in terms of slack variables.

Solution

The linear objective function u is unaffected:

$$u = 3x + 4y \quad \text{(to be maximized)}. \tag{1}$$

Letting the nonnegative slack variables be r and s, we have for our two main constraints

$$2x + y - 12 = -r \tag{2}$$

$$x + 2y - 12 = -s. \tag{3}$$

There is now a total of four nonnegativity constraints:

$$x \geq 0, \tag{4}$$

$$y \geq 0, \tag{5}$$

$$r \geq 0, \tag{6}$$

$$s \geq 0. \tag{7}$$

Problem 4.2

Represent Problem 1.4 in terms of slack variables.

Solution

The unchanged equation for the revenue function u is

$$u = \tfrac{1}{4}x + \tfrac{2}{5}y + \tfrac{1}{2}z \quad \text{(to be maximized).} \tag{1}$$

Again, let the nonnegative slack variables be r and s. Thus, we have

$$1 \cdot x + \tfrac{2}{3}y + \tfrac{1}{4}z - 900 = -r \tag{2}$$

$$0 \cdot x + \tfrac{1}{3}y + \tfrac{3}{4}z - 600 = -s \tag{3}$$

and

$$x \geq 0, \tag{4}$$

$$y \geq 0, \tag{5}$$

$$z \geq 0, \tag{6}$$

$$r \geq 0, \tag{7}$$

$$s \geq 0. \tag{8}$$

Problem 4.3

Represent the standard slack maximum problem in matrix-vector form.

Solution

All that is involved here is an extension of the material in Section 5 of Chapter 2. As before, we have

$$u = \bar{c}^t \bar{x} \quad \text{(to be maximized).} \tag{1}$$

The vector inequality $A\bar{x} \leq \bar{b}$ becomes the vector equality

$$A\bar{x} - \bar{b} = -\bar{t}, \tag{2}$$

if we introduce the m-component slack vector $\bar{t}$, where $\bar{t} = [t_1, t_2, \ldots, t_m]$. Most of our vectors will be column vectors, and these are denoted as *rows* of ordered numbers enclosed by brackets [ ]. Row vectors are denoted as rows enclosed by parentheses ( ). For more information on matrix-vector notation, see Sections 3–5 of Chapter 2.

For the nonnegativity statements we have

$$\bar{x} \geq \bar{0} \quad \text{and} \quad \bar{t} \geq \bar{0}. \tag{3}$$

The first $\bar{0}$ has n components, while the latter $\bar{0}$ has m components.

Problem 4.4

Refer to Figure 1.5 of Problems 1.1 and 1.2. Analyze the extreme point candidates, extreme points, and constraint set in terms of the variables x, y, r, and s of Problem 4.1.

Solution

The slack variables r and s were defined in Problem 4.1, to which the reader is referred at this time. We see that the equations $2x + y = 12$ and $x + 2y = 12$ are also given by $r = 0$ and $s = 0$, respectively. In Figure 1.5 we see that *each* of the bounding edges can thus be described by an equation of the form

$$\text{Variable} = 0.$$

This is indicated in Figure 4.1.

Now we notice that each of the six extreme point candidates O, A, B, C, D, and E is characterized by the fact that for each such point, two of

FIGURE 4.1 The constraint set of Problem 1.1 revisited. *Notes:* The constraint set S_c, is shown shaded. The line $r = 0$ is also $2x + y = 12$. The line $s = 0$ is also $x + 2y = 12$.

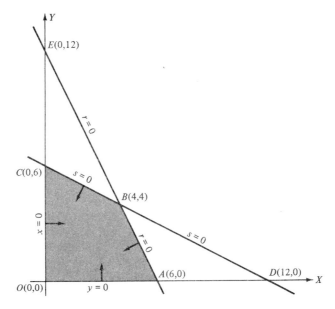

the variables x, y, r, and s are equal to zero. This is so because each such point is the intersection point of two bounding edges.

For example, for point O we have $x = 0$ and $y = 0$; for point A we have $y = 0$ and $r = 0$; and for point B we have $r = 0$ and $s = 0$.

Now let us determine the values of all four variables for each of the extreme point candidates. For this we refer to Equations (2) and (3) of Problem 4.1. Our job is fairly easy, since we already know x and y at these points, as can be seen from Figure 4.1 (or 1.5). For point B, we know that $x = 4$ and $y = 4$, and also that $r = 0$ and $s = 0$, since the point lies at the intersection of these latter two lines.

For point O, where we have $x = 0$ and $y = 0$, Equations (2) and (3) lead us to

$$r = +12 \quad \text{and} \quad s = +12.$$

There is a general approach, based on graphical considerations, that is recommended. We shall illustrate this approach for points A and E. For point A, we see that the point lies at the intersection of $y = 0$ and $r = 0$. We solve Equations (2) and (3) for the remaining variables x and s after setting y and r equal to zero:

$$2x - 12 = 0 \quad \text{and} \quad x - 12 = -s.$$

This leads us to $x = +6$ (from the first equation) and then $s = +6$ after substitution into the second equation. The calculation of $x = 6$ is a check on what we already know.

The point E lies at the intersection of $x = 0$ and $r = 0$. Setting $x = r = 0$ in Equations (2) and (3) leads us to the following two equations in y and s:

$$y - 12 = 0 \quad \text{and} \quad 2y - 12 = -s.$$

From these we obtain $y = +12$ and $s = -12$. We summarize our calculations for all six points in the following table:

Point	x	y	r	s
O	0	0	12	12
A	6	0	0	6
B	4	4	0	0
C	0	6	6	0
D	12	0	-12	0
E	0	12	0	-12

We see that of our six candidates for extreme points, the two candidates that are not extreme points are characterized by the fact that one of the nonzero variables has a value *that is negative*.

In summary, for this problem with a two-dimensional constraint set, we see that two variables are equal to zero for each of the extreme point candidates. Those candidates that turn out to be actual extreme points are those for which the remaining two variables are nonnegative (equivalently, positive or zero) in value. In the above situation, the four extreme points are O, A, B, and C.

Another point is worthy of mention. In Figure 4.1, the arrows point in the directions of increase of the different variables. For example, the arrow on the line segment AE ($r = 0$) points in the direction of increasing r, while the arrow on CD points in the direction of increasing s. Thus, we see that we can characterize all points in the constraint set S_c by the observation that for every such point, each of the variables x, y, r, and s is nonnegative, i.e., for each point in S_c, $x \geq 0$, $y \geq 0$, $r \geq 0$, and $s \geq 0$.

In addition, the variables x, y, r, and s, for the points in S_c, are related through Equations (2) and (3) or through an equivalent system of equations. We shall learn more about the phrase "equivalent system" when we develop the simplex algorithm.

For points on the interior of S_c, each of the four variables x, y, r, and s is strictly positive. On the four bounding edges (except at extreme points), we see that one of the four variables becomes zero in value, while the other three remain positive. At a point of S_c that is an intersection point of two bounding edges, we see that two variables are zero in value, while the other two are positive. Such a point is, of course, an extreme point.

Similar characteristics hold for the constraint sets of many linear programming problems in standard form.

Problem 4.5

Refer to Figure 1.8 of Problem 1.4, the so-called "nut problem." Analyze the extreme point candidates, extreme points, and constraint set in terms of the variables x, y, z, r, and s of Problem 4.2.

Solution

In Problem 4.2, the slack variables r and s were defined by Equations (2) and (3) of that problem as follows:

$$x + \tfrac{2}{3}y + \tfrac{1}{4}z - 900 = -r \tag{2}$$

and

$$\tfrac{1}{3}y + \tfrac{3}{4}z - 600 = -s. \tag{3}$$

We see that all of the bounding planes in Figure 1.8 can be described in the form "Variable $= 0$." For example, we see the following equivalences if we refer to (2) and (3):

$$x + \tfrac{2}{3}y + \tfrac{1}{4}z = 900 \text{ is equivalent to } r = 0$$

and

$$\tfrac{1}{3}y + \tfrac{3}{4}z = 600 \text{ is equivalent to } s = 0.$$

Thus, we can develop Figure 4.2, where the bounding planes are now given in the "Variable = 0" form. This diagram should be contrasted to Figure 1.8.

Several observations are in order. Each of the extreme point candidates (all nine are listed in the table of Problem 1.4) is characterized by the fact that for each such point, three of the variables x, y, z, r, and s are equal to zero. This is true because each such point is the intersection point of three bounding planes. For example, we have the following variables equal to zero for the typical points O, D, A, B, and F:

For point O, $x = y = z = 0$

For point D, $x = z = r = 0$

For point A, $x = r = s = 0$

For point B, $y = r = s = 0$

For point F, $x = z = s = 0.$

FIGURE 4.2 The constraint set of Problem 1.4 revisited. Coordinates of labeled points: $O(0,0,0)$, $A(0,1260,240)$, $B(700,0,800)$, $C(0,0,800)$, $D(0,1350,0)$, $E(900,0,0)$, and $F(0,1800,0)$.

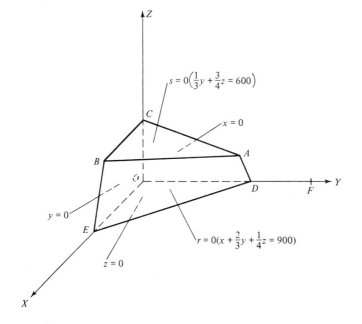

Note that point F is the point where the plane $\frac{1}{3}y + \frac{3}{4}z = 600$ (or $s = 0$) intersects the coordinate planes $x = 0$ and $z = 0$.

Now let us determine the values of all five variables for each of these extreme point candidates. This is fairly easy, since we already know x, y, and z for these points (see Figure 1.8 or the table of Problem 1.4). We shall also employ Equations (2) and (3), which give the relationships among the five variables. A general procedure is to set three variables equal to zero in Equations (2) and (3) and then solve for the remaining two variables (if that is possible). This is what we shall do. Thus, for point O we have

$$r = 900 \quad \text{and} \quad s = 600$$

after setting $x = y = z = 0$ in Equations (2) and (3).

For point D, we set $x = z = r = 0$ in Equations (2) and (3), which leads to

$$\tfrac{2}{3}y - 900 = 0 \quad \text{and} \quad \tfrac{1}{3}y - 600 = -s.$$

We find $y = 1350$ from the first equation and then $s = 150$ from the second equation.

For point A, we set $x = r = s = 0$ in Equations (2) and (3); we obtain

$$\tfrac{2}{3}y + \tfrac{1}{4}z - 900 = 0 \quad \text{and} \quad \tfrac{1}{3}y + \tfrac{3}{4}z - 600 = 0.$$

Solving these two equations results in

$$y = 1260 \quad \text{and} \quad z = 240.$$

This checks with what we already know (see list in Figure 1.8).

For point B, we set $y = r = s = 0$ in Equations (2) and (3); we obtain

$$x + \tfrac{1}{4}z - 900 = 0 \quad \text{and} \quad \tfrac{3}{4}z - 600 = 0.$$

From the second equation we have $z = 800$ and substitution into the first equation yields $x = 700$. This also checks with previously obtained results (see, e.g., Figure 1.8).

For point F, we set $x = z = s = 0$ in Equations (2) and (3); we obtain

$$\tfrac{2}{3}y - 900 = -r \quad \text{and} \quad \tfrac{1}{3}y - 600 = 0.$$

This leads to $y = 1800$ and $r = -300$. We summarize our calculations for the five points in the following table:

Point	x	y	z	r	s
O	0	0	0	900	600
D	0	1350	0	0	150
A	0	1260	240	0	0
B	700	0	800	0	0
F	0	1800	0	-300	0

We see that of the above five extreme point candidates, the one candidate that is not an extreme point (namely, F) has a *negative* value for one of its nonzero variables.

In summary, for this problem with a three-dimensional constraint set, we see that three variables are equal to zero for each of the extreme point candidates. Those candidates that turn out to be actual extreme points are those for which the remaining two variables are nonnegative.

Note that in the above analysis, we did not do calculations for four of the candidates listed in the table of Problem 1.4.

The constraint set S_c is the bounded, three-dimensional region shown in Figure 4.2. We have the following result: *For each point* in S_c, $x \geq 0$, $y \geq 0$, $z \geq 0$, $r \geq 0$, and $s \geq 0$.

In addition, for the points in S_c, the five variables are related through Equations (2) and (3) or through an equivalent system of equations.

For points in the interior of the polyhedral set S_c, each of the five variables x, y, z, r, and s is greater than zero. As we go outward from the interior of S_c, we reach a bounding plane where one of the variables becomes zero. On the edges, where two bounding planes intersect, two variables are equal to zero. For example, on the edge DA, both x and r equal zero. This is confirmed by examining our table; we see that for points D and A we have $x = r = 0$.

Finally, we have the extreme points of S_c. At each of these, three bounding planes meet, and this results in three variables being equal to zero, while the remaining two variables are positive. At an extreme point candidate that is not an extreme point, we have three bounding planes meeting at a point outside the constraint set. This is manifested algebraically by at least one of the nonzero variables being negative (see, e.g., point F).

Problem 4.6

Convert the standard minimum linear programming problem (Problem 1.7) into "standard slack" form by introducing nonnegative slack variables into the main inequalities.

Solution

The statement involving the function to be minimized is unaffected:

$$\text{Minimize} \quad u = c_1 x_1 + c_2 x_2 + \cdots + c_n x_n. \tag{1}$$

We illustrate the introduction of slack variables by focusing on Inequality (2.1):

$$a_{11} x_1 + a_{12} x_2 + \cdots + a_{1n} x_n \geq b_1.$$

We let the slack variable t_1 equal the difference between the left-hand side and the right-hand side. Sometimes in this type of situation, the slack variable is referred to as a *surplus variable*. Thus,

$$a_{11}x_1 + a_{12}x_2 + \cdots + a_{1n}x_n - b_1 = t_1.$$

Equivalent forms for this equation are

$$-a_{11}x_1 - a_{12}x_2 - \cdots - a_{1n}x_n + b_1 = -t_1$$

and

$$a_{11}x_1 + a_{12}x_2 + \cdots + a_{1n}x_n - t_1 = b_1.$$

Proceeding in the same way with the remaining inequalities, we can represent our standard minimum problem in what we shall subsequently call "standard slack" form:

$$\text{Minimize} \quad u = c_1x_1 + c_2x_2 + \cdots + c_nx_n \tag{1}$$

$$\text{subject to} \quad a_{11}x_1 + a_{12}x_2 + \cdots + a_{1n}x_n - b_1 = t_1 \tag{2.1}$$

$$a_{21}x_1 + a_{22}x_2 + \cdots + a_{2n}x_n - b_2 = t_2 \tag{2.2}$$

$$\cdots\cdots\cdots\cdots\cdots\cdots\cdots\cdots\cdots\cdots\cdots\cdots$$

$$\cdots\cdots\cdots\cdots\cdots\cdots\cdots\cdots\cdots\cdots\cdots\cdots$$

$$a_{m1}x_1 + a_{m2}x_2 + \cdots + a_{mn}x_n - b_m = t_m, \tag{2.m}$$

$$x_1 \geq 0 \tag{3.1}$$

$$x_2 \geq 0 \tag{3.2}$$

$$\cdots\cdots\cdots$$

$$\cdots\cdots\cdots$$

$$x_n \geq 0, \tag{3.n}$$

and

$$t_1 \geq 0 \tag{3.n+1}$$

$$t_2 \geq 0 \tag{3.n+2}$$

$$\cdots\cdots\cdots$$

$$\cdots\cdots\cdots$$

$$t_m \geq 0. \tag{3.n+m}$$

We now have a total of $(n + m)$ nonnegativity constraints, m of which involve the slack variables t_1 through t_m.

Problem 4.7

Represent Problem 1.3 in terms of slack variables.

Solution

The linear objective function u is unaffected:

$$u = 200x + 300y \quad \text{(to be minimized).} \tag{1}$$

Letting the nonnegative slack variables be r, s, and t, we have the following for our three main constraints:

$$x + 2y - 60 = r \tag{2}$$

$$4x + 2y - 120 = s \tag{3}$$

$$6x + 2y - 150 = t. \tag{4}$$

There is now a total of five nonnegativity constraints:

$$x \geq 0 \tag{5}$$

$$y \geq 0 \tag{6}$$

$$r \geq 0 \tag{7}$$

$$s \geq 0 \tag{8}$$

$$t \geq 0. \tag{9}$$

When we solve Problem 4.7 by means of the simplex algorithm, Equations (2), (3), and (4) will play a very important part.

Problem 4.8

Refer to Problem 1.3 and the accompanying diagram, Figure 1.7, which is reproduced here for convenience as Figure 4.2a. For this typical minimum problem, analyze bounding edges, extreme points, and the constraint set in terms of the variables x, y, r, s, and t of Problem 4.7.

Solution

In Figure 4.2a, the equation parts of the constraints are indicated by encircled numbers, the equation parts being, of course, equations for straight lines. For example, ② represents $x + 2y = 60$. From the equation for r (Problem 4.7), we see that ② also could be represented by $r = 0$. Proceeding in the same way, we develop the following table, which indicates the correspondence:

Equality		Variable equal to 0
②	$x + 2y = 60$	$r = 0$
③	$4x + 2y = 120$	$s = 0$
④	$6x + 2y = 150$	$t = 0$
⑤	$x = 0$	$x = 0$
⑥	$y = 0$	$y = 0$

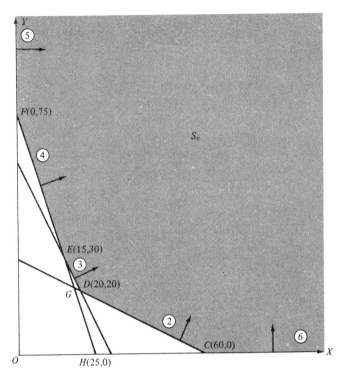

FIGURE 4.2a The set (shown shaded) for Problem 1.3. Encircled numbers indicate equality parts of constraints: Thus, ②represents $x + 2y = 60$, ③represents $4x + 2y = 120$, ④represents $6x + 2y = 150$, ⑤represents $x = 0$, and ⑥represents $y = 0$.

Thus, we see that each bounding edge can be described by an equation in the form "Variable = 0."

From this it is not hard to see why each of the extreme points for this two-dimensional constraint set problem has the characteristic that two variables are zero while the other three are positive. For example, for point D we have $r = 0$ and $s = 0$, since D is at the intersection point of the two lines having those equations. The other three variables are positive, as can be determined by solving Equations (2), (3), and (4) of Problem 4.7. That calculation leads to the following:

For point D, $x = 20$, $y = 20$,   and $t = 10$.

The first two numbers check with what we already know, namely, the x and y coordinates of the extreme point (see, e.g., Figure 4.2a).

The constraint set S_c is that *unbounded,* two-dimensional region indicated in Figure 4.2a by shading. We have the following result: *For each point* in S_c, $x \geq 0$, $y \geq 0$, $r \geq 0$, $s \geq 0$, and $t \geq 0$.

In addition, for the points in S_c, the five variables are related by Equations (2), (3), and (4) of Problem 4.7 or an equivalent system of equations.

We shall now introduce some important definitions. It is useful to think of our linear programming problem as being in "standard slack" form. For the standard maximum problem this form is presented in the discussion preceding Problem 4.1, while the form for the standard minimum problem is given in the solution to Problem 4.6.

The collective set of equations (2.1) through (2.m) will be referred to simply as (2) or Equations (2), while the collection of $(m + n)$ nonnegativity constraints (3.1) through (3.$n + m$) will be referred to as (3) or Inequalities (3). It is useful now to recall our definitions of constraint set and feasible point as given by Definitions 1.2 and 1.3. We defined the constraint set as being the set of points which satisfied the main inequalities (2.1) through (2.m) and also the nonnegativity constraints (3.1) through (3.n) of the linear programming problem in standard form. A typical point in the constraint set was called a feasible point. When we deal with the standard slack form, it is useful to refer to the extended vector that contains the m slack variables as well as the n original variables.

Definition 4.1. The extended vector $\bar{x}_e$ is the $(n + m)$-component vector given by

$$\bar{x}_e = [x_1, x_2, \ldots, x_n, t_1, t_2, \ldots, t_m].$$

Sometimes we shall refer to such vectors as *points*.

Definition 4.2. The set of vectors (or points) with components x_1, $x_2, \ldots, x_n, t_1, t_2, \ldots, t_m$ that satisfy all the equations (2) and all the inequalities (3) of the linear programming problem in standard slack form is called the constraint set. This set is denoted by S_c.

Note: Technically, this constraint set is different from the one discussed previously because of the inclusion of the slack variables t_1 through t_m in typical points of this set. We shall virtually ignore this distinction, unless the discussion calls for a clarification.

Definition 4.3. Those points that lie in the constraint set are called *feasible points*. Here our feasible points have $(m + n)$ components. Thus, a feasible point is a point for which $x_1 \geq 0, x_2 \geq 0, \ldots, x_n \geq 0, t_1 \geq 0, t_2 \geq 0, \ldots, t_m \geq 0$ (i.e., (3) is satisfied) and for which the $(n + m)$ variables are related through Equations (2) or some equivalent system of equations.

In many works dealing with linear programming, the term *feasible solution* is used instead of feasible point.

For the linear programming problem in standard slack form, Equations (2) consisted of m equations in $(n + m)$ unknowns. In the preceding problems of this chapter we saw that we could solve for extreme point candidates (and then the extreme points) by setting n variables equal to zero and then solving (if this is possible) for the remaining m variables. If we think of our points as having $(n + m)$ components, then there are some special definitions which pertain to such points.

Definition 4.4. A *basic point* is a solution to the system of equations (2) (in standard slack form) obtained by setting n variables equal to zero and solving for the remaining m variables. In addition, the solution for the m variables must be *unique*.

Notes:
a. Some authors use *basic solution* in place of basic point.
b. We see that we can identify basic points with extreme point candidates, at least based upon the problems solved thus far in the current chapter.
c. The use of the word *unique* in Definition 4.4 is necessary; there are certain systems of m linear equations in m unknowns that cannot be solved uniquely. We saw an example of this in Problem 1.4. In that problem a case existed (the inconsistency) where we had a system of three equations in three unknowns that led to no solution at all for the variables x, y, and z.

Definition 4.5. A *basic feasible point* is a *basic point* for which the m variables solved for are all greater than or equal to zero (≥ 0).

Notes:
a. Some authors use *basic feasible solution* in place of basic feasible point.
b. Based upon the problems solved thus far in the current chapter, we can see a correspondence between basic feasible points and extreme points. In fact, we shall soon state two theorems that indicate that this correspondence holds in general.
c. Definition 4.5 really is logical. All we are saying is that we shall call a basic point which happens to be feasible a basic feasible point. Perhaps feasible basic point would be better, but such terminology is not common.

Problem 4.9

Identify the basic points and basic feasible points in Problems 4.1 and 4.4.

Solution

Problems 4.1 and 4.4 both dealt with the same situation. The system of $m = 2$ equations in the $m + n = 4$ unknowns was given as follows:

$$2x + y - 12 = -r \tag{2.1}$$

$$x + 2y - 12 = -s. \tag{2.2}$$

Here we have labeled the equations (2.1) and (2.2) and not (2) and (3). Thus, by Equations (2) we mean Equations (2.1) and (2.2). We now set two variables equal to zero and attempt to solve for the remaining two. We have already done all the necessary work in Problem 4.4. In fact, the results of these calculations are given in the table of that problem.

The six points O, A, B, C, D, and E are all basic points, while the points O, A, B, and C are the basic feasible points. (Here we *"extend"* the number of components so that each point has four components: x, y, r, and s.) In this case we clearly see the correspondence between extreme point candidates and basic points on the one hand and between extreme points and basic feasible points on the other hand.

The proofs of the following two theorems require a level of mathematical knowledge (in the area of linear algebra) higher than that which has been assumed in this book thus far. Consequently, no proofs will be given here. Proofs of these theorems appear in [1], [4], and [6]. However, the reader is again warned that the mathematical level is fairly sophisticated; in addition, the notation and terminology vary somewhat from book to book.

Theorem 4.1. *Every basic feasible point is an extreme point.*

Theorem 4.2. *Every extreme point is a basic feasible point.*

In the previous chapters we learned that the solution points for maximum or minimum problems occur at extreme points if they occur at all. Because of the correspondence between extreme points and basic feasible points, as indicated by Theorems 4.1 and 4.2, we may confine our search to basic feasible points. The simplex algorithm is a method where one initially starts at a basic point (such a point is easy to determine in a standard problem—all one has to do is set n variables equal to zero) and then works toward arriving at a basic feasible point. After the initial basic feasible point is reached, the simplex method proceeds in a systematic way to other basic feasible points and finally reaches, in a finite number of steps, the *solution* basic feasible point of the problem. The solution point is the point where u is maximized in a maximum problem or where u is minimized in a minimum problem.

Note: In many cases, it is easy to initially determine a basic feasible point. If this is the case, the total effort required is decreased.

In a few more pages we shall be ready for the simplex algorithm.

At this point it will be useful to recall Theorem 3.1, which deals with extreme points.

Theorem 3.1: Extreme Point. *A point $\bar{e}$ in S_c is an extreme point of S_c if it is the intersection point of n bounding hyperplanes of the set.*

Recalling that hyperplanes are given by the *equation* parts of the constraints, it follows that bounding hyperplanes for our standard problems will have one of the two following forms:

$$a_{i1}x_1 + a_{i2}x_2 + \cdots + a_{in}x_n = b_i \quad \text{or} \quad x_j = 0.$$

Based on the above discussion, we can see the essential equivalence between a basic feasible point and an extreme point. As an example, let us refer to Problem 4.5. The basic feasible point D was characterized by $x = z = r = 0$ and $y > 0$, $s > 0$. The first three are equivalent to the equations of three bounding hyperplanes, namely, $x = 0$, $z = 0$, and

$$x + \tfrac{2}{3}y + \tfrac{1}{4}z = 900.$$

Thus, D lies at the intersection point of three (i.e., $n = 3$) bounding hyperplanes. That the point D is in the constraint set is revealed by the fact that all five inequalities of the original problem are satisfied. For example, $s > 0$ is equivalent to

$$\tfrac{1}{3}y + \tfrac{3}{4}z < 600.$$

Problem 4.10

Show that an upper bound for the number of basic points (and, thus, basic feasible points) is given by

$$\text{Upper bound} = \frac{(m + n)!}{m!n!}.$$

Solution

The reader is referred to the standard slack form of linear programming problems (e.g., refer to the discussion preceding Problem 4.1 (maximum problem) and to the solution of Problem 4.6 (minimum problem)). In both maximum and minimum problems, the relevant equations are the m equations labeled (2.1) through (2.m). The equations contain $(m + n)$ unknowns (or variables). To find a possible basic point, we set n variables equal to zero and then attempt to solve the resulting system of m equations in m unknowns. If a unique solution results, we have found a basic point. Thus, we see that an upper bound for the number of basic points is just the number of ways of choosing n variables (and setting them equal to zero) from a total of $(m + n)$ variables. Thus, the upper bound is none other than $C(m + n, n)$—the number of combinations of $(m +$

n) things taken n at a time; the latter is equal to $(m + n)!/m!n!$. Thus,

$$\text{Upper bound} = C(m + n, n) = \frac{(m + n)!}{m!n!} .$$

Since the number of basic points is at least as great as the number of basic feasible points, the above number is also an upper bound for the number of basic feasible points.

The reader is advised to refer to Problem 1.9, which is very similar.

We note that, with our new terminology, examples of basic points were provided by Problems 4.4 and 4.5. Recall that for basic points we first set n variables equal to zero in Equations (2.1) through (2.m) and then attempt to solve the m equations for the remaining m variables. In Problems 4.4 and 4.5 it turned out that for each basic point, the m solved-for variables are all not equal to zero. Such basic points are called *nondegenerate* basic points. In practice, there do occur basic points for which some of the m solved-for variables turn out to be equal to zero. Such basic points are called *degenerate basic points*. Problem 4.11 will illustrate the preceding discussion, but first we present two definitions.

Definition 4.6. A *nondegenerate* basic point is a basic point such that each of the m solved-for variables is not equal to zero. A *degenerate* basic point is a basic point for which *at least one* of the m solved-for variables is equal to zero.

Definition 4.7. A *nondegenerate* basic feasible point is a basic feasible point such that each of the m solved-for variables is *positive*. A *degenerate* basic feasible point is a basic feasible point for which *at least one* of the m solved-for variables is equal to zero; the rest are, of course, positive.

Problem 4.11

Consider the following standard maximum problem:

$$\text{Maximize} \quad u = 4x + 2y + z \tag{1}$$

$$\text{subject to} \quad x + y \leq 1 \tag{2.1}$$

$$x + z \leq 1 \tag{2.2}$$

and

$$x \geq 0 \tag{3.1}$$

$$y \geq 0 \tag{3.2}$$

$$z \geq 0. \tag{3.3}$$

Identify the basic feasible points (extreme points) of the constraint set. Determine which ones, if any, are degenerate.

Solution

The function to be maximized is really irrelevant to the solution of the problem, i.e., the constraint set is determined solely from (2) and (3). However, later on we shall attempt to solve this problem, with the u as given above, by the simplex algorithm. That is why the equation for u is included.

If we introduce slack variables r and s, we can rewrite (2) and (3) as

$$x + y - 1 = -r \tag{2.1}$$

$$x + z - 1 = -s \tag{2.2}$$

and

$$x \geq 0 \tag{3.1}$$

$$y \geq 0 \tag{3.2}$$

$$z \geq 0 \tag{3.3}$$

$$r \geq 0 \tag{3.4}$$

$$s \geq 0. \tag{3.5}$$

The sketch of the constraint set for this problem is given in Figure 4.3. We plot the bounding planes $x + y = 1$ (or $r = 0$) and $x + z = 1$ (or $s = 0$) as well as $x = 0$, $y = 0$, and $z = 0$. The constraint set consists of those points within and on the surface of the *pyramid* shown. We have a case here of a bounded constraint set.

Note that $x + y = 1$ is parallel to the Z axis because the variable z is missing from it, while $x + z = 1$ is parallel to the Y axis because here the variable y is missing. The plane labeled $x = 0$ is none other than the yz plane. Similarly, $y = 0$ and $z = 0$ are coordinate planes.

At any rate, we can determine the extreme points (or basic feasible points) by inspection of Figure 4.3 and by using Equations (2.1) and (2.2). For example, point B is located at the intersection of $x + z = 1$ ($s = 0$), $y = 0$, and $x = 0$. Thus, setting $x = 0$, $y = 0$, and $s = 0$ in Equations (2.1) and (2.2) yields

$$-1 = -r \quad \text{and} \quad z - 1 = 0.$$

This results in $r = 1$ and $z = 1$ for point B. Thus, B is a nondegenerate basic feasible point because the solved-for variables are both *positive*. Notice that here we set $n = 3$ variables equal to zero and solved for the remaining $m = 2$ variables from Equations (2).

As another example, point C is located at the intersection of $x + z = 1$ ($s = 0$), $x + y = 1$ ($r = 0$), and $x = 0$. Thus, setting $x = 0$, $r = 0$, and

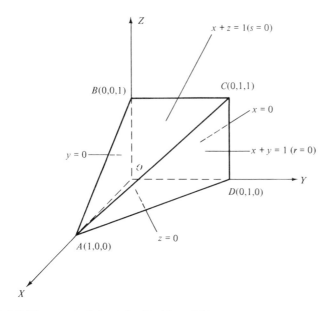

FIGURE 4.3 The constraint set for Problem 4.11.

$s = 0$ in Equations (2.1) and (2.2) yields

$$y - 1 = 0 \quad \text{and} \quad z - 1 = 0.$$

This results in $y = 1$ and $z = 1$ for point C. Thus, point C is also a nondegenerate basic feasible point.

As a final sample calculation, consider point A, which lies at the intersection of $x + y = 1$ $(r = 0)$, $y = 0$, and $z = 0$. Thus, setting $r = 0$, $y = 0$, and $z = 0$ in Equations (2.1) and (2.2) yields

$$x - 1 = 0 \quad \text{and} \quad x - 1 = -s.$$

This results in $x = 1$ and $s = 0$ for point A. Thus, we see that point A is a *degenerate* basic feasible point because one of the solved-for variables is equal to zero.

The following table indicates the values of the five variables (three original plus two slack variables) for the five basic feasible points of this problem. Only one—point A—is a degenerate basic feasible point:

Basic feasible point	x	y	z	r	s	u
O	0	0	0	1	1	0
A (degenerate)	1	0	0	0	0	4
B	0	0	1	1	0	1
C	0	1	1	0	0	3
D	0	1	0	0	1	2

Notes:

a. Since we know that all the basic feasible points O, A, B, C, and D are also extreme points, we may employ Theorem 1.1 (or 3.3) to solve the problem for maximum u. From the above tabulated values for u, we see that u has its maximum value of 4 at point A.

b. There are very simple geometrical interpretations to nondegenerate and degenerate basic feasible points. Recall that for a standard problem a bounding hyperplane (in the case just considered, an ordinary plane) can be represented by an equation of the form

$$\text{Variable} = 0.$$

In a three-dimensional problem such as Problem 4.11, (see Figure 4.3), a nondegenerate basic feasible point occurs at a point where exactly three of the bounding planes intersect, i.e., at a point where exactly three of the variables equal zero, the other two variables being positive at this point. This is what happens at points O, B, C, and D of Problem 4.11.

On the other hand, basic feasible point A occurs at the intersection of four bounding planes, i.e., four variables equal zero, the remaining variable being positive. Thus, for the case where $n = 3$, we may say that a degenerate basic feasible point is a point of intersection of more than three bounding planes.

c. For a standard linear programming problem, we may interpret a degenerate basic feasible point as a point in the constraint set where more than n bounding hyperplanes intersect. If exactly n bounding hyperplanes intersect at a point in the constraint set, then such a point would be a nondegenerate basic feasible point.

3. PRELIMINARY VIEW OF THE SIMPLEX ALGORITHM

In the simplex algorithm, one examines basic points and attempts to arrive at a basic feasible point where the optimum value (i.e., maximum or minimum value) of the objective function u is reached. Thus, it is

FIGURE 4.4 The starting (initial) tableau.

x_1	x_2	$\cdots$	x_j	$\cdots$	x_n	1	
a_{11}	a_{12}	$\cdots$	a_{1j}	$\cdots$	a_{1n}	$-b_1$	$= -t_1$
a_{21}	a_{22}	$\cdots$	a_{2j}	$\cdots$	a_{2n}	$-b_2$	$= -t_2$
$\cdots$						$\cdots$	$\cdots$
a_{i1}	a_{i2}	$\cdots$	a_{ij}	$\cdots$	a_{in}	$-b_i$	$= -t_i$
$\cdots$						$\cdots$	$\cdots$
a_{m1}	a_{m2}	$\cdots$	a_{mj}	$\cdots$	a_{mn}	$-b_m$	$= -t_m$
c_1	c_2	$\cdots$	c_j	$\cdots$	c_n	0	$= u$

clear that it would be useful to determine the variable values at basic points in a fairly simple manner. If we refer to Equations (2.1) through (2.m) that appear in the "standard slack" form for a maximum problem, we can see that it is very easy to determine the coordinates of one such basic point without much effort. (The reader is referred to the development preceding Problem 4.1.) It should again be noted that we shall abbreviate Equations (2.1) through (2.m) by writing merely Equations (2):

$$a_{11}x_1 + a_{12}x_2 + \cdots + a_{1n}x_n - b_1 = -t_1 \qquad (2.1)$$

$$a_{21}x_1 + a_{22}x_2 + \cdots + a_{2n}x_n - b_2 = -t_2 \qquad (2.2)$$

$$\cdots\cdots\cdots\cdots\cdots\cdots\cdots\cdots\cdots\cdots\cdots\cdots\cdots$$

$$\cdots\cdots\cdots\cdots\cdots\cdots\cdots\cdots\cdots\cdots\cdots\cdots\cdots$$

$$a_{m1}x_1 + a_{m2}x_2 + \cdots + a_{mn}x_n - b_m = -t_m. \qquad (2.m)$$

The simplest basic point to determine is the one for which we set all the variables $x_1, x_2, \ldots, x_n$ equal to zero in Equations (2). This leads immediately to the following values for the remaining variables:

$$t_1 = b_1, t_2 = b_2, \ldots, t_m = b_m.$$

If it turns out that the b_i's are all greater than or equal to zero (≥ 0), then the basic point is a basic feasible point. This is the case in many typical maximum problems.

At this point, it is appropriate to introduce the so-called initial (or starting) tableau for a linear programming problem (the words schema and table are also used in place of tableau).

The correspondence between Figure 4.4 and Equations (2) is clear. All we did was detach the coefficients $x_1, x_2, \ldots, x_n$ and 1 and put them in the top margin. We also included in the bottom row items that correspond to the starting equation for u:

$$u = c_1x_1 + c_2x_2 + \cdots + c_nx_n + 0 \quad \text{(to be maximized).} \qquad (1)$$

Thus, the tableau of Figure 4.4 is just a *shorthand notation* for the starting system of equations.

For practice, let us "read off" the first row from the tableau:

$$a_{11}x_1 + a_{12}x_2 + \cdots + a_{1n}x_n - b_1 = -t_1.$$

This is just Equation (2.1).

We are looking for a simple method of determining *other* basic points. Such a method would be available if we could somehow devise a way for solving the system of equations (2.1) through (2.m) for any m variables in terms of the remaining n variables. (Note that there may be certain situations where this is not possible.)

Suppose that we have been able to do this and that we have solved for the variables $r_1, r_2, \ldots, r_m$ in terms of the variables $s_1, s_2, \ldots, s_n$. The labeling $r_1, r_2, \ldots, r_m, s_1, s_2, \ldots, s_n$ is just a relabeling of the original $(m + n)$ x and t variables. In tableau form, this would result in the intermediate tableau of Figure 4.5.

The entries in the tableau of Figure 4.5 are naturally different than those in Figure 4.4, even though we use the same symbols (namely, a_{11}, $a_{12}, \ldots, -b_1, -b_2, \ldots, c_1, c_2, \ldots$, etc.). Here we "read off" the first row as

$$a_{11}s_1 + a_{12}s_2 + \cdots + a_{1n}s_n - b_1 = -r_1.$$

The bottom row is read off as

$$c_1s_1 + c_2s_2 + \cdots + c_ns_n + d = u.$$

Given the tableau of Figure 4.5 (or the equations that the rows of the tableau represent), we may now easily determine a basic point by merely setting all of $s_1, s_2, \ldots, s_n$ equal to zero. We thus obtain

$$r_1 = b_1, r_2 = b_2, \ldots, r_m = b_m.$$

We also see from the bottom row that $u = d$ at this basic point.

So much for an introduction. Our job now is to develop a method for converting from one tableau form to another. (In each form, our corresponding system of equations consists of m variables expressed in terms of the remaining n variables.) The *pivot transformation* (or pivot step or pivot algorithm) provides such a method.

We shall often refer to Figure 4.5 in our subsequent work. Thus, at this point, it is appropriate to give some definitions that are related to the format of Figure 4.5.

Definition 4.8. For a typical intermediate tableau (refer to Figure 4.5), the m variables $r_1, r_2, \ldots, r_m$ for which we have solved are called

FIGURE 4.5 A typical intermediate tableau. *Notes:* The s_j's are the nonbasic variables. The r_i's are the basic variables.

s_1	s_2	$\cdots$	s_j	$\cdots$	s_n	1	
a_{11}	a_{12}	$\cdots$	a_{1j}	$\cdots$	a_{1n}	$-b_1$	$= -r_1$
a_{21}	a_{22}	$\cdots$	a_{2j}	$\cdots$	a_{2n}	$-b_2$	$= -r_2$
$\cdots$						$\cdots$	$\cdots$
a_{i1}	a_{i2}	$\cdots$	a_{ij}	$\cdots$	a_{in}	$-b_i$	$= -r_i$
$\cdots$						$\cdots$	$\cdots$
a_{m1}	a_{m2}	$\cdots$	a_{mj}	$\cdots$	a_{mn}	$-b_m$	$= -r_m$
c_1	c_2	$\cdots$	c_j	$\cdots$	c_n	d	$= u$

basic variables. On occasion, we shall refer to the r_i's as the "solved-for" variables.

Definition 4.9. For a typical intermediate tableau, the n variables s_1, $s_2, \ldots, s_n$ in terms of which we have solved are called *nonbasic variables*.

In a typical transformation from one tableau to another, what we do is interchange (exchange) one of the basic variables and one of the nonbasic variables. The interchange is effected through the pivot transformation. Before we get into the mechanics of the pivot transformation, let us apply some of this new tableau symbolism to some practical problems.

Problem 4.12

Set up the starting tableau corresponding to Problem 4.2. Determine the basic point that corresponds to setting the nonbasic variables equal to zero.

Solution

Recall that in Problem 4.2 we merely introduced slack variables into the starting form of Problem 1.4 (the "nut problem"). If we refer to Equations (1), (2), and (3) of Problem 4.2 and represent them in the shorthand form of Figure 4.4, we arrive at the following starting tableau, which we label Tableau 1 (T.1):

$$
\begin{array}{cccc|cl}
 x & y & z & 1 & \\
\hline
1 & \frac{2}{3} & \frac{1}{4} & -900 & = -r \\
0 & \frac{1}{3} & \frac{3}{4} & -600 & = -s \\
\hline
\frac{1}{4} & \frac{2}{5} & \frac{1}{2} & 0 & = u
\end{array}
\qquad (\text{T.1})
$$

For this tableau, the $n = 3$ nonbasic variables are x, y, and z, while the $m = 2$ basic variables are r and s (the slack variables). The quantity u, although not a basic variable, will always stay in the position to the right of the equals sign in the bottom row. Recall that u is special: It is the objective function.

It is easy to work with Tableau 1. A basic point is easily arrived at by setting the $n = 3$ variables x, y, and z on the top margin equal to zero. We thus obtain

$$-900 = -r, \quad -600 = -s, \quad \text{and} \quad 0 = u,$$

i.e.,

$$r = +900, \quad s = +600, \quad \text{and} \quad u = 0.$$

Referring to Figure 4.2, we see that these calculations correspond to the basic (feasible) point $O(0,0,0)$ since $x = y = z = 0$. We say that this is the basic point *corresponding* to Tableau 1.

Note: Observe that here we list three coordinates for the point O instead of five coordinates. We shall often follow this practice of listing only the coordinates of the *original* variables when referring to a particular point.

In Problem 4.5, we attempted to solve for the basic points (extreme point candidates) and then for basic feasible points (extreme points) of Problem 4.2. We worked with Equations (2) and (3) of Problem 4.2, which are given in shorthand form by the first two rows of Tableau 1 of Problem 4.12. In Problem 4.2, we first set three variables equal to zero and then attempted to solve for the remaining two variables. We note from looking at a typical tableau that this calculation would be easy if the three variables we are setting equal to zero happen to be in the positions of nonbasic variables. For example, if we could interchange the variables y and r in Tableau 1, we could easily solve for the basic point corresponding to setting x, r, and z equal to zero.

The operation of interchanging y and r of Tableau 1 will be carried out later in this section (see Problem 4.13) as a way of illustrating the *pivot transformation*. The tableau entry in the row of r and the column of y of Tableau 1 is called the *pivot entry*; it is usually denoted by the letter p. In the case being discussed, we have $p = \frac{2}{3}$.

We shall now indicate the key aspects of the pivot transformation in schematic form. The letter p denotes the pivot entry (called pivot for short), the letter q denotes any other entry in the pivot row, the letter r denotes any other entry (besides p) in the pivot column, and the letter s denotes an entry not in the same row or column as p; it is convenient to think of s as being an entry in the row of r and column of q.

Method 4.1

The *pivot transformation* is indicated by the following diagram:

$$
\begin{array}{|cc|}
\hline
p & q \\
\\
r & s \\
\hline
\end{array}
\quad \rightarrow \quad
\begin{array}{|cc|}
\hline
1/p & q/p \\
\\
-r/p & s - qr/p \\
\hline
\end{array}
$$

The calculations indicated on the diagram are as follows:

a. The pivot p is replaced by its reciprocal. Thus, p goes to $1/p$. (Note that $p \neq 0$.)
b. The remaining entries in the pivot row are divided by the pivot. Thus, q goes to q/p.

c. The remaining entries in the pivot column are divided by the pivot and then the sign is changed. Thus, r goes to $-r/p$ or $r/(-p)$.

d. Each entry of s type goes to $s - qr/p$, where q is the entry in the row of p and column of s and r is the entry in the column of p and row of s. Note that here q and r are the two entries that "form a rectangle" with p and s.

e. Finally, we must remember to interchange the marginal labels of the pivot's row and column. All other marginal labels are unchanged.

Problem 4.13

Interchange the variables y and r in Tableau 1 of Problem 4.12 by means of the pivot transformation. Interpret the resulting tableau. Determine what basic point corresponds to the resulting tableau. Do a check calculation.

Solution

Recall that Tableau 1, reproduced below, has a strong connection with Problems 4.2 and 4.5 (these two problems are themselves related to the "nut problem," i.e., Problem 1.4):

$$
\begin{array}{cccc|cl}
x & y & z & 1 & & \\
\hline
1 & \frac{2}{3}* & \frac{1}{4} & -900 & = & -r \\
0 & \frac{1}{3} & \frac{3}{4} & -600 & = & -s \\
\hline
\frac{1}{4} & \frac{2}{5} & \frac{1}{2} & 0 & = & u \\
\end{array}
\qquad (\text{T.1})
$$

Since we wish to interchange the nonbasic variable y and the basic variable r, the pivot entry is thus $p = \frac{2}{3}$, i.e., the entry with label a_{12}. We put an asterisk next to it (on occasion, we shall encircle the pivot entry).

The resulting tableau—Tableau 2—is as follows:

$$
\begin{array}{cccc|cl}
x & r & z & 1 & & \\
\hline
\frac{3}{2} & \frac{3}{2} & \frac{3}{8} & -1350 & = & -y \\
-\frac{1}{2} & -\frac{1}{2} & \frac{5}{8} & -150 & = & -s \\
\hline
-\frac{7}{20} & -\frac{3}{5} & \frac{7}{20} & 540 & = & u \\
\end{array}
\qquad (\text{T.2})
$$

Notice that the variables y and r have been interchanged, but that all other variables remained unchanged. Now let us go through the calculations in detail.

Pivot entry: $\frac{2}{3}$ to $\frac{3}{2}$.

Other entries in pivot row (row 1):

$$1 \text{ to } \frac{1}{\frac{2}{3}} = \tfrac{3}{2}, \quad \tfrac{1}{4} \text{ to } \frac{\tfrac{1}{4}}{\frac{2}{3}} = \tfrac{3}{8}, \quad \text{and} \quad -900 \text{ to } \frac{-900}{\frac{2}{3}} = \frac{-2700}{2} = -1350.$$

Note that dividing by $\tfrac{2}{3}$ is equivalent to multiplying by $\tfrac{3}{2}$. Thus, for example, we have $\tfrac{1}{4} \cdot \tfrac{3}{2} = \tfrac{3}{8}$.

Other entries in pivot column (column 2):

$$\tfrac{1}{3} \text{ to } \frac{\tfrac{1}{3}}{-\frac{2}{3}} = -\tfrac{1}{2} \quad \text{and} \quad \tfrac{2}{5} \text{ to } \frac{\tfrac{2}{5}}{-\frac{2}{3}} = -\tfrac{3}{5}.$$

If instead of dividing by $-\tfrac{2}{3}$ we multiplied by $-\tfrac{3}{2}$, the calculations would give the same results. For example, with respect to the second calculation we would have

$$\frac{2}{5} \text{ to } \frac{2}{5} \cdot \frac{-3}{2} = \frac{-3}{5}.$$

Now let us work on the remaining entries, which are all type s entries:

Row 2:

$$0 \text{ to } 0 - \frac{1 \cdot \tfrac{1}{3}}{\frac{2}{3}} = -1 \cdot \tfrac{1}{3} \cdot \tfrac{3}{2} = -\tfrac{1}{2}.$$

$$\tfrac{3}{4} \text{ to } \tfrac{3}{4} - \frac{\tfrac{1}{4} \cdot \tfrac{1}{3}}{\frac{2}{3}} = \tfrac{3}{4} - \tfrac{1}{4} \cdot \tfrac{1}{3} \cdot \tfrac{3}{2} = \tfrac{6}{8} - \tfrac{1}{8} = \tfrac{5}{8}.$$

$$-600 \text{ to } -600 - \frac{(-900) \cdot \tfrac{1}{3}}{\frac{2}{3}} = -600 + 450 = -150.$$

Bottom Row:

$$\tfrac{1}{4} \text{ to } \tfrac{1}{4} - \frac{1 \cdot \tfrac{2}{5}}{\frac{2}{3}} = \tfrac{1}{4} - 1 \cdot \tfrac{2}{5} \cdot \tfrac{3}{2} = \tfrac{1}{4} - \tfrac{3}{5} = \frac{5 - 12}{20} = \frac{-7}{20}.$$

$$\tfrac{1}{2} \text{ to } \tfrac{1}{2} - \frac{\tfrac{1}{4} \cdot \tfrac{2}{5}}{\frac{2}{3}} = \tfrac{1}{2} - \tfrac{1}{4} \cdot \tfrac{2}{5} \cdot \tfrac{3}{2} = \tfrac{1}{2} - \tfrac{3}{20} = \tfrac{7}{20}.$$

$$0 \text{ to } 0 - \frac{(-900) \cdot \tfrac{2}{5}}{\frac{2}{3}} = +900 \cdot \tfrac{2}{5} \cdot \tfrac{3}{2} = 90 \cdot 2 \cdot 3 = 540.$$

Now let us translate the shorthand of Tableau 2 into equations:

$$\tfrac{3}{2}x + \tfrac{3}{8}r + \tfrac{3}{8}z - 1350 = -y \tag{2'}$$

$$-\tfrac{1}{2}x - \tfrac{1}{2}r + \tfrac{5}{8}z - 150 = -s \tag{3'}$$

$$-\tfrac{7}{20}x - \tfrac{3}{5}r + \tfrac{7}{20}z + 540 = u. \tag{1'}$$

Note that we indicate the Tableau 2 equations by primes.

It is now easy to determine coordinates of another basic point by working with the Tableau 2 equations. Let us set the nonbasic variables x, r, and z equal to zero. We immediately obtain

$$y = 1350, \quad s = 150, \quad \text{and} \quad u = 540.$$

Referring to the table of Problem 4.5, we see that we have redone the calculations for point D of Figure 4.2. A new piece of information from the above equations is that $u = 540$ at this point. Recall again that Figure 4.2 clearly illustrates that D is the intersection point of the three planes $x = 0$, $r = 0$, and $z = 0$. The point D is a basic feasible point, since the two variables y and s are positive. We say that point D, with original variables $x = 0$, $y = 1350$, and $z = 0$, is the basic feasible point *corresponding* to Tableau 2.

Tableau 2 (or the equations it stands for) is just an algebraically equivalent way of representing the original problem. The systems of equations corresponding to Tableaus 1 and 2 are referred to as "equivalent systems." Note that we have a convenient new equation for u, where it is given in terms of x, r, and z.

The reader has probably observed by now that it is all too easy to make arithmetical errors in carrying out the pivot transformation calculations. Thus, a check procedure would be useful. One such check is obtained by seeing if the coordinates of the basic point corresponding to Tableau 1 satisfy the equations of Tableau 2. From Tableau 1 we had

$$x = y = z = 0, \quad r = 900, \quad s = 600, \quad \text{and} \quad u = 0.$$

Substituting these into the equations of Tableau 2, we obtain the following:

$$\tfrac{3}{2}{\cdot}0 + \tfrac{3}{2}{\cdot}900 + \tfrac{3}{8}{\cdot}0 - 1350 = 1350 - 1350 = 0 = -y. \qquad (2')$$

Thus, $y = 0$.

$$-\tfrac{1}{2}{\cdot}0 - \tfrac{1}{2}{\cdot}900 + \tfrac{5}{8}{\cdot}0 - 150 = -450 - 150 = -600 = -s. \qquad (3')$$

Thus, $s = 600$.

$$-\tfrac{7}{20}{\cdot}0 - \tfrac{3}{5}{\cdot}900 + \tfrac{7}{20}{\cdot}0 + 540 = -3{\cdot}180 + 540 = 0 = u. \qquad (1')$$

Thus, $u = 0$.

Hence, we see that all the equations of Tableau 2 check out, and this is a partial confirmation that the numbers in Tableau 2 are correct.

Problem 4.14

a. For the two main constraints of Problem 4.1 use the pivot transformation to obtain two equations in which r and y are solved for in terms of x and s. Obtain the corresponding tableaus.

b. Do the same calculations in a purely algebraic way.

Solution

a. Recall that Problem 4.1 is related to Problems 1.1 and 1.2. In Section 4 (Problem 4.18) we shall solve Problem 1.1 by means of the simplex algorithm. The current work is in preparation for that task.

The two main constraints of Problem 4.1 have the two slack variables r and s expressed in terms of x and y:

$$2x + y - 12 = -r \tag{2}$$

$$x + 2y - 12 = -s. \tag{3}$$

In tableau form, the representation is as follows:

x	y	1	
2	1	-12	$= -r$
1	2^*	-12	$= -s$

$$\text{(T.1)}$$

The nonbasic variables are thus x and y, while the basic variables are r and s.

Now we wish to interchange the positions of the variables y and s, as indicated by the asterisk. The pivot entry is thus $a_{22} = 2$, i.e., the 2 in the second row and second column. The calculations are as follows:

Pivot entry: 2 to $\frac{1}{2}$.

Other entries in pivot row (row 2):

$$1 \text{ to } \tfrac{1}{2} \quad \text{and} \quad -12 \text{ to } \frac{-12}{2} = -6.$$

Other entry in pivot column:

$$1 \text{ to } \frac{-1}{2}.$$

Remaining entries (row 1):

$$2 \text{ to } 2 - \frac{1 \cdot 1}{2} = \tfrac{3}{2}$$

and

$$-12 \text{ to } -12 - \frac{(-12) \cdot (1)}{2} = -12 + 6 = -6.$$

Thus, we end up with the following tableau:

x	s	1	
$\frac{3}{2}$	$-\frac{1}{2}$	-6	$= -r$
$\frac{1}{2}$	$\frac{1}{2}$	-6	$= -y$

    (T.2)

The equations corresponding to this tableau are

$$\tfrac{3}{2}x - \tfrac{1}{2}s - 6 = -r \tag{2'}$$

$$\tfrac{1}{2}x + \tfrac{1}{2}s - 6 = -y. \tag{3'}$$

The system of equations (2') and (3') is *equivalent* to the system of equations (2) and (3).

    b. We wish to go from Equations (2) and (3) to Equations (2') and (3') using algebraic means.

First, let us solve Equation (3) for $-y$ in terms of x and s:

$$x + s - 12 = -2y$$

$$\therefore \tfrac{1}{2}x + \tfrac{1}{2}s - 6 = -y.$$

The latter equation is Equation (3'). Let us now multiply (3') by -1 to obtain

$$-\tfrac{1}{2}x - \tfrac{1}{2}s + 6 = y.$$

We now substitute this equation for y into Equation (2):

$$2x + (-\tfrac{1}{2}x - \tfrac{1}{2}s + 6) - 12 = -r.$$

Simplifying, we obtain

$$\tfrac{3}{2}x - \tfrac{1}{2}s - 6 = -r.$$

This is none other than Equation (2').

Problem 4.15

Prove the validity of the pivot transformation for the case where there are two nonbasic and two basic variables.

Solution

Refer to Figure 4.5 for the picture of a typical intermediate tableau. For our problem, it becomes

s_1	s_2	1	
a_{11}	a_{12}	$-b_1$	$= -r_1$
a_{21}	a_{22}	$-b_2$	$= -r_2$
c_1	c_2	d	$= u$

    (T.1)

The equations corresponding to Tableau 1 are

$$a_{11}s_1 + a_{12}s_2 - b_1 = -r_1 \tag{1}$$

$$a_{21}s_1 + a_{22}s_2 - b_2 = -r_2 \tag{2}$$

$$c_1s_1 + c_2s_2 + d = u. \tag{3}$$

Thus, the original nonbasic variables are s_1 and s_2, while the original basic variables are r_1 and r_2. Suppose that we wish to interchange r_1 and s_1 in Tableau 1. Then a_{11} is the pivot entry. We assume that $a_{11} \neq 0$. Our approach will merely be a generalization of the procedure used in Problem 4.14.

First, we solve Equation (1) for $-s_1$ in terms of r_1 and s_2:

$$r_1 + a_{12}s_2 - b_1 = -a_{11}s_1$$

$$\frac{1}{a_{11}}r_1 + \frac{a_{12}}{a_{11}}s_2 - \frac{b_1}{a_{11}} = -s_1. \tag{1'}$$

If we multiply Equation (1') by -1, we obtain

$$-\frac{1}{a_{11}}r_1 - \frac{a_{12}}{a_{11}}s_2 - \frac{(-b_1)}{a_{11}} = s_1.$$

We now substitute this expression for s_1 in terms of r_1 and s_2 into Equations (2) and (3):

$$a_{21}\left(-\frac{1}{a_{11}}r_1 - \frac{a_{12}}{a_{11}}s_2 + \frac{b_1}{a_{11}}\right) + a_{22}s_2 - b_2 = -r_2$$

$$c_1\left(-\frac{1}{a_{11}}r_1 - \frac{a_{12}}{a_{11}}s_2 + \frac{b_1}{a_{11}}\right) + c_2s_2 + d = u.$$

Collecting terms, we obtain

$$-\frac{a_{21}}{a_{11}}r_1 + \left(a_{22} - \frac{a_{12}a_{21}}{a_{11}}\right)s_2 - \left(b_2 - \frac{b_1a_{12}}{a_{11}}\right) = -r_2 \tag{2'}$$

$$-\frac{c_1}{a_{11}}r_1 + \left(c_2 - \frac{a_{12}c_1}{a_{11}}\right)s_2 + \left(d - \frac{(-b_1)c_1}{a_{11}}\right) = u. \tag{3'}$$

If we carefully inspect Equations (1'), (2'), and (3'), we see how the simplex tableau is to be changed. Let us first relabel the coefficients of Equations (1'), (2'), and (3') so that they conform to the standard labeling in a typical tableau:

$$a'_{11}r_1 + a'_{12}s_2 - b'_1 = -s_1 \tag{1a'}$$

$$a'_{21}r_1 + a'_{22}s_2 - b'_2 = -r_2 \tag{2a'}$$

$$c'_1r_1 + c'_2s_2 + d' = u. \tag{3a'}$$

Thus, the tableau corresponding to Equations (1a'), (2a'), and (3a') is as follows:

	r_1	s_2	1	
	a'_{11}	a'_{12}	$-b'_1$	$= -s_1$
	a'_{21}	a'_{22}	$-b'_2$	$= -r_2$
	c'_1	c'_2	d'	$= u$

(T.2)

The superscript prime (') on the coefficients indicates "new" values. If we compare Equations (1'), (2'), and (3') with Equations (1), (2), and (3), we see that r_1 and s_1 have been interchanged and that the respective coefficients are related as follows:

$$a'_{11} = \frac{1}{a_{11}} \text{ (pivot entry), i.e., } p' = \frac{1}{p} \left(p \text{ goes to } \frac{1}{p} \right)$$

$$a'_{12} = \frac{a_{12}}{a_{11}} \text{ (entry in pivot row), i.e., } q' = \frac{q}{p} \left(q \text{ goes to } \frac{q}{p} \right)$$

$$-b'_1 = \frac{-b_1}{a_{11}} \text{ (entry in pivot row), i.e., } q' = \frac{q}{p}$$

$$a'_{21} = -\frac{a_{21}}{a_{11}} \text{ (entry in pivot column), i.e., } r' = -\frac{r}{p}$$

$$c'_1 = -\frac{c_1}{a_{11}} \text{ (entry in pivot column), i.e., } r' = -\frac{r}{p}.$$

Thus, we see that the pivot entry and the remaining entries in the pivot row and pivot column obey the rules of the pivot transformation (Method 4.1). The four remaining entries, namely, a'_{22}, b'_2, c'_2, and d', are s-type entries, and their computation conforms to the transformation rule for such entries. Let us illustrate for a'_{22} and d':

$$a'_{22} = a_{22} - \frac{a_{12}a_{21}}{a_{11}}, \quad \text{i.e., } s' = s - \frac{qr}{p}$$

$$d' = d - \frac{(-b_1)c_1}{a_{11}}, \quad \text{i.e., } s' = s - \frac{qr}{p}.$$

The general proof of the validity of Method 4.1 involves only a more careful bookkeeping with regard to subscripts; in other respects, it does not differ from the procedure just considered.

It will be convenient for the later work if we now define what is meant by a basic point *corresponding* to a given tableau. The reader should refer to the general intermediate tableau of Figure 4.5.

Definition 4.10. The basic point *corresponding* to a typical tableau is that point obtained by setting all the n nonbasic variables equal to zero. Thus, referring to Figure 4.5, we see that the basic point is described by the following equations:

$$s_1 = s_2 = \cdots = s_n = 0 \qquad \text{(nonbasic variables)}$$

$$r_1 = b_1, r_2 = b_2, \ldots, r_m = b_m \quad \text{(basic variables)}.$$

Note: The reader should observe that the symbols r_1, r_2, etc., and s_1, s_2, etc., refer to the basic and nonbasic variables, respectively, while r and s, when used with respect to Method 4.1 (pivot transformation), are labels for locations relative to the pivot entry. The two different usages for the pair of letters r and s should not be confused.

The basic point is feasible if all the r_i's are nonnegative (a quantity is said to be nonnegative if it is positive or zero; it is said to be nonpositive if it is negative or zero). Thus, *a tableau corresponds to a basic feasible point* if all the b_i's (which equal the r_i's) are nonnegative or, equivalently, *if all the* $-b_i$'s *are nonpositive*.

Problem 4.16

Identify the basic points that correspond to the tableaus of Problems 4.13 and 4.14.

Solution

Referring to Tableaus 1 and 2 of Problem 4.13, we respectively set the nonbasic variables equal to zero. Thus, we have the following:

$$\text{Tableau 1:} \quad x = y = z = 0, \quad r = 900, \quad \text{and} \quad s = 600$$

$$\text{Tableau 2:} \quad x = r = z = 0, \quad y = 1350, \quad \text{and} \quad s = 150.$$

If we look at Figure 4.2, we see that Tableau 1 corresponds to basic feasible point O and that Tableau 2 corresponds to basic feasible point D. We note, therefore, the important fact that the pivot transformation from Tableau 1 to Tableau 2 corresponds to movement from point O to point D along the edge connecting both points.

Similarly, let us refer to Tableaus 1 and 2 of Problem 4.14. After setting the nonbasic variables equal to zero, we have the following:

$$\text{Tableau 1:} \quad x = y = 0, \quad r = 12, \quad \text{and} \quad s = 12$$

$$\text{Tableau 2:} \quad x = s = 0, \quad r = 6, \quad \text{and} \quad y = 6.$$

Referring to Figure 4.1, we see that the first tableau corresponds to basic feasible point O and that the second corresponds to basic feasible point C. Thus, the pivot transformation from Tableau 1 to Tableau 2 *corresponds* to movement from point O to point C.

In a previous discussion we distinguished between nondegenerate and degenerate basic points. If we apply the correspondence principle here, we have a convenient way of distinguishing between tableaus representing nondegenerate and degenerate basic points.

A tableau corresponds to a degenerate basic point if one or more of the $-b_i$'s (b_i's) are zero. If all the $-b_i$'s are nonzero, then the tableau corresponds to a nondegenerate basic point.

Definition 4.11. A tableau is called a *degenerate tableau* if one or more of the $-b_i$'s are equal to zero.

If all the $-b_i$'s are nonpositive, the tableau corresponds to a basic *feasible* point. Thus, in this case, we can say that a tableau corresponds to a *degenerate basic feasible point if one or more of the $-b_i$'s is zero and the rest are negative.* If all the $-b_i$'s are negative (all the b_i's are positive), then the tableau corresponds to a nondegenerate basic feasible point.

Problem 4.17

Develop the initial tableau corresponding to Problem 4.11. Then use the pivot transformation to interchange the variables x and s. Determine the basic points corresponding to the tableaus and discuss them with respect to feasibility and degeneracy.

Solution

The initial tableau corresponding to Equations (2.1), (2.2), and (1) of Problem 4.11 is given as follows:

x	y	z	1	
1	1	0	-1	$= -r$
1^*	0	1	-1	$= -s$
4	2	1	0	$= u$

The initial nonbasic variables are the original variables x, y, and z, while the initial basic variables are the slack variables r and s. The bottom row, which represents the equation for u in terms of x, y, and z, is not really relevant to this problem. We include it with a view toward what will come later—we shall need the bottom row when we solve Problem 4.11 by the simplex algorithm (Problem 5.1).

If we set the nonbasic variables equal to zero, we see that Tableau 1 corresponds to point O of Figure 4.3, where $x = y = z = 0$ and $r = 1$ and $s = 1$. In addition, $u = 0$ at this point.

To interchange variables x and s, we pivot on $a_{21} = 1$. Going through the pivot step calculations, we obtain Tableau 2:

s	y	z	1	
-1	1	-1	0	$= -r$
1	0	1	-1	$= -x$
-4	2	-3	4	$= u$

(T.2)

If we set the nonbasic variables s, y, and z equal to zero, we see that $r = 0$ and $x = 1$. Thus, Tableau 2 corresponds to point A of Figure 4.3. We also note from Tableau 2 that $u = 4$ at point A.

Since both $-b_1$ and $-b_2$ are negative in Tableau 1, we see that point O is a nondegenerate basic feasible point. From Tableau 2 we see that point A is a degenerate basic feasible point, since $-b_2$ is negative while $-b_1$ is zero.

The calculation of variables for basic points as well as a discussion of degeneracy were carried out from a somewhat different point of view in Problem 4.11.

In Sections 4 and 5 of this chapter we shall develop the simplex algorithm. For this it is important to keep in mind some points about notation. The key symbolism with regard to tableaus is given in Figure 4.5 (a typical intermediate tableau). The a_{ij}'s given there refer to entries in the body of the tableau. Thus, a_{11} is the entry in row 1 and column 1 *regardless of which tableau* is being analyzed in a problem having more than one tableau. When we speak of *right-hand column* entries, we shall mean the $-b_i$'s in that column, *not* the right-hand corner entry d. When we speak of the *bottom row* entries, we shall mean the c_j's in that row; again, we exclude entry d. Finally, the entry d will be referred to as the *corner entry* or *right-hand corner entry*.

4. STAGE TWO OF THE SIMPLEX ALGORITHM

It is important to keep in mind the correspondence between tableaus and basic points (Definition 4.10). In Stage One of the simplex algorithm one starts with a tableau corresponding to a *nonfeasible* basic point and uses a sequence of pivot steps (following a logical pattern) until a tableau corresponding to a basic feasible point is attained. We shall be concerned with Stage One later on. In Stage Two one starts with a tableau corresponding to a basic feasible point (thus, in this tableau all the $-b_i$'s are nonpositive). Let us tentatively assume that we are dealing with a maximum problem. The Stage Two strategy is to execute a sequence of pivot steps in which the objective function improves (i.e., increases) in each step until finally a tableau is reached where the objective function is maximized. The increase in the objective function is indicated by an increase in the value of the corner entry d. (After all, u equals d at the basic point corresponding to a tableau.) Note that in situations where degeneracy exists it is possible that u *may not change* from one tableau to the next. This exceptional case will be treated later.

The geometrical situation for Stage Two may be described in the following way. In each pivot step from one tableau to the next, in effect there is movement along an edge from one basic feasible point to another. Such basic feasible points are called adjacent. In general, we say that two basic points are *adjacent* if they correspond to two tableaus such that one can be obtained from the other by a single pivot step. During the movement along the edge, the objective function is (usually) increased. In fact, the structure of the Stage Two algorithm is designed in such a way that the objective function *will not decrease* as the pivot transformation is made from one tableau to the next.

The word *edge* was used above. An edge is determined by setting ($n - 1$) variables equal to zero. Two basic feasible points have an edge in common if their corresponding tableaus have ($n - 1$) nonbasic variables in common. This is equivalent to saying that one of their tableaus can be obtained from the other by a single pivot step.

In Problem 4.13, transformation from Tableau 1 to Tableau 2 was equivalent to moving from point O to point D along the edge OD (refer to Figure 4.2). This edge may be described in terms of the nonbasic variables common to both tableaus (x and z in this case). Thus, edge OD is determined by the equations $x = 0$ and $z = 0$. Here $n - 1 = 3 - 1 = 2$.

In Problem 4.14, the pivot transformation from Tableau 1 to Tableau 2 was equivalent to moving from point O to adjacent point C along the edge OC of Figure 4.1. This edge is determined by the equation $x = 0$; the variable x is the only nonbasic variable common to both tableaus. Here $n - 1 = 2 - 1 = 1$.

As a way of introducing the Stage Two algorithm, let us again refer to Problem 4.14. We rewrite Tableau 1, but with a bottom row added for the objective function u of Problem 1.1.

x	y	1	
2	1	-12	$= -r$
1	2*	-12	$= -s$
3	4	0	$= u$

$$(\text{T.1})$$

This tableau corresponds to the basic feasible point $O(0,0)$, where u equals zero while r and s both equal 12. Our goal is to maximize u, the equation for which is

$$u = 3x + 4y. \tag{1}$$

We would like to employ a pivot step in such a way that u increases from the Tableau 1 basic feasible point to the basic feasible point corresponding to the next tableau. This increase in u is manifested as an increase in d—the corner entry.

Equation (1) expresses u in terms of the nonbasic variables of Tableau 1. One of these will remain nonbasic as we go from Tableau 1 to Tableau 2. Suppose that we keep x nonbasic (in other words, x will equal zero at the basic feasible points corresponding to both tableaus). Since the coefficient of y is positive, we would want to increase y as much as possible from its starting value of zero. To see how much we can increase y, let us write out the other equations corresponding to Tableau 1. After transposing a little, we obtain

$$2x + \ y + r = 12 \tag{2}$$

$$x + 2y + s = 12. \tag{3}$$

In addition to these main constraints, we also have the nonnegativity constraints

$$x \geq 0,\ y \geq 0,\ r \geq 0, \quad \text{and} \quad s \geq 0.$$

Since x is to be equal to zero at both points, we can rewrite Equations (2) and (3) as

$$y + r = 12 \tag{2'}$$

$$2y + s = 12. \tag{3'}$$

In Equation (2′) we can increase y to 12, corresponding to which $r = 0$. If y increases beyond 12, then r would become negative, and this is disallowed by the $r \geq 0$ constraint. In Equation (3′) we can increase y to 6, at which point $s = 0$. Increasing y beyond 6 would cause s to become negative, which is also disallowed.

Because all the constraints must be satisfied, it is clear that we can only increase y from 0 to 6. This tells us that *in our pivot transformation we should interchange y and s*. These are the two variables in Equation (3′)—the equation that determined the largest allowable increase in y. The effect of this transformation will be to increase y from 0 to 6 and to decrease s from 12 to 0. Equivalently, y goes from nonbasic to basic, while s takes the reverse path. From Equation (1) we see that u will increase to $4 \cdot 6 = 24$ from its initial value of zero.

Thus, let us pivot on $a_{22} = 2$ in Tableau 1. Following the usual pivot step rules, we obtain Tableau 2:

x	s	1	
$\frac{3}{2}*$	$-\frac{1}{2}$	-6	$= -r$
$\frac{1}{2}$	$\frac{1}{2}$	-6	$= -y$
1	-2	24	$= u$

(T.2)

If we set the nonbasic variables equal to zero, we have $x = s = 0$, $r = 6$, $y = 6$, and $u = 24$. Thus, Tableau 2 corresponds to point $C(0,6)$ in Figure 4.1. Let us write out the equations corresponding to Tableau 2, where the first corresponds to the bottom row:

$$x - 2s + 24 = u \tag{1a}$$

$$\tfrac{3}{2}x - \tfrac{1}{2}s + r = 6 \tag{2a}$$

$$\tfrac{1}{2}x + \tfrac{1}{2}s + y = 6. \tag{3a}$$

Equation (1a) indicates that we would like to increase x from zero in the next pivot step because the coefficient of x is positive, namely, $+1$. This increase in x would cause u to increase, and that is desirable. Increasing s from zero would only cause u to decrease, since the coefficient of s is negative. A decrease in u is to be avoided. Thus, there is no doubt that x is the nonbasic variable to be interchanged. The other nonbasic variable—s—will remain nonbasic. Thus, at the basic feasible points corresponding to Tableaus 2 and 3, we have $s = 0$. Using this information, we rewrite Equations (2a) and (3a) as follows:

$$\tfrac{3}{2}x + r = 6 \tag{2a'}$$

$$\tfrac{1}{2}x + y = 6. \tag{3a'}$$

In Equation (2a'), we can increase x to the value $\tfrac{2}{3} \cdot 6 = 4$, i.e., x increases from 0 to 4, while r goes from 6 to 0. In Equation (3a'), we can increase x to the value $2 \cdot 6 = 12$. Thus, the above calculations mean that x can only increase to 4, the lower of the values 4 and 12. Increasing x beyond 4 would violate the constraint $r \geq 0$. Our pivot transformation is determined by the equation giving the largest allowable increase in x. This is Equation (2a'), and the pivot transformation thus interchanges x and r. Therefore, an asterisk is placed next to $a_{11} = \tfrac{3}{2}$, which is the pivot entry. Carrying out the pivot step leads to Tableau 3, in which r and s are nonbasic variables, while x and y are basic variables:

r	s	1	
$\tfrac{2}{3}$	$-\tfrac{1}{3}$	-4	$= -x$
$-\tfrac{1}{3}$	$\tfrac{2}{3}$	-4	$= -y$
$-\tfrac{2}{3}$	$-\tfrac{5}{3}$	28	$= u$

(T.3)

To determine the corresponding basic feasible point, we set the nonbasic variables equal to zero:

$$r = s = 0, \ x = 4, \ y = 4, \quad \text{and} \quad u = 28.$$

Thus, Tableau 3 corresponds to point $B(4,4)$ in Figure 4.1. Let us write the equations corresponding to Tableau 3, again putting the equation for u first:

$$-\tfrac{2}{3}r - \tfrac{5}{3}s + 28 = u \tag{1b}$$

$$\tfrac{2}{3}r - \tfrac{1}{3}s + x = 4 \tag{2b}$$

$$-\tfrac{1}{3}r + \tfrac{2}{3}s + y = 4. \tag{3b}$$

Notice that the coefficients of the variables r and s in the equation for u are both negative. This fact leads us to conclude that the maximum value of u on the constraint set S_c is 28 and that this occurs when r and s are both zero. In proving this, what we first have to show is that $u \leq 28$ for any point in S_c. The demonstration follows:

a. Pick any point $[x,y,r,s]$ in S_c.
b. Thus, the original constraints (2) and (3) and all the nonnegativity constraints are satisfied.
c. In particular, $r \geq 0$ and $s \geq 0$.
d. In Equation (1b), we thus have that

$$-\tfrac{2}{3}r - \tfrac{5}{3}s \leq 0.$$

e. However,

$$u = -\tfrac{2}{3}r - \tfrac{5}{3}s + 28.$$

f. Thus, we see that u equals a number that is less than or equal to 28, i.e.,

$$u \leq 28.$$

Thus, we have shown that u is less than or equal to 28 for all points in the constraint set. One point in the constraint set where u equals 28 *exactly* is given by the equations corresponding to Tableau 3. Setting r and s equal to zero gives $x = 4$, $y = 4$, and $u = 28$. These x, y, r, and s values satisfy Equations (2b) and (3b), which are equivalent to the original main constraints (2) and (3). Thus, the main constraints and, clearly, also the nonnegativity constraints are satisfied. It then follows that the point labeled $B(4,4)$ in Figure 4.1 is the solution point for the problem. In addition, Max $u = 28$.

The preceding discussion illustrates the key aspects of Stage Two of the simplex algorithm. Note that in going from Tableau 1 to the next tableau we could have chosen x as the nonbasic variable to be interchanged. This is so because the x coefficient in the equation for u (Equation (1)) was positive (namely, 3), meaning that u could have been increased through an increase in x.

We now state the Stage Two rules with respect to a typical intermediate tableau (see Figure 4.5). Remember that we assume that the first tableau under consideration corresponds to a basic feasible point, i.e., all the

$-b_i$'s are nonpositive. The Stage Two rules prescribe a way to choose a pivot entry. The resulting interchange of basic and nonbasic variables will lead to a new tableau in which the new $-b_i$'s will all be nonpositive. The new d value is not less than the old d value in the case of a maximum problem ($d_{\text{new}} \geq d_{\text{old}}$), while the opposite holds in a minimum problem.

Method 4.2: Stage Two of Simplex Algorithm

Part (a), Maximum Case. Let c_j be any *positive* entry in the bottom row. Thus, column j is an acceptable pivot column. For each *positive* a_{ij} in this column, calculate the ratio b_i/a_{ij} by dividing the negative of the entry in the right-hand column by a_{ij} (such b_i/a_{ij} ratios will all be nonnegative). Suppose that b_k/a_{kj} is the smallest of the ratios. Then row k is the pivot row and a_{kj} *is the pivot entry.* (If there is a tie for the smallest ratio, the pivot entry may be chosen to be *any* of those corresponding to the tie.) In the usual *terminal* tableau, all the c_j's will be nonpositive. The basic feasible point corresponding to this tableau is the solution point of the problem.

Part (b), Minimum Case. Let c_j be any *negative* entry in the bottom row. Thus, column j is an acceptable pivot column. For each *positive* a_{ij} in this column, calculate the ratio b_i/a_{ij} by dividing the negative of the entry in the right-hand column by a_{ij} (such b_i/a_{ij} ratios will all be nonnegative). Suppose that b_k/a_{kj} is the smallest of the ratios. Then row k is the pivot row and a_{kj} *is the pivot entry.* (For the procedure in the case of ties, see Part (a).) In the usual *terminal* tableau, all the c_j's will be nonnegative. The basic feasible point corresponding to this tableau is the solution point of the problem.

Figure 4.6 illustrates the various terms involved in the Stage Two rules.

FIGURE 4.6 Schematic diagram for Stage Two of the simplex algorithm. b_k/a_{kj} is the smallest of the b_i/a_{ij} ratios for which a_{ij} is positive. Thus, a_{kj} is the pivot entry.

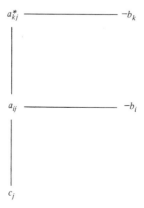

Proofs that application of the Stage Two rules leads to a new tableau which is an improvement over the old tableau will be given later, but it is instructive to first work through some problems. The proof dealing with the usual terminal tableau for a maximum problem is presented in Problem 4.21.

Problem 4.18

Solve Problem 4.1 (or 1.1) by using Stage Two of the simplex algorithm. Interpret the results graphically.

Solution

First we develop the starting tableau from Equations (1), (2), and (3) of Problem 4.1 (this is also the starting tableau in the discussion immediately preceding Method 4.2):

x	y	1		
2	1	-12	$= -r$	
1	2^*	-12	$= -s$	$O(0,0)$
3	4	0	$= u$	

(T.1)

The basic point corresponding to this tableau is the origin $O(0,0)$ of Figure 4.7. Henceforth, the corresponding point will be written to the left of the tableau label.

Stage Two clearly applies, since $-b_1 = -12$ and $-b_2 = -12$ ($b_1 = +12$ and $b_2 = +12$). Both c_1 and c_2 are positive, so that either column 1 or column 2 is an acceptable pivot column (we have a maximum problem). Suppose we let column 2 be the pivot column (in Problem 4.19, we shall let column 1 be the pivot column). The relevant b_i/a_{i2} ratios are as follows:

$$\frac{b_1}{a_{12}} = -\frac{(-12)}{1} = \frac{12}{1} = 12$$

$$\frac{b_2}{a_{22}} = -\frac{(-12)}{2} = \frac{12}{2} = \underline{6}.$$

(Underlining indicates the smallest b/a ratio.) Since the latter value is smaller, it follows that $a_{22} = 2$ is the pivot element (marked with an asterisk in Tableau 1). We carry through the pivot step and arrive at Tableau 2. Note that this is the same second tableau as in the discussion preceding Method 4.2. Tableau 2 corresponds to the basic feasible point C of Figure 4.7:

x	s	1		
$\frac{3}{2}^*$	$-\frac{1}{2}$	-6	$= -r$	
$\frac{1}{2}$	$\frac{1}{2}$	-6	$= -y$	$C(0,6)$
1	-2	24	$= u$	

(T.2)

We note that the new $-b_i$'s are nonpositive, as predicted, and that the new d value (24) is larger than the old value (0). The Stage Two rules apply to Tableau 2 because one of the c_j's is positive, namely, $c_1 = +1$. Forming the b_i/a_{i1} ratios, we have

$$\frac{b_1}{a_{11}} = \frac{-(-6)}{(\frac{3}{2})} = 4$$

$$\frac{b_2}{a_{21}} = \frac{-(-6)}{(\frac{1}{2})} = 12.$$

Thus, $a_{11} = \frac{3}{2}$ is the pivot entry in Tableau 2, since b_1/a_{11} is the smaller quantity. Again carrying through the pivot step, we obtain Tableau 3, which is the same third tableau as in the discussion preceding Method 4.2:

r	s	1			
$\frac{2}{3}$	$-\frac{1}{3}$	-4	$= -x$	$B(4,4)$	(T.3)
$-\frac{1}{3}$	$\frac{2}{3}$	-4	$= -y$		
$-\frac{2}{3}$	$-\frac{5}{3}$	28	$=\ u$		

We cannot apply the Stage Two rules to Tableau 3 because both c_j values are negative.

Thus, Tableau 3 corresponds to the solution point of the maximum problem and is called a terminal tableau. Setting the nonbasic variables equal to zero leads to

$$r = s = 0,\ x = 4,\ y = 4, \quad \text{and Max } u = u(B) = 28.$$

The sequence of Tableaus 1, 2, and 3 corresponds to the arrow path from O to C to B in Figure 4.7.

Problem 4.19

Choose column 1 of Tableau 1 of Problem 4.18 as the initial pivot column. Rework the problem and interpret the results graphically.

Solution

The starting tableau—Tableau 1—is the same as in Problem 4.18:

x	y	1			
2*	1	-12	$= -r$	$O(0,0)$	(T.1)
1	2	-12	$= -s$		
3	4	0	$=\ u$		

Since $c_1 = 3$ is positive, column one is an acceptable pivot column. The

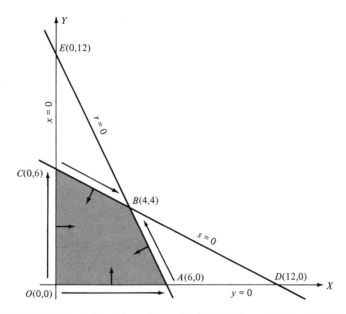

FIGURE 4.7 The basic feasible point paths for Problems 4.18 and 4.19.

calculations for pivot entry are as follows:

$$\frac{b_1}{a_{11}} = \frac{12}{2} = \underline{6}$$

$$\frac{b_2}{a_{21}} = \frac{12}{1} = 12.$$

Thus, $a_{11} = 2$ is the pivot entry. We exchange x and r and obtain Tableau $2'$ by the usual pivot step calculations:

r	y	1	
$\frac{1}{2}$	$\frac{1}{2}$	-6	$= -x$
$-\frac{1}{2}$	$\frac{3}{2}*$	-6	$= -s$
$-\frac{3}{2}$	$\frac{5}{2}$	18	$= u$

$A(6,0)$ (T.2′)

To find the point corresponding to Tableau $2'$, we set the nonbasic variables r and y equal to zero:

$$r = y = 0, \ x = 6, \ s = 6, \quad \text{and} \quad u = 18.$$

Thus, Tableau $2'$ corresponds to basic feasible point A of Figure 4.7, and u has improved from 0 to 18 in going from the origin to point A. Tableau $2'$ is not terminal because c_2 is positive. Thus, we choose column 2 as the pivot column. The calculations for pivot entry are as follows:

$$\frac{b_1}{a_{12}} = \frac{6}{(\frac{1}{2})} = 12$$

$$\frac{b_2}{a_{22}} = \frac{6}{(\frac{3}{2})} = 4.$$

Thus, $a_{22} = \frac{3}{2}$ is the pivot entry; the pivot step from Tableau 2' to Tableau 3' results in the interchange of y and s:

r	s	1		
$\frac{2}{3}$	$-\frac{1}{3}$	-4	$= -x$	
$-\frac{1}{3}$	$\frac{2}{3}$	-4	$= -y$	$B(4,4)$
$-\frac{2}{3}$	$-\frac{5}{3}$	28	$= u$	

$$(\text{T.3}')$$

We notice that Tableau 3' is identical to Tableau 3 of Problem 4.18. Tableau 3' is *terminal*, and it corresponds to the solution point $B(4,4)$ of Figure 4.7, where $u(B) = \text{Max } u = 28$, as before.

The transition from Tableau 1 to Tableau 2' was equivalent to moving from the origin to point A. The nonbasic variable that was not interchanged, namely, y, corresponded to movement along the edge $y = 0$. In the transition from Tableau 2' to Tableau 3', the unchanging nonbasic variable was r. The tableau transition corresponded to movement from point A to point B along the edge $r = 0$. Thus, the sequence of Tableaus 1, 2', and 3' corresponds to the path (indicated by arrows) from O to A to B in Figure 4.7.

Problem 4.20

Solve Problem 1.4 (the ''nut problem'') by Stage Two of the simplex algorithm. Describe the path corresponding to the sequence of tableaus.

Solution

The main constraints in terms of slack variables are given in Problem 4.2, and the corresponding initial tableau appears in Problem 4.13. The relevant diagram—Figure 4.2—is reproduced here for convenience as Figure 4.8. Tableau 1 is as follows:

x	y	z	1		
1	$\frac{2}{3}*$	$\frac{1}{4}$	-900	$= -r$	
0	$\frac{1}{3}$	$\frac{3}{4}$	-600	$= -s$	$O(0,0,0)$
$\frac{1}{4}$	$\frac{2}{5}$	$\frac{1}{2}$	0	$= u$	

$$(\text{T.1})$$

Stage Two applies to Tableau 1 because both $-b_1$ and $-b_2$ are nonpositive (here both are negative). In other words, Tableau 1 corresponds to

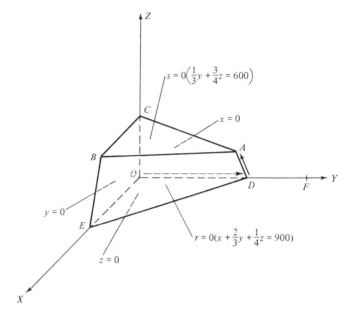

FIGURE 4.8 The basic feasible point path for Problem 4.20. Coordinates of labeled points: O(0,0,0), A(0,1260,240), B(700,0,800), C(0,0,800), D(0,1350,0), E(900,0,0), and F(0,1800,0).

a basic feasible point. Since c_1, c_2, and c_3 are all positive, the pivot column could be any of the three columns. Suppose we choose column 2—the y column—to be the pivot column. The calculations for pivot entry are as follows:

$$\frac{b_1}{a_{12}} = \frac{900}{(\frac{2}{3})} = \frac{2700}{2} = \underline{1350}$$

$$\frac{b_2}{a_{22}} = \frac{600}{(\frac{1}{3})} = 1800.$$

Thus, $a_{12} = \frac{2}{3}$ is the pivot entry; the pivot step from Tableau 1 to Tableau 2 will thus interchange y and r. The detailed calculation of the entries in Tableau 2 was already carried out in Problem 4.13:

x	r	z	1	
$\frac{3}{2}$	$\frac{3}{2}$	$\frac{3}{8}$	-1350	$= -y$
$-\frac{1}{2}$	$-\frac{1}{2}$	$\frac{5}{8}*$	-150	$= -s$
$-\frac{7}{20}$	$-\frac{3}{5}$	$\frac{7}{20}$	540	$= u$

D(0,1350,0)   (T.2)

The basic point corresponding to Tableau 2 is obtained by setting the nonbasic variables equal to zero:

$$x = r = z = 0, \; y = 1350, \; s = 150, \quad \text{and} \quad u = 540.$$

Thus, Tableau 2 corresponds to basic feasible point D of Figure 4.8. Notice that in going from Tableau 1 to Tableau 2, the variables x and z remained nonbasic. Thus, the edge corresponding to the Tableau 1 to Tableau 2 transition is the edge described by $x = 0$ and $z = 0$. This is just another way of identifying the edge OD.

Tableau 2 is not terminal because c_3 is positive. Since this is the only positive c_j, it follows that column 3 is the pivot column. The calculations for pivot entry are as follows:

$$\frac{b_1}{a_{13}} = \frac{1350}{\left(\frac{3}{8}\right)} = \frac{1350 \cdot 8}{3} = 3600$$

$$\frac{b_2}{a_{23}} = \frac{150}{\left(\frac{5}{8}\right)} = \frac{150 \cdot 8}{5} = \underline{240}.$$

Thus, $a_{23} = \frac{5}{8}$ is the pivot entry in the Tableau 2 to Tableau 3 transition. The variables z and s will be interchanged by the associated pivot step. The resulting tableau—namely, Tableau 3—is as follows:

x	r	s	1		
$\frac{9}{5}$	$\frac{9}{5}$	$-\frac{3}{5}$	-1260	$= -y$	$A(0,1260,240)$
$-\frac{4}{5}$	$-\frac{4}{5}$	$\frac{8}{5}$	-240	$= -z$	(solution point)
$-\frac{7}{100}$	$-\frac{8}{25}$	$-\frac{14}{25}$	624	$= u$	

(T.3)

Tableau 3 is a terminal tableau because all the c_j's are negative. The solution point is the point corresponding to Tableau 3. This is obtained by setting the nonbasic variables equal to zero:

$$x = r = s = 0, \; y = 1260, \; z = 240, \quad \text{and Max } u = 624.$$

In terms of the original variables x, y, and z, we see that this point is the basic feasible point labeled A in Figure 4.8. It can be shown that the maximum value of u is 624 as a result of the fact that all the c_j's in Tableau 3 are nonpositive, and the general proof of this will be carried out in Problem 4.21. For now, let us write the equation for u from Tableau 3:

$$u = -\tfrac{7}{100}x - \tfrac{8}{25}r - \tfrac{14}{25}s + 624.$$

We know that $u = 624$ at the basic feasible point A. To prove that the maximum value of u is 624, all we have to show is that $u \le 624$ for any feasible point (i.e., any point in the constraint set). To begin with, for any feasible point, all of the variables x, y, z, r, and s are greater than or equal to zero. Thus, the right-hand side of the equation for u is the sum of three nonpositive terms plus 624. In other words, $u \le 624$, as was to be shown.

In going from Tableau 2 to Tableau 3, the unchanging nonbasic variables were x and r. Thus, the corresponding edge is given by $x = 0$ and $r = 0$; this is none other than the edge DA. Therefore, the basic feasible path for Problem 4.20 is from O to D to A.

Problem 4.21

Refer to the format of the general intermediate tableau (Figure 4.5). Show that a tableau in which all the $-b_i$'s and c_j's are nonpositive is a *terminal* tableau for a maximum problem (i.e., show that the tableau *corresponds* to a solution point of a maximum problem and that the corner entry d of the tableau is the value of Max u).

Solution

Let us find the basic point corresponding to the tableau. We set all the nonbasic variables equal to zero:

$$s_1 = s_2 = \cdots = s_n = 0. \tag{1}$$

This yields the following values for the r_i's and u:

$$r_1 = b_1, r_2 = b_2, \ldots, r_m = b_m, \quad \text{and} \quad u = d. \tag{2}$$

Thus, the r_i's are all nonnegative, since the b_i's are all nonnegative. In other words, corresponding to the tableau, we have a *basic feasible* point for which $u = d$. Let us write the equation for u corresponding to the bottom row and d entries:

$$u = c_1 s_1 + c_2 s_2 + \cdots + c_n s_n + d. \tag{3}$$

Let us now pick *any* feasible point, i.e., a point whose coordinates satisfy the m main constraints and the n nonnegativity constraints. For such a point, all of the $c_j s_j$ terms of the u equation would be nonpositive (nonpositive times nonnegative). Thus, from (3) we see that $u \le d$ for such a point. In other words, the basic feasible point above for which u exactly equals d is a solution point of the problem, and the maximum value of u (Max $u = d$) occurs at that point.

Note: For a minimum problem, a terminal tableau that results from applying the Stage Two rules is one in which all the $-b_i$'s are nonpositive (as in the maximum case), but where *all the c_j's are nonnegative*. The argument is similar to that of Problem 4.21. Here, for a typical feasible point, all of the $c_j s_j$ terms of Equation (3) would be nonnegative (nonnegative times nonnegative), and thus $u \ge d$ for such a point. We conclude that a solution point of the minimum problem is the basic feasible point corresponding to the terminal tableau and that Min u equals the corner entry d of that tableau. We will now illustrate the situation for a minimum problem.

Problem 4.22

Given the following tableau for a minimum problem, apply the Stage Two rules until a terminal tableau is reached; cite the values of the variables and Min u at the solution point:

	x	y	s	1		
	2	1^*	2	-10	$= -r$	
	-4	5	-4	-80	$= -z$	(T.1)
	3	-9	6	600	$=\ \ u$	

Solution

The Stage Two rules are applicable, since the $-b_i$'s in Tableau 1 are both negative, i.e., Tableau 1 corresponds to the following basic feasible point:

$$x = y = s = 0, r = 10, \quad \text{and} \quad z = 80.$$

In addition, $u = 600$ at this point. Since we have a minimum problem, our goal is to reach a tableau in which all the c_j's are nonnegative. Thus, we follow the procedure suggested by Part (b) of Stage Two (Method 4.2). Since only c_2 is negative, we focus on column 2:

$$\frac{b_1}{a_{12}} = \frac{10}{1} = \underline{10}$$

$$\frac{b_2}{a_{22}} = \frac{80}{5} = 16.$$

(Recall that underlining indicates the smallest b/a ratio.) Thus, $a_{12} = 1$ is the pivot entry, and the resulting pivot step will interchange y and r. Tableau 2 is as follows:

	x	r	s	1		
	2	1	2	-10	$= -y$	
	-14	-5	-14	-30	$= -z$	(T.2)
	21	9	24	510	$=\ \ u$	

Tableau 2 is a terminal tableau, since all the c_j's are positive. The solution point is the basic feasible point corresponding to Tableau 2:

$$x = r = s = 0, y = 10, z = 30, \quad \text{and} \quad \text{Min } u = 510.$$

Thus far, in our worked-out problems on the Stage Two rules, the terminal (i.e., last) tableau was one in which all the c_j's were nonpositive (maximum case) or nonnegative (minimum case). By terminal, we mean that it *is not possible* to apply the Stage Two rules. We saw that such a tableau corresponded to a *solution point* for the problem and that the optimum value for u (maximum or minimum) was the d value of that terminal tableau.

There exists another case where it is not possible to apply the Stage Two rules. This occurs in a maximum problem when we arrive at a tableau for which we have a column (not the right-hand column) with a positive entry in the bottom row but no other positive entries; i.e., for some j, c_j is positive but all the a_{ij}'s in the column are nonpositive. Such a tableau corresponds to an *unbounded* linear programming problem.

Definition 4.12. An *unbounded* linear programming problem is one in which the constraint set is an unbounded set and the objective function is an unbounded function; i.e., u can be made as large as one wishes in the maximum case and as small as one desires in the minimum case.

Problem 4.23: Unbounded Linear Programming Problem— A Particular Case

Attempt to solve the following maximum problem by applying the Stage Two rules; draw the relevant sketch and interpret it:

$$\text{Maximize} \quad u = x + y \tag{1}$$

$$\text{subject to} \quad -2x + y \le 2 \tag{2}$$

$$-x + y \le 4 \tag{3}$$

$$x \ge 0 \tag{4}$$

$$y \ge 0. \tag{5}$$

Solution

If we introduce nonnegative slack variables r and s into (2) and (3), we obtain the following starting tableau:

x	y	1	
-2	1^*	-2	$= -r$
-1	1	-4	$= -s$
1	1	0	$= u$

(T.1)

This corresponds to basic feasible point $O(0,0)$ in Figure 4.9. Let us choose column 2 to be the pivot column in Tableau 1. In the usual way,

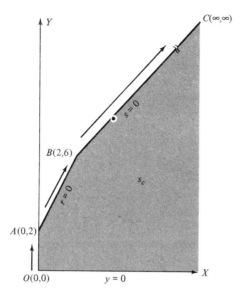

FIGURE 4.9 An unbounded linear programming problem. $r = 0$ is also $-2x + y = 2$. $s = 0$ is also $-x + y = 4$.

we see that $a_{12} = 1$ is the pivot entry:

$$\frac{b_1}{a_{12}} = \frac{2}{1} = 2,$$

$$\frac{b_2}{a_{22}} = \frac{4}{1} = 4.$$

The pivot transformation leads us to Tableau 2, which corresponds to basic feasible point A in Figure 4.9:

x	r	1	
-2	1	-2	$= -y$
1^*	-1	-2	$= -s$
3	-1	$+2$	$= u$

(T.2)

The pivot column is column 1, since $c_1 = +3$. Also, $a_{21} = 1$ is the pivot entry, since this is the only positive a_{ij} in the first column. Pivoting now leads us to Tableau 3, which corresponds to basic feasible point $B(2,6)$ in Figure 4.9:

s	r	1	
2	-1	-6	$= -y$
1	-1	-2	$= -x$
-3	$+2$	8	$= u$

(T.3)

We cannot apply the Stage Two rules to this tableau because while c_2 is positive, both a_{12} and a_{22} are negative. Looking at Figure 4.9, we see that we would like to move along the edge $s = 0$. It is easy to see from the sketch that the edge $s = 0$ starting from B is of infinite extent and that the constraint set is also unbounded. We can see this algebraically by writing the equations corresponding to Tableau 3, after setting $s = 0$:

$$-3(0) + 2r + 8 = \quad u$$
$$2(0) - \quad r - 6 = -y$$
$$1(0) - \quad r - 2 = -x.$$

Simplifying and transposing leads to

$$u = 2r + 8 \tag{6}$$
$$y = \quad r + 6 \tag{7}$$
$$x = \quad r + 2. \tag{8}$$

From (7) and (8) we see that for any positive value of r we shall have a point that satisfies the constraints (i.e., a feasible point). Furthermore, u can be made as large as we desire simply by making r large enough. For example, if we wanted to locate a feasible point where u is 2,000,008, all we would have to do is let $r = 1,000,000$.

In conclusion, the constraint set is such that we can make u as large as we wish merely by making r large enough (and holding s equal to zero). This means that we have an unbounded linear programming problem.

Note: In Tableau 1 of Problem 4.23 we notice that the first column indicates that the problem is unbounded: c_1 is positive there, but a_{11} and a_{21} are both negative. Using the same approach as was used for Tableau 3, we could show that the edge $y = 0$ (the X axis) extends indefinitely in the constraint set and that u is unbounded on that edge.

Problem 4.24: Unbounded Maximum Linear Programming Problem—General Case

Suppose that in the course of applying Stage Two rules, we arrive at a tableau in which there is some column j for which the c_j entry is positive but all the a_{ij} entries in the column are nonpositive. Show that this implies an unbounded linear programming problem.

Solution

Refer to Figure 4.5 for an example of a typical intermediate tableau. Suppose, without loss of generality, that column 1 is the column where

c_1 is positive and all the a_{i1} entries are nonpositive:

$$c_1 > 0, \quad \text{but} \quad a_{i1} \le 0 \quad \text{for} \quad i = 1, 2, \ldots, m.$$

Let us set all the nonbasic variables, except s_1, equal to zero (i.e., we are focusing on the edge given by $s_2 = 0$, $s_3 = 0$, ..., $s_n = 0$). The equations corresponding to the tableau thus become

$$u = c_1 s_1 + d \tag{1}$$

$$r_1 = -a_{11}s_1 + b_1 \tag{2.1}$$

$$r_2 = -a_{21}s_1 + b_2 \tag{2.2}$$

$$\ldots\ldots\ldots\ldots\ldots$$

$$\ldots\ldots\ldots\ldots\ldots$$

$$r_m = -a_{m1}s_1 + b_m. \tag{2.m}$$

What we shall now show is that the edge $s_2 = s_3 = \cdots = s_n = 0$ is an indefinitely extending edge on which u increases without bound.

For any positive value of s_1, each of the r_i's will be nonnegative. We will demonstrate this for r_1, but the same reasoning applies for all the r_i's.

Since $a_{11} \le 0$ and $s_1 > 0$, we have

$$-a_{11}s_1 \ge 0. \tag{a}$$

Because we have a Stage Two tableau, it follows that

$$b_1 \ge 0. \tag{b}$$

Thus,

$$r_1 = -a_{11}s_1 + b_1 \ge 0. \tag{c}$$

In other words, we have a feasible point (a point that satisfies all constraints) for any positive value of s_1. We see from (1) that we can make u as large as we wish simply by taking s_1 large enough (recall that $c_1 > 0$).

Thus, the edge given by $s_2 = s_3 = \cdots = s_n = 0$ is an indefinitely extending (unbounded) edge on which u is unbounded from above. In other words, we have an unbounded linear programming problem.

Notes:
a. In Problem 4.24, the constraint set is clearly unbounded because it contains an unbounded edge.
b. The unbounded situation arises in a minimum linear programming problem if there occurs some column j where c_j is negative, but all the a_{ij}'s in the column are nonpositive.

c. We shall refer to an unbounded linear programming problem as a *problem that has no solution*. This is not a general practice of all authors!

d. The phenomenon of an unbounded linear programming problem arising in Stage Two is considered to be a pathological situation; most real-life linear programming problems, if properly posed, will have well-defined solutions. Unbounded linear programming problems seldom arise in the real world.

We shall now prove that applying the Stage Two rules to a tableau corresponding to a basic feasible point will lead us to a terminal tableau *after a finite number of pivot steps* (or tableaus). To save space, we shall henceforth use the abbreviations b.p. for basic point and b.f.p. for basic feasible point. Also, it should be noted that many authors use the word *solution* in place of point (e.g., basic feasible point is called basic feasible solution).

The reader who is more interested in numerical techniques than in theory may wish to omit reading the rest of this section and jump ahead to Section 5.

In our proof we shall make the assumption of nondegeneracy. Geometrically, this means that for each b.f.p. *not more than n* of the bounding hyperplanes pass through it. Because of our correspondence principle, this means that for *each* Stage Two tableau encountered, all the $-b_i$'s will be negative (see Definition 4.11). We shall have more to say later about the theoretical and computational factors associated with degeneracy (Section 1 of Chapter 5).

It is useful to introduce some new symbols. If a number is positive, we indicate this by the letter p. If a number is positive or zero (i.e., nonnegative), we indicate this by $p.z$. The following table indicates the four possibilities:

Number	Symbol
Positive	p
Negative	n
Positive or zero (nonnegative)	$p.z.$
Negative or zero (nonpositive)	$n.z.$

The rules of multiplication and division for these symbols follow the corresponding rules that hold for the real number system. For example,

$$p \cdot (p.z.) = p.z., \qquad p \cdot (n.z.) = n.z.$$

$$\frac{n.z.}{p} = n.z., \qquad \frac{n.z.}{n} = p.z.$$

$$p \cdot p = p, \quad p \cdot n = n, \quad n \cdot n = p.$$

Problem 4.25

Given a nonterminal b.f.p. tableau for a maximum linear programming problem. Assume nondegeneracy. Show that applying the Stage Two rules will lead to a new tableau such that (a) it also corresponds to a b.f.p., and (b) the objective function improves (increases) in value. (c) Discuss why eventually a tableau will be reached where the Stage Two rules (Method 4.2) cannot be applied. Interpret what this means.

Solution

A nondegenerate, b.f.p. tableau is one in which all the $-b_i$'s are negative (Definitions 4.10, 4.11, and accompanying discussions). Figure 4.10 is a schematic diagram indicating the relevant entries.

The fact that the tableau is nonterminal means that there is a column (which we label j) where c_j is positive and for which there are some positive a_{ij} entries. Thus, column j is an acceptable pivot column. We let a_{hj} be a typical nonpositive entry (assuming such exists) and a_{ij} be a typical positive entry. We label a_{kj} as the pivot entry. This means that a_{kj} is positive and also that b_k/a_{kj} is the *smallest* of those b_i/a_{ij} ratios for which the a_{ij} terms are positive (see Method 4.2 for the Stage Two rules). The terms $-b_h$, $-b_k$, and $-b_i$ in Figure 4.10 are all negative because of the nondegeneracy assumption. Thus, b_h, b_k and b_i are all positive.

Part (a)

Here we have to show that all the "new" $-b$ values will be negative. By "new," we refer to the *next* tableau that will be calculated by applying the pivot transformation to the current tableau. The new values of entries

FIGURE 4.10 Schematic diagram for Stage Two proofs (maximum case). The pivot entry is a_{kj}. c_j is positive. a_{ij} is positive. a_{hj} is negative or zero.

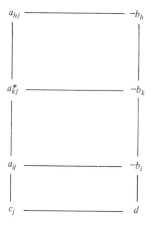

are indicated by *primes*. For example, $-b'_k$ is the value that replaces $-b_k$ in the next tableau.

Part (a), Case 1: Pivot Row k

Row k is the pivot row. We apply the pivot transformation (Method 4.1) to the row entry $-b_k$:

$$-b'_k = \frac{-b_k}{a_{kj}} .$$

Using the symbols mentioned in the discussion preceding this problem, we have

$$-b'_k = \frac{n}{p} = n.$$

Part (a), Case 2: Row i, where $a_{ij} > 0, i \neq k$

Row i is a row *other than* the pivot row where the a term is positive. By the rules for selecting a pivot in Stage Two, we have

$$\frac{b_k}{a_{kj}} \leq \frac{b_i}{a_{ij}} .$$

Thus, adding $-b_i/a_{ij}$ to both sides, we obtain

$$-\frac{b_i}{a_{ij}} + \frac{b_k}{a_{kj}} \leq 0. \tag{1}$$

Applying the pivot transformation to $-b_i$, which is neither in the pivot row nor in the pivot column (see Method 4.1), leads to

$$-b'_i = -b_i - \frac{(-b_k)a_{ij}}{a_{kj}}$$

or

$$-b'_i = -b_i + \frac{b_k a_{ij}}{a_{kj}} . \tag{2}$$

We now multiply the $-b_i$ term by $a_{ij}/a_{ij} = 1$ and then factor a_{ij} out of both terms:

$$-b'_i = a_{ij}\left(-\frac{b_i}{a_{ij}} + \frac{b_k}{a_{kj}}\right) . \tag{3}$$

If we now make use of (1) and the fact that a_{ij} is positive, we obtain (in our new symbols)

$$-b'_i = p \cdot (n.z.) = n.z. \tag{4}$$

By the nondegeneracy *assumption* we then have

$$-b'_i = n. \tag{5}$$

Part (a), Case 3: Row h, where $a_{hj} \leq 0$

By applying the pivot transformation to $-b_h$, we obtain

$$-b'_h = -b_h + \frac{b_k a_{hj}}{a_{kj}}. \tag{1}$$

We have

$$\frac{b_k a_{hj}}{a_{kj}} = \frac{p \cdot (n.z.)}{p} = n.z. \tag{2}$$

Thus, (1) becomes

$$-b'_h = n + n.z. = n. \tag{3}$$

This last step says that a negative number plus a negative or zero number equals a negative number.

Part (b)

By the value of the objective function, we mean the value at the b.p. *corresponding* to the tableau. Thus, we mean the value of the corner entry d (see Definition 4.10).

If we now apply the pivot transformation to d, which is neither in the pivot row nor in the pivot column (see Method 4.1), we obtain

$$d' = d + \frac{b_k c_j}{a_{kj}}. \tag{1}$$

We have

$$\frac{b_k c_j}{a_{kj}} = \frac{p \cdot p}{p} = p \tag{2}$$

Thus, d' equals d plus a positive number. In other words,

$$d' > d. \tag{3}$$

For a maximum problem, this means improvement, because the u value increases as we go from the current to the next tableau.

Part (c)

We see that following the Stage Two rules leads to a new tableau that also corresponds to a b.f.p. In addition, the objective function will have an improved value for the new tableau. Thus, each time we apply the rules we arrive at a b.f.p. (extreme point) that is an improvement over the prior one (the nondegeneracy assumption is important here). Hence, we cannot obtain the same b.f.p. (or tableau) more than once. The results of Problem 4.10 (or 1.9) indicate that there are a finite number of b.f.p.'s Thus, there are a finite number of distinct b.f.p. tableaus. It follows that we shall eventually reach a tableau where the Stage Two rules cannot be applied (terminal tableau).

There are two cases for which this occurs. One case is when we reach a tableau where all the c_j's are nonpositive. Such a tableau is called a *solution tableau.*

The other case occurs when there is some column j such that c_j is positive but *all* the a_{ij}'s in the column are nonpositive. We have shown that such an occurrence implies that the linear programming problem is *unbounded.*

Thus, if we start with a b.f.p. tableau, successive application of the Stage Two rules will eventually either lead to a *solution* of the linear programming problem or to a tableau that indicates *unboundedness.*

The above Stage Two demonstration also applies to minimum problems. The only change involves the sign of c_j, i.e., we apply the rules to a column for which the c_j is negative. Improvement in a tableau is manifested here by a decrease in d value from one tableau to the next.

5. STAGE ONE OF THE SIMPLEX ALGORITHM

We now turn our attention to Stage One of the simplex algorithm. For a standard linear programming problem it is an easy matter to develop an initial tableau that corresponds to a basic point, although not necessarily a basic feasible point. The purpose of Stage One is to start with a *nonfeasible* basic point tableau and proceed, by pivot steps, until we reach a b.f.p. tableau. In other words, we start with a tableau for which some of the $-b_i$'s are positive, and we try to reach a tableau for which all the $-b_i$'s are nonpositive. Once we reach a b.f.p. tableau (assuming that it is possible), we then switch over to Stage Two. In the following discussion of Stage One, $-b_\ell$ will denote the physically lowest (i.e., lowest in physical position) positive entry in the right-hand column (we exclude the corner entry here); i.e., all the $-b_i$'s, that are physically below $-b_\ell$ (if any exist) are nonpositive. We shall divide the statement of the Stage One rules into two parts; in Part (a) there do exist $-b_i$'s that are physically below $-b_\ell$.

Method 4.3: Stage One of Simplex Algorithm

Part (a)

In this case, $-b_\ell$ is the physically lowest positive entry in the right-hand column and there *do exist* nonpositive $-b_i$'s that are physically below $-b_\ell$. Suppose that $a_{\ell j}$ is a *negative* entry in row ℓ. Then column j is an acceptable *pivot column.* For $a_{\ell j}$ we now compute the ratio $b_\ell / a_{\ell j}$. In addition, for each *positive* entry a_{ij} physically below $a_{\ell j}$ in column j (if any exist), we compute the ratio b_i / a_{ij}. Let b_k / a_{kj} be the smallest of all the computed b/a ratios. Then row k is the pivot row, and a_{kj} is the pivot entry.

Notes:

a. If there are no positive a_{ij} entries physically below $a_{\ell j}$, then $a_{\ell j}$ is automatically the pivot entry. If there are positive a_{ij} entries physically below $a_{\ell j}$, then $a_{\ell j}$ *may* turn out to be the pivot entry. Here we have a situation where the pivot entry is negative.

b. In the usual numbering scheme for rows, the numbers increase as we go from top to bottom. Thus, if row i is physically below row ℓ, then i is greater than ℓ $(i > \ell)$.

Part (b)

In this case, the positive entry $-b_\ell$ is the *physically lowest* entry in the right-hand column and there *do not exist* $-b_i$'s that are physically below $-b_\ell$. Let $a_{\ell j}$ be any *negative* entry in row ℓ (if any exist). Then $a_{\ell j}$ is the pivot entry.

Stage One is designed to generate a new tableau that is "closer" to a b.f.p. tableau. It can be shown that following the Stage One, Part (a) rules will, in the nondegenerate case, lead to a tableau such that $-b_\ell$ is decreased and the $-b_i$'s physically below $-b_\ell$ will stay nonpositive. The process of applying Stage One rules is best illustrated by working out some problems.

Problem 4.26

Solve Problem 1.3 by applying Stage One and (if necessary) Stage Two of the simplex algorithm. Show the basic point (b.p.) path on a constraint set diagram.

Solution

The problem is restated as follows:

$$\text{Minimize} \quad u = 200x + 300y \tag{1}$$

$$\text{subject to} \quad x + 2y \geq 60 \tag{2}$$

$$4x + 2y \geq 120 \tag{3}$$

$$6x + 2y \geq 150 \tag{4}$$

$$x \geq 0 \tag{5}$$

$$y \geq 0. \tag{6}$$

Introducing nonnegative slack variables r, s, and t (see also Problem 4.7), the above main inequalities are converted to the following three equations:

$$-x - 2y + 60 = -r \tag{2a}$$

$$-4x - 2y + 120 = -s \tag{3a}$$

$$-6x - 2y + 150 = -t. \tag{4a}$$

From these equations and Equation (1), we obtain the following starting tableau:

x	y	1	
-1	-2	60	$= -r$
-4	-2	120	$= -s$
-6^*	-2	150	$= -t$
200	300	0	$= u$

b.p. $O(0,0)$ (T.1)

To determine the b.p. that corresponds to Tableau 1, we set $x = y = 0$. We obtain

$$r = -60, \ s = -120, \quad \text{and} \quad t = -150.$$

Thus, we see that this b.p., which is the origin (see $O(0,0)$ in Figure 4.11), is nonfeasible because r, s, and t are all negative.

Note that our starting tableau conforms to the general diagram of Figure 4.4, i.e., we solved for $-r$, $-s$, and $-t$ (negatives of slack variables) in terms of the original variables x and y. In accordance with the labeling in Figure 4.4, we have

$$-b_1 = 60, \quad -b_2 = 120, \quad \text{and} \quad -b_3 = 150$$

FIGURE 4.11 Basic point path (indicated by arrows) for Problem 4.26. $r = 0$ is also $x + 2y = 60$. $s = 0$ is also $4x + 2y = 120$. $t = 0$ is also $6x + 2y = 150$.

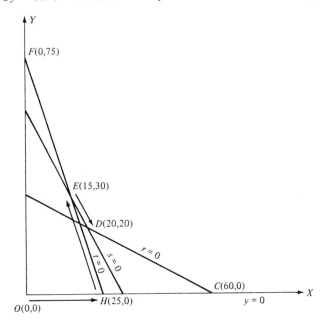

or

$$b_1 = -60, \qquad b_2 = -120, \quad \text{and} \qquad b_3 = -150$$

for our starting tableau.

Thus, Stage One is called for. The physically lowest positive $-b_i$ is $-b_3$, and we see that there are no $-b$ entries physically below it. Consequently, Part (b) of Stage One is applicable.

We now proceed. According to the rules, either $a_{31} = -6$ or $a_{32} = -2$ is acceptable as a pivot entry. Let us choose a_{31}. This means that variables x and t will be interchanged. Applying the pivot transformation (Method 4.1) to Tableau 1 results in Tableau 2:

	t	y	1			
	$-\frac{1}{6}$	$-\frac{5}{3}$	35	$= -r$		
	$-\frac{2}{3}$	$-\frac{2}{3}*$	20	$= -s$	b.p. $H(25,0)$	(T.2)
	$-\frac{1}{6}$	$\frac{1}{3}$	-25	$= -x$		
	$\frac{100}{3}$	$\frac{700}{3}$	5000	$= u$		

We see that Tableau 2 does not correspond to a b.f.p. either, since some of the $-b_i$'s are positive. The b.p. corresponding to Tableau 2 is obtained by setting $t = y = 0$. Thus, for this point (labeled H in Figure 4.11) we have $r = -35$, $s = -20$, $x = 25$, and $u = 5000$.

The transition from Tableau 1 to Tableau 2 corresponds to movement from point O to point H along the edge $y = 0$ (here and subsequently, refer to the path indicated by the arrows in Figure 4.11). Note that y is the unchanging nonbasic variable of Tableaus 1 and 2. From Figure 4.11 it is clear that points O and H are outside the constraint set, which extends above and to the right of the boundary $CDEF$, and includes the boundary points (see also Figure 1.7 for a clear demarcation of the unbounded constraint set).

Thus, Part (a) of Stage One is applicable to Tableau 2, and the $-b_\ell$ term is $-b_2 = +20$. We now examine row 2 for negative a_{2j} entries. Let us focus on $a_{22} = -\frac{2}{3}$, although a_{21} also qualifies. Thus, we let column 2 be the pivot column. We compute b/a ratios according to the Part (a) rules of Method 4.3:

$$\frac{b_2}{a_{22}} = \frac{-20}{\left(-\frac{2}{3}\right)} = \frac{60}{2} = \underline{30},$$

$$\frac{b_3}{a_{32}} = \frac{25}{\left(\frac{1}{3}\right)} = 75.$$

Note that the last term is computed because a_{32} is a positive entry that is *physically below* a_{22} in column 2. We see that a_{22} is the pivot entry for Tableau 2, since its b/a ratio is the smaller of the two ratios computed. Carrying out the pivot step leads to Tableau 3, in which y and s of

Tableau 2 have been interchanged:

t	s	1		
$\frac{3}{2}*$	$-\frac{5}{2}$	-15	$= -r$	
1	$-\frac{3}{2}$	-30	$= -y$	b.f.p. $E(15,30)$
$-\frac{1}{2}$	$\frac{1}{2}$	-15	$= -x$	
-200	350	$12,000$	$= \quad u$	

(T.3)

Setting $t = s = 0$ in Tableau 3, we see that $r = 15$, $y = 30$, $x = 15$, and $u = 12,000$. Thus, Tableau 3 corresponds to point E in Figure 4.11, which is a b.f.p. The path corresponding to the Tableau 2 to Tableau 3 transition is from point H to point E along the edge $t = 0$; t is the unchanging nonbasic variable in Tableaus 2 and 3.

Since E is a b.f.p. (equivalently, since all the $-b_i$'s of Tableau 3 are nonpositive), we see that Stage Two is now applicable. Our problem is to minimize u so the pivot column is column 1, where $c_1 = -200$. Computing the eligible b/a ratios in column 1 (recall that the a's must be positive), we have

$$\frac{b_1}{a_{11}} = \frac{15}{(\frac{3}{2})} = \frac{30}{3} = \underline{10},$$

$$\frac{b_2}{a_{12}} = \frac{30}{1} = 30.$$

Thus, a_{11} is the pivot entry of Tableau 3. Pivoting then leads to Tableau 4:

r	s	1		
$\frac{2}{3}$	$-\frac{5}{3}$	-10	$= -t$	
$-\frac{2}{3}$	$\frac{1}{6}$	-20	$= -y$	b.f.p. $D(20,20)$
$\frac{1}{3}$	$-\frac{1}{3}$	-20	$= -x$	
$\frac{400}{3}$	$\frac{50}{3}$	$10,000$	$= \quad u$	

(T.4)

Tableau 4 is a terminal tableau, since both c_j's are positive. Setting the nonbasic variables equal to zero, we have

$$r = s = 0, \, t = 10, \, y = 20, \, x = 20, \quad \text{and} \quad u = 10,000.$$

Thus, Tableau 4 corresponds to the solution point D, where $x = y = 20$ and Min $u = 10,000$. The transition from Tableau 3 to Tableau 4 corresponds to movement along the edge ED ($s = 0$). Note that s is the unchanging nonbasic variable in Tableaus 3 and 4.

It is instructive to compare the above simplex algorithm approach to the "scanning the extreme points" approach in Problem 1.3. Although the former approach is faster for this problem, the simplex algorithm is much better suited to real-world problems, in which m and n are much larger. This is particularly true because very efficient computer programs on the simplex algorithm have been written.

Problem 4.27

Use the Stage One rules on the following tableau to determine the pivot entry:

x	y	z	1	
3	20	-2	10	$= -r$
14	-20	4	-10	$= -s$
6	3	3^*	-6	$= -t$
1	10	2	15	$= \quad u$

Solution

In the right-hand column, $-b_1$ is the lowest (and the only) positive entry. Part (a) of Stage One applies. In row 1, a_{13} is the only negative entry; thus, column 3 is the pivot column. The positive a_{i3}'s physically below a_{13} in column 3 are a_{23} and a_{33}. The rules call for computing the b/a ratios for a_{13}, a_{23}, and a_{33}:

$$\frac{b_1}{a_{13}} = \frac{-10}{(-2)} = 5$$

$$\frac{b_2}{a_{23}} = \frac{10}{4} = 2.5$$

$$\frac{b_3}{a_{33}} = \frac{6}{3} = 2.$$

Since b_3/a_{33} is the smallest, the pivot entry is a_{33}.

Recall that in Stage Two there were two situations in which we could no longer apply the Stage Two rules. One was the *normal* situation, in which the terminal tableau corresponded to a solution b.f.p. The other was the *pathological* situation, in which the linear programming problem was *unbounded*.

Similar phenomena occur with respect to Stage One. On the one hand, we may end up with a so-called normal terminal tableau in which all the $-b_i$'s are nonpositive. In this case, we can no longer apply the Stage One rules. However, such a tableau means that we have located a b.f.p. and that henceforth Stage Two applies.

The other situation in which Stage One rules cannot be applied is where we encounter a row, e.g., row i, such that $-b_i$ is positive but *all* the a_{ij}'s in that row are nonnegative (i.e., positive or zero). We shall prove that this means there are no feasible points for the linear programming problem, i.e., there is no point for which all the constraints are satisfied. In other words, the constraint set has no points in it, i.e., it is the empty (null) set. We say that such a linear programming problem is *infeasible* and refer to this case as a Stage One *pathological* situation. This sort of thing should not happen in well-posed problems dealing with real-world situations.

Definition 4.13. A linear programming problem is *infeasible* if there are no points for which all the main and nonnegativity constraints are satisfied.

Let us now work on a simple numerical problem where this infeasibility situation occurs. Then we shall consider a general infeasible linear programming problem.

Problem 4.28

$$\text{Maximize} \quad u = 2x + y \tag{1}$$

$$\text{subject to} \quad x + y \leq 1 \tag{2}$$

$$x + 2y \geq 4 \tag{3}$$

$$x \geq 0 \tag{4}$$

$$y \geq 0. \tag{5}$$

Solution

We introduce nonnegative slack variables r and s into (2) and (3) and obtain

$$x + y - 1 = -r \tag{2'}$$

$$-x - 2y + 4 = -s \tag{3'}$$

We use Equations (1), (2'), and (3') to form the following starting tableau:

x	y	1		
1	1	-1	$= -r$	
-1	-2^*	$+4$	$= -s$	b.p. $O(0,0)$
2	1	0	$= u$	

(T.1)

So far, nothing unusual has appeared, and the Stage One, Part (b) rules are applicable. Either $a_{21} = -1$ or $a_{22} = -2$ is an acceptable pivot entry.

We choose the latter and thereby obtain Tableau 2:

x	s	1	
$\frac{1}{2}$	$\frac{1}{2}$	$+1$	$= -r$
$\frac{1}{2}$	$-\frac{1}{2}$	-2	$= -y$
$\frac{3}{2}$	$\frac{1}{2}$	2	$= u$

b.p. $A(0,2)$ (T.2)

It is clear that Part (a) of Stage One applies, since $-b_1 = +1$ is positive. However, we cannot follow the relevant rules because both a_{11} and a_{12} are positive!

We shall now prove, by contradiction, that row 1 of Tableau 2 implies that the problem is infeasible. The equation corresponding to row 1 is

$$-r = \tfrac{1}{2}x + \tfrac{1}{2}s + 1$$

or

$$r = -\frac{x}{2} - \frac{s}{2} - 1. \tag{6}$$

Assume there is a point such that all the constraints are satisfied. In particular, for this point we have $x \geq 0$, $y \geq 0$, $r \geq 0$, and $s \geq 0$. However, the coordinates x, s, and r are related by Equation (6). Furthermore, the first two terms on the right-hand side are nonpositive, while the third term (-1) is negative. Thus, the right-hand side of (6) is negative. However, this contradicts the fact that the left-hand side (namely, r) has to be nonnegative. This contradiction means that there are no points for which all the constraints are satisfied, i.e., the problem is infeasible.

We shall now present an interesting graphical analysis of this problem. In Figure 4.12, the set S_3 is the set of points which satisfy Inequality (3). The set $S_2 \cap S_4 \cap S_5$ (the triangle by the origin) is the set of points for which Inequalities (2), (4), and (5) are satisfied. The constraint set S_c is the intersection set of sets S_2, S_3, S_4, and S_5. As a result of the commutative and associative properties of set intersection, we can express S_c as follows:

$$S_c = S_3 \cap (S_2 \cap S_4 \cap S_5). \tag{7}$$

It is clear from Figure 4.12 that this set is the *empty set*, i.e., there is no point that is simultaneously in the region S_3 and the region $S_2 \cap S_4 \cap S_5$. This is just another way of saying that the problem is infeasible.

Note that the transition from Tableau 1 to Tableau 2 corresponds to movement from basic point $O(0,0)$ to basic point $A(0,2)$.

The following problem is a generalized version of Problem 4.28.

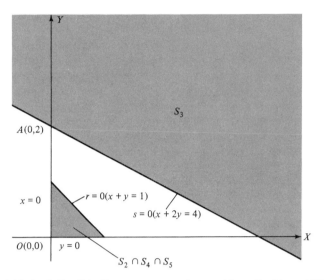

FIGURE 4.12 An infeasible linear programming problem (Problem 4.28).

Problem 4.29

Suppose that in Stage One we encounter a row, e.g., row i, for which $-b_i$ is positive, but all the a_{ij}'s in the row are nonnegative:

$$-b_i > 0, \, a_{ij} \geq 0 \quad \text{for} \quad j = 1, 2, \ldots, n.$$

Show that this implies that the linear programming problem is infeasible.

Solution

We assume that we have a standard linear programming problem, i.e., in addition to m main constraints there are nonnegativity requirements on all $m + n$ variables (n original and m slack).

Let us focus on the ith row, which represents the following equation (refer to Figure 4.5):

$$a_{i1}s_1 + a_{i2}s_2 + \cdots + a_{in}s_n - b_i = -r_i$$

or

$$r_i = b_i - (a_{i1}s_1 + a_{i2}s_2 + \cdots + a_{in}s_n). \tag{2.i}$$

Let us suppose that there is a point such that all the main and nonnegativity constraints are satisfied. Thus, for this point $r_1, r_2, \ldots, r_m$ and $s_1, s_2, \ldots, s_n$ are all nonnegative. In addition, the variables are related by the m equations corresponding to a typical tableau. Equation (2.i) is one such equation.

Each product of the form $a_{ij}s_j$ is nonnegative because each a_{ij} term and also each s_j term is nonnegative. Thus, the expression in parentheses is nonnegative, and minus that quantity (note the minus sign in front of the parenthetical expression in Equation $(2.i)$) yields a nonpositive quantity. The term b_i is negative because $-b_i$ is positive.

Thus, Equation $(2.i)$ says that r_i equals the sum of a negative quantity and a nonpositive quantity, i.e., a negative quantity. However, this is a contradiction because r_i must be nonnegative. This means that the problem is infeasible: There is no point for which all the constraints are satisfied.

Note that in the preceding argument the $-b_i$ did not have to be the physically lowest positive quantity in the right-hand column. Thus, in Stage One, if there is any row for which the $-b$ entry is positive and all the a entries are nonnegative, then it follows that the problem is infeasible.

It is possible to prove that applying the Stage One rules to a tableau corresponding to a nonfeasible b.p. will lead to a b.f.p. tableau or an infeasible tableau in a finite number of pivot steps. However, it would again be necessary to invoke the nondegeneracy assumption. The techniques involved in the proofs are similar to those used in the Stage Two proofs (Problem 4.25).

6. CONCLUDING REMARKS

In the preceding sections of this chapter we have developed the simplex algorithm. We have seen how it can be used efficiently to solve linear programming problems in standard form. In addition, we have shown by means of proofs (for Stage Two) that repeated use of the algorithm would ultimately lead to a terminal tableau (e.g., a solution tableau) in a finite number of steps. We assumed nondegeneracy in our proofs; however, the conclusions of these proofs (concerning the attainment of various terminal tableaus) hold even if degeneracy occurs. We shall have more to say about degeneracy in the coming pages (e.g., in Chapter 5).

From the various problems solved in the text, we have learned about the different types of possibilities that can arise when employing the simplex algorithm on standard linear programming problems. One such possibility is an *infeasible* linear programming problem; this is clearly revealed in Stage One. The existence of a feasible problem (i.e., there are points that satisfy all of the constraints) is also revealed in Stage One.

The subsequent use of Stage Two reveals whether the problem is *unbounded* or not. In the latter case, there ultimately occurs a solution tableau in which a solution point and the maximum or minimum value of the objective function are clearly revealed. We can separate the case in which a solution exists into two subcases. One is where the constraint set is bounded, and the other is where it is unbounded (e.g., Problem

4.26). An important feature of the Stage Two part of the simplex algorithm is that it will reach a solution tableau if one exists, irrespective of the boundedness of the constraint set. Thus, in Problem 4.26, the fact that the constraint set was unbounded offered no special difficulties.

The approach we have used, which involves a compact tableau, is being adopted by more and more authors and practitioners in the field of linear programming. Unfortunately, there is a lack of standardization in format. Our symbolism and format are closest to that of Owen [10, 11], although that used by Kemeny et al. in [8] is quite similar. Singleton and Tyndall [12] employ a type of tableau where the various equations are lined up vertically instead of horizontally as in the current text. In other respects, their approach is very close to that found here. The dual tableau, to be developed in the next chapter, is fairly similar to that of the authors mentioned above with respect to format, symbolism, and approach. Of course, there is a slight variation in dual tableaus among the different authors.

In the literature, there appear several order of magnitude estimates for the number of pivot steps needed by the simplex algorithm to solve a standard linear programming problem with m main inequality constraints and n original variables. Owen [10] cites the estimate $(m + n)$. Thus, one would expect between $(m + n)/2$ and $2(m + n)$ pivot steps in most problems. It is interesting to compare the respective efforts involved in using the simplex algorithm and the "scanning the extreme points" approach of Chapter 1. For example, if $m = 5$ and $n = 10$, the simplex algorithm would require between about 8 and 30 pivot steps. The scanning approach would involve solving $15!/(10! \cdot 5!) = 3003$ systems of equations, where each system has 10 equations in 10 unknowns (see Sections 2 and 3 of Chapter 1). It is clear that the simplex algorithm is preferable here (and for most problems where m and n are not small). Another reason for the simplex method being so widely used is that it is fairly easy to write computer programs for it. This has been done for most digital computer systems currently in existence.

The reader now has a fairly good acquaintance with some of the key characteristics of many linear programming problems. He may now wish to jump ahead to portions of Chapter 8, in which several real-world applications involving the linear programming approach are considered.

REFERENCES

1. Cooper, L., and Steinberg, D. I. *Introduction to Methods of Optimization.* Philadelphia, Pennsylvania: W. B. Saunders Company, 1970.
2. Dantzig, G. B. "Maximization of a Linear Function of Variables Subject to Linear Inequalities." In *Activity Analysis of Production and Allocation,* edited by T. C. Koopmans. New York: John Wiley and Sons, 1951.
3. Dantzig, G. B. *Linear Programs and Extensions.* Princeton, New Jersey: Princeton University Press, 1963.

4. Gass, S. I. *Linear Programming*. Third edition. New York: McGraw-Hill, 1969.
5. Hadley, G. *Linear Algebra*. Reading, Massachusetts: Addison-Wesley, 1961.
6. Hadley, G. *Linear Programming*. Reading, Massachusetts: Addison-Wesley, 1962.
7. Kemeny, J. G., Mirkil, H., Snell, J. L., and Thompson, G. L. *Finite Mathematics with Business Applications*. Englewood Cliffs, New Jersey: Prentice-Hall, 1962.
8. Kemeny, J. G., Mirkil, H., Snell, J. L., and Thompson, G. L. *Finite Mathematics with Business Applications*. Second edition. Englewood Cliffs, New Jersey: Prentice-Hall, 1972.
9. Nemhauser, G. L., and Garfinkel, R. *Integer Programming*. New York: John Wiley and Sons, 1972.
10. Owen, G. *Finite Mathematics*. Philadelphia, Pennsylvania: W. B. Saunders Company, 1970.
11. Owen, G. *Game Theory*. Philadelphia, Pennsylvania: W. B. Saunders Company, 1968.
12. Singleton, R. R., and Tyndall, W. F. *Games and Programs*. San Francisco, California: W. H. Freeman and Company, 1974.
13. Tucker, A. W. "Combinatorial Algebra of Matrix Games and Linear Programming." In *Applied Combinatorial Mathematics*, edited by E. F. Beckenbach. New York: John Wiley and Sons, 1964.
14. Tucker, A. W., and Balinski, M. L. "Duality Theory of Linear Programs: A Constructive Approach with Applications." *SIAM Review*, vol. 11, 1969.

SUPPLEMENTARY PROBLEMS

Slack Variables

S.P. 4.1: Represent the constraint set of S.P. 1.1 (S.P. means Supplementary Problem) in terms of nonnegative slack variables.

S.P. 4.2: Represent the constraint set of S.P. 1.2 in terms of nonnegative slack variables.

S.P. 4.3: Represent the constraint set of S.P. 1.5 in terms of nonnegative slack variables.

S.P. 4.4: Analyze the extreme point candidates of S.P. 1.1 in terms of the variables x, y, r, and s. (Refer to S.P. 4.1.)

S.P. 4.5: Analyze the extreme point candidates of S.P. 1.2 in terms of the variables x, y, r, and s. (Refer to S.P. 4.2.)

S.P. 4.6: Analyze the extreme points of S.P. 1.5 in terms of the variables x, y, z, r, and s. (Refer to S.P. 4.3.)

S.P. 4.7: Analyze the bounding edges of S.P. 1.1 by giving equations for them in terms of x and y and also in "Variable $= 0$" form.

S.P. 4.8: Repeat S.P. 4.7 for S.P. 1.2.

S.P. 4.9: Analyze the bounding planes of S.P. 1.5 by giving equations for them in terms of x, y, and z and also in "Variable $= 0$" form. (Refer to S.P. 4.3.)

S.P. 4.10: Identify the basic points and basic feasible points for S.P. 1.1. (Refer to S.P. 4.4.)

S.P. 4.11: Repeat S.P. 4.10 for S.P. 1.2. (Refer to S.P. 4.5.)

S.P. 4.12: Identify the basic feasible points for S.P. 1.5. (Refer to S.P. 4.6.)

S.P. 4.13: Consider the following linear programming problem, whose constraint set is in 2-space.

$$\text{Maximize} \quad u = 6x + 2y \tag{1}$$

$$\text{subject to} \quad 6x + 4y \leq 48 \tag{2.1}$$

$$x \leq 8 \tag{2.2}$$

$$y \leq 6 \tag{2.3}$$

and

$$x \geq 0. \tag{3.1}$$

$$y \geq 0. \tag{3.2}$$

Introduce nonnegative slack variables. Identify the basic feasible points (extreme points) of the constraint set. Determine which ones, if any, are degenerate.

Preliminary View of the Simplex Algorithm

S.P. 4.14: Set up the starting tableau corresponding to S.P. 1.1. Determine the basic point that corresponds to the tableau.

S.P. 4.15: Refer to S.P. 4.14. Interchange the variables x and s by using the pivot transformation. What basic point corresponds to the resulting tableau. Do a check calculation (see Problem 4.13).

S.P. 4.16: Set up the initial tableau corresponding to problem S.P. 4.13. Then use the pivot transformation to interchange x and s. Determine the basic points corresponding to the tableaus. Discuss with respect to feasibility and degeneracy.

Stage Two of the Simplex Algorithm

S.P. 4.17: Solve S.P. 1.1 by using Stage Two of the simplex algorithm. Use the x column of Tableau 1 as the initial pivot column (refer to the answer to S.P. 4.14 for Tableau 1). Display the tableaus.

S.P. 4.18: Solve S.P. 1.1 by using Stage Two of the simplex algorithm, but this time use the y column of Tableau 1 as the initial pivot column (refer to the answer to S.P. 4.14 for Tableau 1).

S.P. 4.19: Solve S.P. 1.5 by using Stage Two of the simplex algorithm. Display the tableaus and indicate the corresponding basic feasible points in x, y, z coordinates.

S.P. 4.20: Given the following tableau for a minimum problem in which all the variables are required to be nonnegative, apply the Stage Two rules until a terminal tableau is reached and list variable and Min u values at the solution point:

s	t	1	
4	−5	−20	$= -y$
2	−3	−50	$= -r$
−1	1	−30	$= -x$
−20	35	120	$= u$

(T.1)

S.P. 4.21: Given the following tableau for a maximum problem in which all the variables are required to be nonnegative, apply Stage Two rules as far as possible and indicate why the problem is unbounded:

y	r	1	
2	−2	−6	$= -x$
−4	2	−4	$= -s$
4	−2	3	$= u$

(T.1)

S.P. 4.22: Given the following tableau for a minimum problem in which all the variables are required to be nonnegative, apply Stage Two rules as far as possible and indicate why the problem is unbounded.

y	r	1	
2	−2	−6	$= -x$
−4	2	−4	$= -s$
−8	5	30	$= u$

(T.1)

Stage One of the Simplex Algorithm

S.P. 4.23: Solve S.P. 1.2 by applying Stage One and (if necessary) Stage Two of the simplex algorithm. Display the tableaus and indicate the corresponding basic points in x, y coordinates.

S.P. 4.24: Refer to Problem 4.26. In Tableau 1, use column 2 as the pivot column. Obtain successive tableaus until the optimal solution is reached. Trace through the corresponding path in Figure 4.11.

S.P. 4.25: Apply the Stage One rules to the following tableau in order to determine *all* possible pivot entries:

x	r	z	1	
−5	−10	−6	10	$= -y$
8	−7	6	−8	$= -s$
10	−8	10	−12	$= -t$
3	5	−2	17	$= u$

S.P. 4.26: Explain why the following tableau indicates an infeasible linear programming problem. Assume that all variables are required to be nonnegative.

x	y	z	1	
-5	7	8	-10	$= -r$
6	4	5	8	$= -s$
9	2	-3	-6	$= -t$
4	3	7	0	$=\ \ u$

Miscellaneous Problems Apply the simplex algorithm to each of the following problems.

S.P. 4.27: Solve S.P. 1.3.
S.P. 4.28: Solve S.P. 1.4.
S.P. 4.29: Solve S.P. 1.8.
S.P. 4.30: Solve Problem 1.12 of Chapter 1 by the simplex algorithm.
S.P. 4.31: Solve S.P. 1.11.
S.P. 4.32: Solve S.P. 1.12.
S.P. 4.33: Solve S.P. 1.13.
S.P. 4.34: Solve S.P. 1.14.
S.P. 4.35: Solve S.P. 1.15.
S.P. 4.36: Solve S.P. 1.16.
S.P. 4.37: Solve S.P. 1.17.
S.P. 4.38: Solve S.P. 1.18.
S.P. 4.39: Solve S.P. 1.19.
S.P. 4.40: A food processing company has 12,000 lb. of Brazilian coffee, 15,000 lb. of African coffee and 10,000 lb of Colombian coffee in stock. It sells three blends of coffee, labeled I, II, and III. The contents (in ounces of coffee per pound of blend) and the prices (in dollars) per pound are given in the following table:

Coffee Blend	Brazilian	African	Colombian	Price ($)
I	4	0	12	1.50
II	4	4	8	1.20
III	4	8	4	1.00

How many pounds of each blend should be produced so as to maximize sales revenues? (Recall that there are 16 ounces per pound.)

S.P. 4.41: A dietitian is given three foods: two meats and a vegetable. Each pound of the first meat contains 1 unit of carbohydrate, 4 units of vitamins, and 10 units of protein, and costs $1.98. Each pound of the second meat contains 2 units of carbohydrate, 6 units of vitamins, and 8 units of protein, and costs $1.50. Each pound of the vegetable contains 2 units of carbohydrate, 4 units of vitamins, and 2 units of protein, and costs $0.54. The minimum daily requirements for a certain diet are 8 units of carbohydrate, 20 units of vitamins, and 15 units of protein. What is the minimum-cost diet?

S.P. 4.42: A manufacturer of plumbing fixtures produces fixture types A, B and C using lathes, drill presses, and grinders. The machinery requirements (in hours) for one unit of each fixture and the daily capacities (in hours) for the different types of machines are as follows:

Fixture	Lathe	Drill press	Grinder
A	0.2	0.2	0.25
B	0.1	0.4	0.5
C	0.15	0.3	0.4
Daily machine capacity (hr)	8	10	12

The per unit profits for fixtures A, B, and C are \$1.20, \$1.40, and \$1.50, respectively. What should the daily production of each fixture be in order to maximize profits?

ANSWERS TO SUPPLEMENTARY PROBLEMS

S.P. 4.1: We have

$$x + 3y + r = 12,$$
$$2x + y + s = 19,$$

where $x, y, r, s \geq 0$.

S.P. 4.2: We have

$$x + 3y - r = 12,$$
$$2x + y - s = 19,$$

where $x, y, r, s \geq 0$.

S.P. 4.3: We have

$$12x + 9y + 3z + r = 1200$$
$$5y + 10z + s = 1000,$$

where $x, y, z, r, s \geq 0$.

S.P. 4.4:

x	y	r	s	Comment
0	0	12	19	Extreme point
0	4	0	15	Extreme point
0	19	−45	0	Not extreme point
12	0	0	−5	Not extreme point
$\frac{19}{2}$	0	$\frac{5}{2}$	0	Extreme point
9	1	0	0	Extreme point

S.P. 4.5:

x	y	r	s	Comment
0	0	-12	-19	Not extreme point
0	4	0	-15	Not extreme point
0	19	45	0	Extreme point
12	0	0	5	Extreme point
$\frac{19}{2}$	0	$-\frac{5}{2}$	0	Not extreme point
9	1	0	0	Extreme point

S.P. 4.6:

x	y	z	r	s
0	0	0	1200	1000
0	$\frac{400}{3}$	0	0	$\frac{1000}{3}$
100	0	0	0	1000
0	0	100	900	0
0	120	40	0	0
75	0	100	0	0

S.P. 4.7:

Equation in x and y for edge.	Variable $= 0$ form.
$x + 3y = 12$	$r = 0$
$2x + y = 19$	$s = 0$
$x = 0$	$x = 0$
$y = 0$	$y = 0$

S.P. 4.8: See S.P. 4.7 for the answer.

S.P. 4.9:

Equation in x, y, and z for plane.	Variable $= 0$ form.
$12x + 9y + 3z = 1200$	$r = 0$
$5y + 10z = 1000$	$s = 0$
$x = 0$	$x = 0$
$y = 0$	$y = 0$
$z = 0$	$z = 0$

S.P. 4.10: See the answer to S.P. 4.4. The table gives all the basic points. The basic feasible points are equivalent to the extreme points.

S.P. 4.11: See the answer to S.P. 4.5. The table gives all the basic points. The basic feasible points are equivalent to the extreme points.

S.P. 4.12: See the answer to S.P. 4.6. The table gives all the basic feasible points. The basic feasible points are equivalent to the extreme points.

S.P. 4.13:

b.f.p. label	x	y	r	s	t
O	0	0	48	8	6
A (Degenerate)	8	0	0	0	6
B	4	6	0	4	0
C	0	6	24	8	0

S.P. 4.14:

x	y	1	
1	3	-12	$= -r$
2*	1	-19	$= -s$
4	3	0	$= u$

(T.1)

The basic point is $x = 0$, $y = 0$ (feasible).

S.P. 4.15:

s	y	1	
$-\frac{1}{2}$	$\frac{5}{2}$	$-\frac{5}{2}$	$= -r$
$\frac{1}{2}$	$\frac{1}{2}$	$-\frac{19}{2}$	$= -x$
-2	1	38	$= u$

(T.2)

The basic point is $x = 9.5$, $y = 0$ (feasible); $u = 38$.

S.P. 4.16:

x	y	1	
6	4	-48	$= -r$
1*	0	-8	$= -s$
0	1	-6	$= -t$
6	2	0	$= u$

(T.1)

The basic point is $x = 0$, $y = 0$ (feasible).

s	y	1	
-6	4	0	$= -r$
1	0	-8	$= -x$
0	1	-6	$= -t$
-6	2	48	$= u$

(T.2)

The basic point is $x = 8$, $y = 0$ (degenerate and feasible); $u = 48$.

S.P. 4.17: The first two tableaus are given in the answers to S.P. 4.14 and S.P. 4.15. The terminal tableau is as follows:

s	r	1	
$-\frac{1}{5}$	$\frac{2}{5}$	-1	$= -y$
$\frac{3}{5}$	$-\frac{1}{5}$	-9	$= -x$
$-\frac{9}{5}$	$-\frac{2}{5}$	39	$= u$

(T.3)

For the optimal solution point, $s = r = 0$, $x = 9$, $y = 1$, and Max u = 39.

S.P. 4.18: Tableau 1 appears in the answer to S.P. 4.14, while Tableau 3 and the optimal solution data are given in the answer to S.P. 4.17. Tableau 2' is as follows:

x	r	1	
$\frac{1}{3}$	$\frac{1}{3}$	-4	$= -y$
$\frac{5}{3}$	$-\frac{1}{3}$	-15	$= -s$
3	-1	12	$= u$

(T.2')

S.P. 4.19:

x	y	z	1			
12	9	3	-1200	$= -r$		
0	5	10*	-1000	$= -s$	(0,0,0)	(T.1)
2	3	5	0	$= u$		

x	y	s	1			
12*	$\frac{15}{2}$	$-\frac{3}{10}$	-900	$= -r$		
0	$\frac{1}{2}$	$\frac{1}{10}$	-100	$= -z$	(0,0,100)	(T.2)
2	$\frac{1}{2}$	$-\frac{1}{2}$	500	$= u$		

r	y	s	1			
$\frac{1}{12}$	$\frac{5}{8}$	$-\frac{1}{40}$	-75	$= -x$		
0	$\frac{1}{2}$	$\frac{1}{10}$	-100	$= -z$	(75,0,100)	(T.3)
$-\frac{1}{6}$	$-\frac{3}{4}$	$-\frac{9}{20}$	650	$= u$		

Tableau 3 is the terminal tableau. Optimal solution data: $x = 75$, $y = 0$, $z = 100$, $r = s = 0$, and Max $u = 650$.

S.P. 4.20: Terminal Tableau 2 is as follows:

y	t	1	
$\frac{1}{4}$	$-\frac{5}{4}$	-5	$= -s$
$-\frac{1}{2}$	$-\frac{1}{2}$	-40	$= -r$
$\frac{1}{4}$	$-\frac{1}{4}$	-35	$= -x$
5	10	20	$= u$

(T.2)

Optimal solution data: $y = t = 0$, $s = 5$, $r = 40$, $x = 35$, and Min u = 20.

S.P. 4.21: From Tableau 1 obtain Tableau 2, which follows:

x	r	1	
$\frac{1}{2}$	-1	-3	$= -y$
2	-2	-16	$= -s$
-2	2	15	$= u$

(T.2)

Stage Two rules cannot be applied to Tableau 2, since $c_2 > 0$ while a_{12}, $a_{22} < 0$. This indicates an unbounded linear programming problem.

S.P. 4.22: From Tableau 1 obtain Tableau 2, which follows:

x	r	1	
$\frac{1}{2}$	-1	-3	$= -y$
2	-2	-16	$= -s$
4	-3	6	$= u$

(T.2)

Stage Two rules cannot be applied to Tableau 2, since $c_2 < 0$ (recall that this is a minimum problem) while a_{12}, $a_{22} < 0$. This indicates an unbounded linear programming problem.

S.P. 4.23:

x	y	1	
-1	-3	12	$= -r$
-2^*	-1	19	$= -s$
4	5	0	$= u$

(0,0)   (T.1)

s	y	1	
$-\frac{1}{2}$	$-\frac{5}{2}^*$	$\frac{5}{2}$	$= -r$
$-\frac{1}{2}$	$\frac{1}{2}$	$-\frac{19}{2}$	$= -x$
2	3	38	$= u$

(9.5,0)   (T.2)

$$
\begin{array}{ccc|c}
s & r & 1 & \\
\hline
\frac{1}{5} & -\frac{2}{5} & -1 & = -y \\
-\frac{3}{5} & \frac{1}{5} & -9 & = -x \\
\hline
\frac{7}{5} & \frac{6}{5} & 41 & = u
\end{array}
\qquad (9,1) \qquad\qquad (\mathrm{T}.3)
$$

The optimal solution from Tableau 3 is Min $u = 41$ when $x = 9$, $y = 1$, and $r = s = 0$. Stage One is used in Tableaus 1 and 2. Note that other tableau sequences are possible.

S.P. 4.24: The path is from $(0,0)$ to $(0,75)$ to $(15,30)$ to $(20,20)$. Stage One is needed only for the first pivot.

S.P. 4.25: Possible pivot entries are $a_{21} = 8$, $a_{12} = -10$, and $a_{33} = 10$.

S.P. 4.26: Here $-b_2 = 8$ is the physically lowest positive entry in the right-hand column, but $a_{2j} > 0$ for $j = 1, 2$, and 3. Thus (see Problem 4.29 in the text), linear programming problem is infeasible.

S.P. 4.27: Max $u = 27$ at $x = 3$, $y = 3$.

S.P. 4.28: Min $u = 21$ at $x = \frac{1}{3}$, $y = \frac{14}{3}$.

S.P. 4.29: Minimum cost equals \$13,500; operate mine A for 15 days and mine B for 30 days.

S.P. 4.30: Minimum cost equals \$8.3125 when $x = 21.25$ units and $y = 7.5$ units.

S.P. 4.31: Max $u = \$635.71$, $x = 5.71$, and $y = 3.57$. (Rounded-off *integer* solution: $u = \$630.00$, $x = 6$, and $y = 3$. For the latter, all constraints must be satisfied.)

S.P. 4.32: (a) Min $u = \$33,000$, $x = 60$, and $y = 30$. (b) Min $u = \$31,500$, $x = 60$, and $y = 30$.

S.P. 4.33: Min $u = \$12.00$, $x = 30$, and $y = 0$.

S.P. 4.34: Max $u = \$283.33$, $x = 5.10$, $y = 4.83$, and $z = 3.67$. (Rounded-off integer solution: $u = \$265.00$, $x = y = 5$, and $z = 3$.) *Alternate solution:* Max $u = \$283.33$, $x = 0$, $y = 2.92$, and $z = 7.5$. (Rounded-off integer solution: $u = \$270.00$, $x = 0$, $y = 3$, and $z = 7$.)

S.P. 4.35: (a) Maximum profit equals \$42.00; $x_1 = 12$ and $x_2 = 6$. (b) Maximum profit equals \$30.00; attained at both $x_1 = 12$, $x_2 = 6$ and $x_1 = 16$, $x_2 = 3$.

S.P. 4.36: Maximum profit equals \$13,500 per week; $x_1 = 15$ standard cars per week and $x_2 = 25$ compact cars per week.

S.P. 4.37: (a) Minimum cost equals \$3.60, $x_1 = \frac{5}{8}$ lb, and $x_2 = \frac{35}{8}$ lb. (b) Minimum cost equals \$3.852, $x_1 = 0.8$ lb, and $x_2 = 4.2$ lb.

S.P. 4.38: *Hint:* don't forget the overall requirement $x + y + z \leq 50{,}000$. Maximum annual interest equals \$5,400; $x = \$20,000$, $y = \$5,000$, and $z = \$25,000$.

S.P. 4.39: Maximum daily profit equals \$534.00; $x_1 = 3{,}700$ per day and $x_2 = 600$ per day.

S.P. 4.40: Max $u = \$35,000$; $x_1 = 3{,}333.33$ lb, $x_2 = 0$, and $x_3 = 30{,}000$ lb.

S.P. 4.41: Min $u = \$3.39$; $x_1 = 0$, $x_2 = 1.0$ lb, and $x_3 = 3.5$ lb.

S.P. 4.42: Max $u = \$53.647$; $x_1 = 32.94$, $x_2 = 0$, and $x_3 = 9.42$. (Rounded-off integer solution: Max $u = \$53.40$; $x_1 = 32$, $x_2 = 0$, and $x_3 = 10$.)

Chapter Five

The Simplex Algorithm and Duality Theory

1. DEGENERACY

We shall now investigate the difficulties associated with degeneracy. Degenerate basic points and degenerate basic feasible points were defined in Definitions 4.6 and 4.7, respectively, while Problem 4.11 dealt with a degenerate constraint set.

The Stage Two arguments depended upon the $-b_i$'s being nonzero. This ensured that successive tableaus would yield improved values of the objective function. In the case of a degenerate tableau, at least one of the $-b_i$'s is zero. Thus, the next tableau will not necessarily yield an improved value of the objective function, and the improvement of the objective function is what guarantees that the Stage Two simplex algorithm will *terminate*. Thus, in the case of degeneracy, the process of "cycling" may occur. In this process, a sequence of $(p + 1)$ tableaus is encountered, e.g., T.k, T.$(k + 1)$, T.$(k + 2)$, . . . , T.$(k + p)$, where the objective function value d is unchanged and for which Tableau $(k + p)$ is identical to Tableau k. If a computer, using a systematic process to choose the pivot element, were attempting to solve the problem, it would go through such a cycle again and again. A linear programming problem that cycles is discussed on pages 130–133 of [4]; this example was artificially constructed by Hoffman and Beale.

Such difficulties could be avoided if the computer were instructed to choose pivot elements *at random* if more than one element qualified as a pivot element. This is not hard to do if the library of routines (called subroutines) available on the computer being used contains a routine for generating random numbers. Choosing pivot elements at random would eliminate the *systematic* aspect of choosing a pivot element, and cycling would not occur.

Among the many linear programming problems considered by investigators in the field, only a few have been known to cycle, and several of these were artificially constructed [4, pages 130–133]. Up until recently (ca. 1975), it was commonly believed that cycling had never occurred in practical problems [6, pages 174, 175]. However, Kotiah and Steinberg [9] recently encountered a class of nonartificial linear programming problems involving queuing (waiting line) theory which gave rise to cycling. They found that small perturbations in the input data caused the cycling to be eliminated. We shall illustrate the perturbation approach in Problem 5.2.

Degeneracy does not necessarily lead to cycling. From a practical viewpoint, we shall not let ourselves be worried about degeneracy. The normal procedure for degeneracy is to treat it as nothing unusual, i.e., one should use conventional methods. This is illustrated in the following problem.

Problem 5.1

Use the simplex algorithm to solve Problem 4.11. Discuss the degeneracy aspects of the problem.

Solution

The problem, restated in terms of nonnegative slack variables r and s, is as follows:

$$\text{Maximize} \quad u = 4x + 2y + z \tag{1}$$

$$\text{subject to} \quad x + y - 1 = -r \tag{2.1}$$

$$x + z - 1 = -s \tag{2.2}$$

and

$$x \geq 0 \tag{3.1}$$

$$y \geq 0 \tag{3.2}$$

$$z \geq 0 \tag{3.3}$$

$$r \geq 0 \tag{3.4}$$

$$s \geq 0. \tag{3.5}$$

The initial tableau, which corresponds to the origin in Figure 4.3, is as follows:

x	y	z	1		
1	1	0	-1	$= -r$	
1^*	0	1	-1	$= -s$	$O(0,0,0)$ (T.1)
4	2	1	0	$= u$	

The first column is a suitable pivot column. Both a_{11} and a_{21} qualify as the pivot entry; because of this the next tableau will be degenerate. Choosing a_{21} as the pivot entry gives rise to Tableau 2:

s	y	z	1	
-1	1^*	-1	0	$= -r$
1	0	1	-1	$= -x$ $A(1,0,0)$ (T.2)
-4	$+2$	-3	4	$= u$

Tableau 2 is degenerate since $-b_1 = 0$. Graphically, this corresponds to the planes $s = 0$, $y = 0$, $z = 0$, and $r = 0$ intersecting at point A of Figure 4.3. Tableau 2 is not terminal; pivoting on a_{12} produces Tableau 3:

s	r	z	1	
-1	1	-1	0	$= -y$
1	0	1	-1	$= -x$ $A(1,0,0)$ (T.3)
-2	-2	-1	4	$= u$

Tableau 3 is a terminal tableau, since all the c_j's are negative. Note that both Tableau 2 and Tableau 3 correspond to the point A, where the degeneracy occurs. The b.f.p. path is thus from O to A to A. This behavior is typical of what happens when a degenerate b.f.p. is the solution point. Note that Max $u = 4$.

One way of resolving degeneracy from a theoretical point of view is through the *perturbation* approach. We shall illustrate this method by applying it to Problem 5.1. It should be stressed that this method is not usually recommended as a computational device. Discussions of the perturbation approach appear in [3], [4], and [6].

Problem 5.2

Rework Problem 5.1 by using the perturbation approach. Make a graphical interpretation. Indicate how the degeneracy has been removed.

Solution

In Tableau 2 of Problem 5.1, replace the 0 in the right-hand column by $-\epsilon$, where the "perturbation" ϵ denotes a small positive number. We thus obtain Tableau 2':

s	y	z	1	
-1	1^*	-1	$-\epsilon$	$= -r'$
1	0	1	-1	$= -x$ $A(1,0,0)$ (T.2')
-4	$+2$	-3	4	$= u$

We see that the second tableau (in this case, Tableau 2') is no longer degenerate. To obtain a graphical interpretation, let us write out the equation for $-r'$:

$$-r' = -s + y - z - \epsilon.$$

Substituting for $-s$ from Equation (2.2) of Problem 5.1, we obtain

$$-r' = x + y - (1 + \epsilon).$$

The plane $r' = 0$ is thus equivalent to $x + y = 1 + \epsilon$. We can now construct the constraint set $S_{c,\epsilon}$ that applies to the perturbed problem. This appears in Figure 5.1. Notice that the degenerate basic feasible point A of Figure 4.3 has been split into the two nondegenerate basic feasible points A and A'. Both are intersection points of exactly three bounding planes.

Pivoting in Tableau 2' with a_{12} as the pivot entry results in Tableau 3, which is a terminal tableau:

s	r'	z	1	
-1	1	-1	$-\epsilon$	$= -y$
1	0	1	-1	$= -x$ $A'(1,\epsilon,0)$
-2	-2	-1	$4 + 2\epsilon$	$= u$

(T.3')

FIGURE 5.1 The Perturbation approach (Problem 5.2).

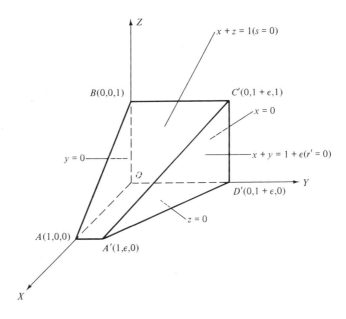

The b.f.p. corresponding to Tableau 3′ is the solution point A' of the perturbed problem:

$$x = 1, y = \epsilon, z = 0, \quad \text{and} \quad \text{Max } u = 4 + 2\epsilon.$$

The solution to the original unperturbed (but degenerate) problem is obtained by letting ϵ approach zero. This results in $x = 1$, $y = 0$, $z = 0$ (point A as before), and Max $u = 4$.

Problem 5.2 illustrates the perturbation approach. When a zero first appears in the right-hand column, it is replaced by $-\epsilon$, which represents a small negative number. If another zero appears in the right-hand column, perhaps of a subsequent tableau, it is replaced by $-\epsilon^2$, etc. In the terminal tableau of the perturbed, but *nondegenerate*, problem, the solution will be in terms of ϵ. The solution of the original unperturbed problem is obtained by letting ϵ approach zero.

The perturbation approach is useful from a theoretical standpoint because it enables one to establish that the Stage One and Two proofs now hold also for problems that are degenerate. One merely replaces a degenerate problem by an equivalent perturbed problem that is nondegenerate.

Kotiah and Steinberg (see discussion preceding Problem 5.1, and [9]) speculate that the primary reason that cycling is not known to occur often in practice is that round-off errors in the computations (done, e.g., by a digital computer) result inadvertently in perturbations. These perturbations apparently prevent cycling or, at least, indefinite cycling from occurring.

2. CLASSIFYING STANDARD LINEAR PROGRAMMING PROBLEMS

As a result of the validity of applying the perturbation approach to degenerate problems, it follows that standard linear programming problems fall into four classes. For the latter three classes, the constraint set S_c is nonempty (the problem is feasible). The four classes are labeled as follows:

A. Infeasible (S_c is the empty set)
B. Unbounded (S_c is an unbounded set and u is unbounded with respect to S_c)
C. S_c is a bounded set; the extreme (i.e., maximum or minimum) value of u is finite, and it occurs at a basic feasible point (extreme point)
D. S_c is an unbounded set; the extreme value of u is finite, and it occurs at a basic feasible point.

Stage One reveals whether or not the problem is infeasible. If a Stage Two tableau is achieved, then subsequent calculations will lead to either a tableau that reveals unboundedness (case (B)) or a solution tableau of the normal type. In this book, problems of the latter type (case (C) or (D)) are called linear programming problems *with solutions*.

Due to the nature of the above classification, we may state the following theorem, which is given for a maximum problem (however, an equivalent statement applies to a minimum problem).

Theorem 5.1. *If the linear programming problem is feasible (i.e., S_c is not the empty set), then either (1) or (2) holds:*

1. *The objective function u is maximized at a basic feasible point (case (C) or (D)).*
2. *The objective function u is unbounded from above on S_c, which itself is an unbounded set (case (B)).*

Corollary: Maximum Case. *Suppose that S_c is an unbounded set. If u is bounded from above on S_c, then the maximum value of u will occur at a basic feasible point of S_c.*

PROOF OF COROLLARY. Either (1) or (2) of Theorem 5.1 holds. Case (2) cannot hold, since by hypothesis u is bounded from above on S_c. Thus, (1) must hold. ☐

Notes:
a. The Corollary of Theorem 5.1 is equivalent to Theorem 1.2. Note also that the Corollary for the *minimum* case reads as follows: "if u is bounded from below on S_c, then the minimum value of u will occur at a basic feasible point of S_c."
b. We see that problems with solutions are those in either class (C) or (D). The following statement is a consequence of this fact: If the objective function u is bounded from above (below) with respect to the constraint set of the problem, then the maximum (minimum) value of u occurs at a b.f.p.
c. The solution class (C) indicates that if the constraint set of a linear programming problem is bounded, then the extreme value of the objective function will definitely occur at a b.f.p. This statement is equivalent to Theorem 1.1.

3. NONSTANDARD LINEAR PROGRAMMING PROBLEMS

By means of examples (Problems 5.3–5.5), we shall now illustrate how two frequently occurring nonstandard types of linear programming problems can be converted into standard slack form.

Problem 5.3

Convert Problem 3.3 into standard slack form and develop the initial tableau.

Solution

In Problem 3.3 the two original variables x_1 and x_2 were both *unrestricted*. The problem was developed in standard inequality form by writing $x_1 = x_1^+ - x_1^-$ and $x_2 = x_2^+ - x_2^-$ (refer to the items labeled (1) through (7) in Problem 3.3). In the usual way, we introduce nonnegative slack variables r and s into Inequalities (2) and (3) of Problem 3.3; the problem is then in standard slack form. We thereby obtain the following tableau, which is ready for the Stage Two algorithm:

x_1^+	x_1^-	x_2^+	x_2^-	1	
3	-3	1	-1	-5	$= -r$
2	-2	4	-4	-10	$= -s$
1	-1	3	-3	0	$= u$

(T.1)

Problem 5.3 indicates how to proceed when confronted with a problem containing unrestricted variables. In certain situations, some of the main constraints of a problem may be equations instead of inequality constraints. We shall call these *equation (equality) constraints*. There are two approaches that can be used to convert the problem into a standard form.

Problem 5.4

Convert the following problem into a standard form:

$$\text{Maximize} \quad u = x + 3y + 2z \tag{1}$$

$$\text{subject to} \quad y + 2z \le 21 \tag{2}$$

$$3x + 2y + z = 24 \tag{3}$$

and

$$x, y, z \ge 0.$$

Solution: One Approach

First, we can replace Equation (3) by the following two inequalities:

$$3x + 2y + z \le 24 \tag{3a}$$

and

$$3x + 2y + z \ge 24. \tag{3b}$$

Note that (3) will be satisfied when *both* (3a) and (3b) are satisfied and vice versa; thus, we have the equivalence between (3) and (3a), (3b). If we now introduce nonnegative slack variables r, s, and t, we obtain the following equations in place of (2), (3a), and (3b):

$$y + 2z - 21 = -r \tag{2'}$$

$$3x + 2y + z - 24 = -s \tag{3a'}$$

$$-3x - 2y - z + 24 = -t. \tag{3b'}$$

If we now place these equations together with (1) into a tableau, we will be ready to apply Stage One of the simplex algorithm ($-b_3 = +24$). The remainder of the computations are straightforward and thus will be omitted here.

In Problem 5.4 we replaced an equation by two inequalities and then introduced two slack variables to convert the inequalities into equations. There is another approach that does not require as much work. This involves leaving equations alone and then attempting to develop a tableau that has the usual (or standard) appearance. In this approach there will be *no* slack variable corresponding to an equation constraint. One then attempts to develop a tableau in standard form; this process is called *Stage Zero*. The approach is similar to that recommended by Owen [11, pp. 135–138]. We will now illustrate the Stage Zero approach by means of an example.

Problem 5.5

Employ the Stage Zero approach on Problem 5.4 and work the problem through to completion.

Solution

We introduce the nonnegative slack variable r into Inequality (2) and then enter (1), (2), and (3) into Tableau 1:

x	y	z	1	
0	1	2	-21	$= -r$
3	2	1^*	-24	$= -0$
1	3	2	0	$= u$

(T.1)

Note the -0 (which equals $+0$) in the margin of the second row; in a standard tableau we would have the negative of a basic variable there. To achieve this, we pivot on *any* nonzero entry in row 2. Thus, pivoting on $a_{23} = 1$, we obtain Tableau 2:

x	y	0	1	
-6	-3^*	-2	$+27$	$= -r$
3	2	1	-24	$= -z$
-5	-1	-2	48	$= u$

(T.2)

We now delete column 3, which has a zero in the top margin, since the elements in that column are all multiplied by zero. Tableau 2 will then have the appearance of a standard tableau with two basic variables solved for in terms of two nonbasic variables. The process of going from Tableau 1 to Tableau 2 is known as Stage Zero. Tableau 2 is now ready for Stage One. Pivoting on $a_{12} = -3$ results in

x	r	1	
$+2$	$-\frac{1}{3}$	-9	$= -y$
-1	$\frac{2}{3}$	-6	$= -z$
-3	$-\frac{1}{3}$	39	$= u$

(T.3)

We see that Tableau 3 is a terminal tableau and that the solution of the problem is

$$x = r = 0, \ y = 9, \ z = 6, \quad \text{and} \quad \text{Max } u = 39.$$

4. TRANSPORTATION PROBLEMS

A particular type of problem, where the main constraints are equations, falls into the general category known as *transportation problems*. Such problems are concerned with shipping at lowest cost a product such as wood, oil, or steel from different source locations (e.g., warehouses) to various markets. We shall begin by illustrating this type of problem with a numerical example.

Problem 5.6

Refer to the following table:

	M_1	M_2		
W_1	5	8	15	Supplies
W_2	4	10	5	(tons)
	12	8		

Demands (tons)

We consider the shipment of steel from two warehouses W_1 and W_2 to two markets M_1 and M_2. The cost of shipping from warehouse W_i to market M_j is given in the ith row and jth column of the table. For example, the cost of shipping from W_1 to M_2 is $c_{12} = 8$ \$/ton. The supplies (a_i) at the warehouses are listed at the right of the table; thus, the supply at W_1 is $a_1 = 15$ tons. The demands (b_j) at the markets are listed at the bottom of the table; thus, the demand at M_1 is $b_1 = 12$ tons. Note that the sum of the supplies equals the sum of the demands:

$$a_1 + a_2 = 20 = b_1 + b_2.$$

Transportation problems in which total supply equals total demand are called *balanced*.

Let x_{ij} be the amount in tons to be shipped from warehouse W_i to market M_j. The problem is to ship the steel in the least expensive (minimum cost) way and in so doing completely exhaust the supplies at the warehouses and exactly satisfy the demands at the markets.

Set up the problem in linear programming form and solve it by employing the simplex algorithm.

Solution

The cost in dollars in shipping x_{11} tons from W_1 to M_1 is five times x_{11}, i.e., $5x_{11}$. We can thus express the total shipping cost u in the following way:

$$u = 5x_{11} + 8x_{12} + 4x_{21} + 10x_{22} \quad \text{(to be minimized)}. \tag{1}$$

The requirements pertaining to the warehouses and markets are equation constraints (see the next-to-last paragraph of the problem statement):

$$x_{11} + x_{12} \qquad\qquad\qquad = 15 \tag{2.1}$$

$$x_{21} + x_{22} = 5 \tag{2.2}$$

$$x_{11} \qquad + x_{21} \qquad = 12 \tag{2.3}$$

$$x_{12} \qquad + x_{22} = 8 \tag{2.4}$$

and

$$x_{11} \geq 0 \tag{3.1}$$

$$x_{12} \geq 0 \tag{3.2}$$

$$x_{21} \geq 0 \tag{3.3}$$

$$x_{22} \geq 0. \tag{3.4}$$

Of the four equation constraints, one is redundant. For example, the sum of (2.1) and (2.2) minus (2.3) yields (2.4). Thus, we may eliminate (2.4) as a constraint (any other would do equally well as the one to be eliminated). This ability to eliminate one of the constraints is a direct consequence of the *balanced nature* of the problem.

The initial tableau corresponding to (1), (2.1), (2.2), and (2.3) is thus

x_{11}	x_{12}	x_{21}	x_{22}	1		
1*	1	0	0	-15	$= -0$	
0	0	1	1	-5	$= -0$	(T.1)
1	0	1	0	-12	$= -0$	
5	8	4	10	0	$= u$	

We now apply Stage Zero three times in succession to replace the zeros in the right-hand margin by basic variables. This is done in Tableaus 2, 3, and 4. In each of these, we delete the column with the zero in the top margin. These tableaus, which are very easy to calculate, are as follows:

$\cancel{0}$	x_{12}	x_{21}	x_{22}	1		
1	1	0	0	-15	$= -x_{11}$	
0	0	1	1	-5	$= -0$	(T.2)
-1	-1	1*	0	3	$= -0$	
-5	3	4	10	75	$= u$	

x_{12}	$\cancel{0}$	x_{22}	1		
1	0	0	-15	$= -x_{11}$	
1	-1	1*	-8	$= -0$	(T.3)
-1	1	0	3	$= -x_{21}$	
7	-4	10	63	$= u$	

x_{12}	$\cancel{0}$	1		
1	0	-15	$= -x_{11}$	
1	1	-8	$= -x_{22}$	(T.4)
$-1*$	0	$+3$	$= -x_{21}$	
-3	-10	143	$= u$	

Tableau 4 has the appearance of a standard tableau; there are three basic variables and one nonbasic variable. Since $-b_3 = +3$, this tableau is ready for Stage One. We obtain Tableau 5:

$$\begin{array}{c} \quad\quad x_{21} \quad\quad 1 \\ \begin{array}{|c|c|} \hline 1 & -12 \\ 1^* & -5 \\ -1 & -3 \\ \hline -3 & 134 \\ \hline \end{array} \begin{array}{l} = -x_{11} \\ = -x_{22} \\ = -x_{12} \\ = u \end{array} \end{array} \quad \text{(T.5)}$$

Tableau 5 corresponds to a b.f.p. tableau because all the entries in the right-hand column are negative. We apply the Stage Two rules. This leads to the solution tableau for this minimum problem:

$$\begin{array}{c} \quad\quad x_{22} \quad\quad 1 \\ \begin{array}{|c|c|} \hline -1 & -7 \\ 1 & -5 \\ 1 & -8 \\ \hline +3 & 119 \\ \hline \end{array} \begin{array}{l} = -x_{11} \\ = -x_{21} \\ = -x_{12} \\ = u \end{array} \end{array} \quad \text{(T.6)}$$

We read off the optimal solution in the usual way:

$$x_{22} = 0, \ x_{11} = 7, \ x_{21} = 5, \ x_{12} = 8, \quad \text{and} \quad \text{Min } u = \$119.$$

We can check out our solution by substituting the above x_{ij} values into the original equation for u:

$$u = 5{\cdot}7 + 8{\cdot}8 + 4{\cdot}5 = 119 \quad \text{(check)}.$$

The supply and demand requirements also check out.

Notes:
a. Transportation problems of the Problem 5.6 type can be solved by a much simpler procedure—the *stepping stone method*. This method will be discussed in Chapter 9.
b. Observe that one of the x_{ij}'s in the optimal solution is equal to zero. The simplex algorithm dictates that at least one of the x_{ij}'s will turn out to be zero. This is clear because the system of constraints contains three equations and four variables; a standard tableau will thus have three basic variables solved in terms of a remaining (nonbasic) variable.

Problem 5.7

Set up the equations and inequalities for a general transportation problem in which there are m' warehouses and n' markets. Suppose that the sum of the supplies equals the sum of the demands (balanced problem) and that all the main constraints are equations (as in Problem 5.6).

Solution

We make reference to the following table:

	M_1	M_2	$\cdots$	$M_{n'}$		
W_1	c_{11}	c_{12}	$\cdots$	$c_{1n'}$	a_1	
W_2	c_{21}	c_{22}	$\cdots$	$c_{2n'}$	a_2	Supplies
$\cdots$					$\cdots$	(units)
$\cdots$					$\cdots$	
$W_{m'}$	$c_{m'1}$	$c_{m'2}$	$\cdots$	$c_{m'n'}$	$a_{m'}$	
	b_1	b_2	$\cdots$	$b_{n'}$		

Demands (units)

Here c_{ij} is the cost in dollars per unit for shipping from warehouse W_i to market M_j; a_i is the number of units supplied by warehouse W_i, while b_j is the number of units demanded by market M_j. The balance of supplies and demands can be written as

$$a_1 + a_2 + \cdots + a_{m'} = b_1 + b_2 + \cdots + b_{n'}.$$

The cost of shipping x_{ij} units from W_i to M_j is $c_{ij}x_{ij}$. The total shipping cost u is the sum of all such products for all i and all j. We can write this as

$$u = c_{11}x_{11} + c_{12}x_{12} + \cdots + c_{m'n'}x_{m'n'}. \tag{1}$$

This expression, which contains m' times n' terms in the sum on the right, is to be *minimized*. The total amount to be shipped from W_i, namely, a_i, is $x_{i1} + x_{i2} + \cdots + x_{in'}$. The total amount to be shipped to M_j, namely, b_j is given by $x_{1j} + x_{2j} + \cdots + x_{m'j}$. The equation constraints are thus given by

$$x_{i1} + x_{i2} + \cdots + x_{in'} = a_i \quad \text{for} \quad i = 1, 2, \ldots, m' \tag{2a}$$

$$x_{1j} + x_{2j} + \cdots + x_{m'j} = b_j \quad \text{for} \quad j = 1, 2, \ldots, n'. \tag{2b}$$

The remaining constraints are of the nonnegative type:

$$x_{ij} \geq 0 \quad \text{for all } i \text{ and } j. \tag{3}$$

In (2a) and (2b) there are $m' + n'$ equations. One of these is redundant because of the balanced nature of the problem; thus, in effect, there are $m' + n' - 1$ equation constraints.

Problem 5.8

Suppose that the system defined in Problem 5.7 is converted (e.g., using Stage Zero) to a standard simplex tableau. Discuss the total number of equations and variables in such a system. What is the minimum number of zero x_{ij}'s in an optimal solution obtained by the simplex algorithm?

Solution

We shall use the usual symbols m and n for the number of basic and nonbasic variables, respectively, in a *standard* linear program tableau. The total number of variables is thus $m + n$; the quantity m also denotes the total number of equations.

We see from Problem 5.7 that the total number of equations (or basic variables) is given by

$$m = m' + n' - 1. \tag{A}$$

The total number of variables is m' times n', i.e.,

$$m + n = m'n'. \tag{B}$$

Thus, the total number of nonbasic variables in a standard tableau is

$$n = m'n' - (m' + n' - 1),$$

or

$$n = (m' - 1)\cdot(n' - 1). \tag{C}$$

To obtain the optimal solution from the terminal simplex tableau we set the nonbasic variables equal to zero. Thus, we see that the optimal solution to the transportation problem will contain at least $(m' - 1)\cdot(n' - 1)$ zero terms among the x_{ij}'s. If we apply this to the information of Problem 5.6, we have

$$m = 2 + 2 - 1 = 3 \text{ basic variables.}$$

and

$$n = (2 - 1)\cdot(2 - 1) = 1\cdot1 = 1 \text{ nonbasic variable.}$$

This checks with Problem 5.6; in that problem the single variable x_{22} was zero in the optimal solution.

5. DUALITY THEORY

The following two linear programming problems are defined as being *dual* problems:

$$\text{Maximize} \quad u = c_1x_1 + c_2x_2 + \cdots + c_nx_n \tag{1L}$$

$$\text{subject to} \quad a_{11}x_1 + a_{12}x_2 + \cdots + a_{1n}x_n \leq b_1 \tag{2.1L}$$

$$a_{21}x_1 + a_{22}x_2 + \cdots + a_{2n}x_n \leq b_2 \tag{2.2L}$$

$$\ldots\ldots\ldots\ldots\ldots\ldots\ldots\ldots\ldots\ldots\ldots\ldots\ldots\ldots$$

$$\ldots\ldots\ldots\ldots\ldots\ldots\ldots\ldots\ldots\ldots\ldots\ldots\ldots\ldots$$

$$a_{m1}x_1 + a_{m2}x_2 + \cdots + a_{mn}x_n \leq b_m \tag{2.mL}$$

and

$$x_1 \geq 0 \qquad (3.1L)$$

$$x_2 \geq 0 \qquad (3.2L)$$

$$\cdots\cdots\cdots$$

$$\cdots\cdots\cdots$$

$$x_n \geq 0. \qquad (3.nL)$$

Minimize $\quad w = b_1 y_1 + b_2 y_2 + \cdots + b_m y_m \qquad\qquad (4L)$

subject to $\quad a_{11} y_1 + a_{21} y_2 + \cdots + a_{m1} y_m \geq c_1 \qquad (5.1L)$

$$a_{12} y_1 + a_{22} y_2 + \cdots + a_{m2} y_m \geq c_2 \qquad (5.2L)$$

$$\cdots\cdots\cdots\cdots\cdots\cdots\cdots\cdots\cdots\cdots$$

$$\cdots\cdots\cdots\cdots\cdots\cdots\cdots\cdots\cdots\cdots$$

$$a_{1n} y_1 + a_{2n} y_2 + \cdots + a_{mn} y_m \geq c_n \qquad (5.nL)$$

and

$$y_1 \geq 0 \qquad (6.1L)$$

$$y_2 \geq 0 \qquad (6.2L)$$

$$\cdots\cdots\cdots$$

$$\cdots\cdots\cdots$$

$$y_m \geq 0. \qquad (6.mL)$$

This can be written more compactly if we use matrix-vector notation. (The reader is referred to Chapter 2 for a review of manipulations involving vectors and matrices.) In the following, $\bar{x}$ and $\bar{c}$ are n-component column vectors and $\bar{y}$ and $\bar{b}$ are m-component column vectors:

Maximize $\qquad u = \bar{c}^t \bar{x} \qquad\qquad\qquad (1L)$

subject to $\qquad A\bar{x} \leq \bar{b} \qquad\qquad\qquad (2L)$

and

$$\bar{x} \geq \bar{0}. \qquad (3L)$$

Minimize $\qquad w = \bar{b}^t \bar{y} \qquad\qquad\qquad (4L)$

subject to $\qquad A^t \bar{y} \geq \bar{c} \qquad\qquad\qquad (5L)$

and

$$\bar{y} \geq \bar{0}. \qquad (6L)$$

The maximum problem given by (1L), (2L), (3L) is said to be the *dual* of the minimum problem given by (4L), (5L), (6L), and vice-versa. In a practical situation, one usually expresses a linear programming problem in a particular way (either maximum or minimum) depending on the situation. This particular problem is called the *primal* problem. The dual of the primal problem is called the *dual* problem.

In practice, it is easy to state the dual problem once the primal problem is known. The data box (Figure 5.2) makes this apparent; the maximum problem is read off "horizontally," while the minimum problem is read off "vertically."

Problem 5.9

Develop the data box and the dual problem of the linear programming problem given in Problem 1.4 (also Problem 4.20).

Solution

We first relabel the original variables from x, y, and z as x_1, x_2, and x_3, respectively. Thus, the primal problem is

$$\text{Maximize} \quad u = \tfrac{1}{4}x_1 + \tfrac{2}{5}x_2 + \tfrac{1}{2}x_3 \tag{1}$$

$$\text{subject to} \quad x_1 + \tfrac{2}{3}x_2 + \tfrac{1}{4}x_3 \le 900 \tag{2.1}$$

$$\tfrac{1}{3}x_2 + \tfrac{3}{4}x_3 \le 600 \tag{2.2}$$

and

$$x_1 \ge 0 \tag{3.1}$$

$$x_2 \ge 0 \tag{3.2}$$

$$x_3 \ge 0. \tag{3.3}$$

We now line this up horizontally in a data box:

	x_1	x_2	x_3	
y_1	1	$\tfrac{2}{3}$	$\tfrac{1}{4}$	900
y_2	0	$\tfrac{1}{3}$	$\tfrac{3}{4}$	600
	$\tfrac{1}{4}$	$\tfrac{2}{5}$	$\tfrac{1}{2}$	

FIGURE 5.2 The data box for dual linear programming problems.

	x_1	x_2	$\cdots$	x_n	
y_1	a_{11}	a_{12}	$\cdots$	a_{1n}	b_1
y_2	a_{21}	a_{22}	$\cdots$	a_{2n}	b_2
	$\cdots\cdots\cdots\cdots\cdots$				$\cdots$
	$\cdots\cdots\cdots\cdots\cdots$				$\cdots$
y_m	a_{m1}	a_{m2}	$\cdots$	a_{mn}	b_m
	c_1	c_2	$\cdots$	c_n	

The dual (minimum) problem is read off vertically as follows:

$$\text{Minimize} \quad w = 900y_1 + 600y_2 \tag{4}$$

$$\text{subject to} \quad y_1 \qquad \geq \tfrac{1}{4} \tag{5.1}$$

$$\tfrac{2}{3}y_1 + \tfrac{1}{3}y_2 \geq \tfrac{2}{5} \tag{5.2}$$

$$\tfrac{1}{4}y_1 + \tfrac{3}{4}y_2 \geq \tfrac{1}{2} \tag{5.3}$$

and

$$y_1 \geq 0 \tag{6.1}$$

$$y_2 \geq 0. \tag{6.2}$$

Problem 5.10

Develop the data box and the dual problem of the linear programming problem given in Problem 1.3.

Solution

The primal problem is

$$\text{Minimize} \quad w = 200y_1 + 300y_2$$

$$\text{subject to} \quad y_1 + 2y_2 \geq 60$$

$$4y_1 + 2y_2 \geq 120$$

$$6y_1 + 2y_2 \geq 150$$

and

$$y_1, y_2 \geq 0.$$

Note how we relabeled the cost function from u to w, and the original variables from x and y to y_1 and y_2, respectively. We now line these statements up vertically in a data box:

	x_1	x_2	x_3	
y_1	1	4	6	200
y_2	2	2	2	300
	60	120	150	

The dual (maximum) problem is read off horizontally as follows:

$$\text{Maximize} \quad u = 60x_1 + 120x_2 + 150x_3$$

$$\text{subject to} \quad x_1 + 4x_2 + 6x_3 \leq 200$$

$$2x_1 + 2x_2 + 2x_3 \leq 300$$

and

$$x_1, x_2, x_3 \geq 0.$$

Problem 5.11

Introduce slack variables into the dual linear programming problems in standard inequality form. Develop the starting (initial) dual tableau.

Solution

If we introduce nonnegative slack variables t_i into (2.1L) through (2.mL), we obtain the following equations:

$$a_{11}x_1 + a_{12}x_2 + \cdots + a_{1n}x_n - b_1 = -t_1 \qquad (2.1\text{L})$$

$$a_{21}x_1 + a_{22}x_2 + \cdots + a_{2n}x_n - b_2 = -t_2 \qquad (2.2\text{L})$$

$$\dots\dots\dots\dots\dots\dots\dots\dots\dots\dots\dots\dots\dots\dots\dots\dots$$

$$\dots\dots\dots\dots\dots\dots\dots\dots\dots\dots\dots\dots\dots\dots\dots\dots$$

$$a_{m1}x_1 + a_{m2}x_2 + \cdots + a_{mn}x_n - b_m = -t_m \qquad (2.m\text{L})$$

where

$x_1 \geq 0$	(3.1L)	and	$t_1 \geq 0$	(3.n+1 L)
$x_2 \geq 0$	(3.2L)		$t_2 \geq 0$	(3.n+2 L)
$\dots\dots$			$\dots\dots$	
$\dots\dots$			$\dots\dots$	
$x_n \geq 0$	(3.nL)		$t_m \geq 0$	(3.n+m L)

Using matrix-vector notation, these take the form

$$A\bar{x} - \bar{b} = -\bar{t} \qquad (2\text{L})$$

$$\bar{x} \geq \bar{0}; \quad \bar{t} \geq \bar{0}. \qquad (3\text{L})$$

Introducing nonnegative slack variables v_j into (5.1L) through (5.nL) yields

$$a_{11}y_1 + a_{21}y_2 + \cdots + a_{m1}y_m - c_1 = v_1 \qquad (5.1\text{L})$$

$$a_{12}y_1 + a_{22}y_2 + \cdots + a_{m2}y_m - c_2 = v_2 \qquad (5.2\text{L})$$

$$\dots\dots\dots\dots\dots\dots\dots\dots\dots\dots\dots\dots\dots\dots\dots\dots$$

$$\dots\dots\dots\dots\dots\dots\dots\dots\dots\dots\dots\dots\dots\dots\dots\dots$$

$$a_{1n}y_1 + a_{2n}y_2 + \cdots + a_{mn}y_m - c_n = v_n \qquad (5.n\text{L})$$

where

$y_1 \geq 0$	(6.1L)	and	$v_1 \geq 0$	(6.m+1 L)
$y_2 \geq 0$	(6.2L)		$v_2 \geq 0$	(6.m+2 L)
......			......	
......			......	
$y_m \geq 0$	(6.mL)		$v_n \geq 0$	(6.m+n L)

In matrix-vector form these become

$$A^t \bar{y} - \bar{c} = \bar{v} \tag{5L}$$

$$\bar{y} \geq \bar{0}; \quad \bar{v} \geq \bar{0}. \tag{6L}$$

Both problems are represented simultaneously in Figure 5.3—the starting dual tableau. The rows represent the maximum problem and the columns the minimum problem. The variables for the latter appear in the left and bottom margins. Note the occurrence of the -1 and $-w$, respectively, in these margins.

Problem 5.12

Develop the starting dual tableau for Problem 5.9.

Solution

We introduce nonnegative slack variables t_1 and t_2 (maximum problem) and v_1, v_2, and v_3 (minimum problem). We thus obtain the following starting dual tableau, in which the maximum problem is lined up horizontally and the minimum problem vertically:

	x_1	x_2	x_3	1	
y_1	1	$\frac{2}{3}$	$\frac{1}{4}$	-900	$= -t_1$
y_2	0	$\frac{1}{3}$	$\frac{3}{4}$	-600	$= -t_2$
-1	$\frac{1}{4}$	$\frac{2}{5}$	$\frac{1}{2}$	0	$= u$
	$= v_1$	$= v_2$	$= v_3$	$= -w$	

For example, row 1 represents

$$x_1 + \tfrac{2}{3}x_2 + \tfrac{1}{4}x_3 - 900 = -t_1,$$

while column 2 represents

$$\tfrac{2}{3}y_1 + \tfrac{1}{3}y_2 - \tfrac{2}{5} = v_2.$$

In this equation, we multiply the numbers in column 2 by the entries in the left-hand margin and set the result equal to v_2. It is instructive to compare the above tableau with the data box of Problem 5.9.

The pivot transformation algorithm (Method 4.1) caused the interchange of a basic and nonbasic variable in such a way that the validity of the row equations was preserved. (Refer to the top and right margins of Figure 5.4 below.) It is important to note that the *same* pivot transformation also effects the interchange of a basic and nonbasic variable (in the column system) in a way such that the validity of the column equations is also preserved. This will be shown in Problem 5.13 for the special case where $m = n = 2$.

Problem 5.13

Consider Figure 5.4 for the case where $m = n = 2$. Develop the algorithm needed to interchange a basic and nonbasic variable for the system defined by the column equations.

Solution

The column version of Figure 5.4 (refer to left and bottom margins) for the $m = n = 2$ case is as follows:

q_1	a_{11}^{*}	a_{12}	$-b_1$
q_2	a_{21}	a_{22}	$-b_2$
-1	c_1	c_2	d
	$= p_1$	$= p_2$	$= -w$

(T.1)

If we read off vertically, we obtain the following equations, which correspond to Tableau 1:

$$a_{11}q_1 + a_{21}q_2 - c_1 = p_1 \qquad (1)$$

$$a_{12}q_1 + a_{22}q_2 - c_2 = p_2 \qquad (2)$$

$$-b_1q_1 - b_2q_2 - d = -w. \qquad (3)$$

Equation (3) is equivalent to

$$b_1q_1 + b_2q_2 + d = w.$$

FIGURE 5.3 The starting dual tableau.

	x_1	x_2	$\cdots$	x_j	$\cdots$	x_n	1	
y_1	a_{11}	a_{12}	$\cdots$	a_{1j}	$\cdots$	a_{1n}	$-b_1$	$=-t_1$
y_2	a_{21}	a_{22}	$\cdots$	a_{2j}	$\cdots$	a_{2n}	$-b_2$	$=-t_2$
$\cdots$								
y_i	a_{i1}	a_{i2}	$\cdots$	a_{ij}	$\cdots$	a_{in}	$-b_i$	$=-t_i$
$\cdots$								
y_m	a_{m1}	a_{m2}	$\cdots$	a_{mj}	$\cdots$	a_{mn}	$-b_m$	$=-t_m$
-1	c_1	c_2	$\cdots$	c_j	$\cdots$	c_n	0	$= u$
	$= v_1$	$= v_2$	$\cdots$	$= v_j$	$\cdots$	$= v_n$	$= -w$	

Suppose that we wish to interchange q_1 and p_1 in Tableau 1 (we assume that $a_{11} \neq 0$). To do this, we first solve Equation (1) for q_1 in terms of p_1 and q_2:

$$-p_1 + a_{21}q_2 - c_1 = -a_{11}q_1$$

$$\frac{1}{a_{11}} p_1 - \frac{a_{21}}{a_{11}} q_2 - \left(-\frac{c_1}{a_{11}} \right) = q_1. \tag{1'}$$

We now substitute this expression for q_1 into Equations (2) and (3). After collecting terms, we obtain

$$\left(\frac{a_{12}}{a_{11}} \right) p_1 + \left(a_{22} - \frac{a_{12}a_{21}}{a_{11}} \right) q_2 - \left(c_2 - \frac{a_{12}c_1}{a_{11}} \right) = p_2 \tag{2'}$$

$$-\frac{b_1}{a_{11}} p_1 + \left(-b_2 - \frac{(-b_1)a_{21}}{a_{11}} \right) q_2 - \left(d - \frac{(-b_1)c_1}{a_{11}} \right) = -w \tag{3'}$$

The new tableau corresponding to Equations (1'), (2'), and (3') is as follows:

p_1	a'_{11}	a'_{12}	$-b'_1$
q_2	a'_{21}	a'_{22}	$-b'_2$
-1	c'_1	c'_2	d'
	$= q_1$	$= p_2$	$= -w$

Note that p_1 is now nonbasic and q_1 is now a basic variable. From Equations (1'), (2'), and (3') we see that

$$a'_{11} = \frac{1}{a_{11}}, \qquad a'_{12} = \frac{a_{12}}{a_{11}},$$

$$a'_{21} = -\frac{a_{21}}{a_{11}}, \qquad a'_{22} = \left(a_{22} - \frac{a_{12}a_{21}}{a_{11}} \right), \text{ etc.,}$$

That is, the changes in *all* the coefficients can be summarized by means of the *pivot transformation rules of Method 4.1*!

The demonstration in Problem 5.13 can be carried out without much more effort for a system involving general m and n. One merely has to be more careful with respect to bookkeeping. From Figure 5.4 we see that if we employ Method 4.1 to interchange s_j and r_i in the maximum problem, we shall simultaneously cause the interchange of variables q_i and p_j in the minimum problem. We summarize the effects of the pivot transformation in Method 5.1.

Method 5.1

Suppose that the pivot transformation of Method 4.1 is applied to a typical dual tableau (Figure 5.4). Then if a_{ij} is the pivot element (it is

	s_1	s_2	$\cdots$	s_j	$\cdots$	s_n	1	
q_1	a_{11}	a_{12}	$\cdots$	a_{1j}	$\cdots$	a_{1n}	$-b_1$	$= -r_1$
q_2	a_{21}	a_{22}	$\cdots$	a_{2j}	$\cdots$	a_{2n}	$-b_2$	$= -r_2$
	. . . .						. . . .	. . . .
q_i	a_{i1}	a_{i2}	$\cdots$	a_{ij}	$\cdots$	a_{in}	$-b_i$	$= -r_i$
	. . . .						. . . .	. . . .
q_m	a_{m1}	a_{m2}	$\cdots$	a_{mj}	$\cdots$	a_{mn}	$-b_m$	$= -r_m$
-1	c_1	c_2	$\cdots$	c_j	$\cdots$	c_n	d	$= u$
	$= p_1$	$= p_2$		$= p_j$		$= p_n$	$= -w$	

	Nonbasic variables	Basic variables
Maximum problem	s_j	r_i
Minimum problem	q_i	p_j

FIGURE 5.4 A typical intermediate dual tableau.

assumed that $a_{ij} \neq 0$), a new tableau is obtained in which:

a. The variables s_j and r_i of the maximum problem (row equations) will be interchanged.
b. The variables q_i and p_j of the minimum problem (column equations) will be interchanged.
c. The validity of both the new row and the new column equations will be maintained.

Some theorems dealing with duality theory will now be stated and proved. They have great significance with respect to the mathematical aspects, computational procedures, and economic interpretations of linear programming. Since the development involves matrix-vector manipulations, the reader is advised to study and/or review the appropriate sections of Chapters 2 and 3.

Note: The reader who is not very concerned with theory may wish merely to *skim* over the proofs and theoretical problems contained in the following pages. However, such a reader is urged to carefully read the statements of the theorems and to observe how some of these theorems relate to the computational problems in the rest of the chapter.

Theorem 5.2. *Suppose that the vectors* $\bar{x} = [x_1, x_2, \ldots, x_n]$ *and* $\bar{y} = [y_1, y_2, \ldots, y_m]$ *satisfy the constraints (2L), (3L) and (5L), (6L), respectively, of the dual linear programming problems (see the beginning of Section 5). Then*

$$\bar{c}^t \bar{x} \leq \bar{b}^t \bar{y} \quad \text{or, equivalently,} \quad u(\bar{x}) \leq w(\bar{y}).$$

Notes:

a. Henceforth, we shall occasionally write "$\bar{x}$ in S_{c-u}" to indicate that $\bar{x}$ satisfies (2L) and (3L). This means that $\bar{x}$ is in the constraint set S_{c-u} of the maximum problem. Also, "$\bar{y}$ in S_{c-w}" will mean that $\bar{y}$ satisfies (5L) and (6L). Here S_{c-w} denotes the constraint set of the minimum problem. It is also common practice to refer to vectors $\bar{x}$ and $\bar{y}$ that satisfy (2L), (3L) and (5L), (6L), respectively, as *feasible vectors*.

b. We shall at times write $u(\bar{x})$ and $w(\bar{y})$ in place of u and w, respectively. (If one writes $u(\bar{x})$, one implies that u is a function of $\bar{x}$.)

c. Recall that the square bracket notation indicates *column* vectors (refer to Chapter 2). The latter are interpreted also as matrices with one column.

The solution of the following problem will be needed to carry out the proof of Theorem 5.2.

Problem 5.14

Given that $\bar{a} \geq \bar{c}$ and $\bar{x} \geq \bar{0}$, where all vectors have the same number of components. Prove that the following is a valid scalar (i.e., real number) inequality:

$$\bar{x}^t \bar{a} \geq \bar{x}^t \bar{c}.$$

Solution

We shall carry out the proof for the case of two-component vectors. By $\bar{a} \geq \bar{c}$ we mean that $a_1 \geq c_1$ and $a_2 \geq c_2$, where the subscripts indicate the vector components. Analogously, $\bar{x} \geq \bar{0}$ indicates that $x_1 \geq 0$ and $x_2 \geq 0$.

Multiplying the $\bar{a} \geq \bar{c}$ inequalities by the nonnegative numbers x_1 and x_2 leads to

$$x_1 a_1 \geq x_1 c_1 \tag{1a}$$

$$x_2 a_2 \geq x_2 c_2. \tag{1b}$$

Adding (1a) and (1b), we obtain

$$x_1 a_1 + x_2 a_2 \geq x_1 c_1 + x_2 c_2. \tag{2}$$

However, we realize that this is none other than

$$\bar{x}^t \bar{a} \geq \bar{x}^t \bar{c} \tag{3}$$

when we recall the rules for matrix multiplication. For example, on either side we have a 1 by 2 matrix times a 2 by 1 matrix yielding a 1 by 1 matrix; the latter is equivalent to a real number.

We are now in a position to prove Theorem 5.2.

PROOF OF THEOREM 5.2. We shall make use of the following matrix properties:

a. $(AB)^t = B^t A^t$ and $(A^t)^t = A$. These results were stated in Theorem 2.3.
b. $K = K^t$ if K is a 1 by 1 matrix (i.e., a real number).

Refer to statements (1L) through (6L) at the beginning of Section 5.

Suppose that $\bar{x}$ and $\bar{y}$ satisfy (2L), (3L) and (5L), (6L), respectively. Inequality (5L) is of the form $\bar{a} \geq \bar{c}$, since $A^t \bar{y}$ is equivalent to an n by 1 matrix (n by m times m by 1). If we premultiply it by the nonnegative vector $\bar{x}^t$, we obtain

$$\bar{x}^t A^t \bar{y} \geq \bar{x}^t \bar{c},$$

when we make use of the result of Problem 5.14. The right-hand side of this inequality is none other than $x_1 c_1 + \cdots + x_n c_n$, which is $u(\bar{x})$. Thus, we have

$$u(\bar{x}) \leq \bar{x}^t A^t \bar{y}. \tag{i}$$

Premultiplication of (2L) by the nonnegative vector $\bar{y}^t$ yields

$$\bar{y}^t A \bar{x} \leq \bar{y}^t \bar{b},$$

if we again consider the result of Problem 5.14. The right-hand side of this inequality is $y_1 b_1 + \cdots + y_m b_m = w(\bar{y})$. Thus, we have

$$\bar{y}^t A \bar{x} \leq w(\bar{y}). \tag{ii}$$

The quantity $\bar{x}^t A^t \bar{y}$ in (i) is a 1 by 1 matrix (a 1 by n matrix times an n by m matrix times an m by 1 matrix). Thus, it equals its transpose by Note (b) above. That is,

$$\bar{x}^t A^t \bar{y} = (\bar{x}^t A^t \bar{y})^t. \tag{iii}$$

Applying Note (a) to the right-hand side of Equation (iii) leads to

$$(\bar{x}^t A^t \bar{y})^t = \bar{y}^t (A^t)^t (\bar{x}^t)^t = \bar{y}^t A \bar{x}. \tag{iv}$$

From (iii) and (iv) we have

$$\bar{x}^t A^t \bar{y} = \bar{y}^t A \bar{x}. \tag{v}$$

Thus, the right-hand side of (i) equals the left side of (ii), and we have

$$u(x) \leq \bar{y}^t A \bar{x} \leq w(\bar{y}) \tag{vi}$$

or

$$u(\bar{x}) \leq w(\bar{y}). \tag{vii}$$

$\square$

Notice that the conclusion of the above theorem also holds for the respective solution vectors $\bar{x}^*$ and $\bar{y}^*$, i.e., $u(\bar{x}^*) \leq w(\bar{y}^*)$. In other words, Max $u \leq$ Min w. It then follows from the definition of solution vector that the following inequality relates $\bar{x}$, $\bar{x}^*$, $\bar{y}$, and $\bar{y}^*$, where $\bar{x}$ satisfies (2L), and (3L) and $\bar{y}$ satisfies (5L) and (6L):

$$u(\bar{x}) \leq u(\bar{x}^*) \leq w(\bar{y}^*) \leq w(\bar{y}).$$

The following theorem follows directly from Theorem 5.2.

Theorem 5.3. *Suppose that $\bar{x}^\circ$ and $\bar{y}^\circ$ satisfy (2L), (3L) and (5L), (6L), respectively, and, in addition, that $u(\bar{x}^\circ) = w(\bar{y}^\circ)$. Then $\bar{x}^\circ$ and $\bar{y}^\circ$ are solutions of the dual linear programming problems.*

Note that the equality given in Theorem 5.3 may be written as $\bar{c}^t \bar{x}^\circ = \bar{b}^t \bar{y}^\circ$.

PROOF OF THEOREM 5.3. We have to show that $u(\bar{x}) \leq u(\bar{x}^\circ)$ for any $\bar{x}$ in S_{c-u} and $w(\bar{y}^\circ) \leq w(\bar{y})$ for any $\bar{y}$ in S_{c-w}. Recall what is meant by a solution (vector). Let $\bar{x}$ be any vector in S_{c-u}. Since $\bar{y}^\circ$ is in S_{c-w}, it follows from Theorem 5.2 that

$$u(\bar{x}) \leq w(\bar{y}^\circ). \tag{1}$$

Using the hypothesis that $u(\bar{x}^\circ) = w(\bar{y}^\circ)$, we obtain

$$u(\bar{x}) \leq u(\bar{x}^\circ). \tag{2}$$

Now let $\bar{y}$ be any vector in S_{c-w}. Since $\bar{x}^\circ$ is in S_{c-u}, it follows from Theorem 5.2 that

$$u(\bar{x}^\circ) \leq w(\bar{y}). \tag{3}$$

Using the hypothesis that $u(\bar{x}^\circ) = w(\bar{y}^\circ)$, we obtain

$$w(\bar{y}^\circ) \leq w(\bar{y}). \tag{4}$$

$\square$

Notes:
a. From Inequality (2) it is clear that $u(\bar{x}^\circ) =$ Max u, while from Inequality (4) it is clear that $w(\bar{y}^\circ) =$ Min w. Thus, we may also write $u(\bar{x}^\circ) = w(\bar{y}^\circ)$ as Max $u =$ Min w.
b. If we examine Equation (vi) in the proof of Theorem 5.2, we see that both the maximum and minimum values can be expressed as $(\bar{y}^\circ)^t A(\bar{x}^\circ)$, since this would have to equal $u(\bar{x}^\circ) = w(\bar{y}^\circ)$. Thus, Max $u =$ Min $w = (\bar{y}^\circ)^t A(\bar{x}^\circ)$.

The following very important theorem is dependent on Method 5.1.

Theorem 5.4. *Suppose that one of the dual linear programming problems (1L), (2L), (3L) and (4L), (5L), (6L) has a solution. Then the other dual* *problem has a solution. Moreover, both solutions give the same value to the respective objective functions.*

PROOF OF THEOREM 5.4. Suppose that the maximum problem has a solution. Refer to Figure 5.4 for the picture of a typical intermediate tableau for both dual programming problems. Since the maximum problem has a solution, a tableau will be reached such that all the entries in the right-hand column and bottom row (with the exception of d) will be nonpositive, i.e.,

$$-b_i \le 0 \quad \text{for} \quad i = 1, 2, \ldots, m \tag{1}$$

$$c_j \le 0 \quad \text{for} \quad j = 1, 2, \ldots, n \tag{2}$$

Recall that the solution to the maximum problem is obtained from the basic feasible point *corresponding* to such a tableau. That is, for the solution, we set the nonbasic variables s_j equal to zero for $j = 1, 2, \ldots, n$. Thus, the basic variables and maximum value are given by

$$r_i = b_i \ge 0 \quad \text{for} \quad i = 1, 2, \ldots, m, \quad \text{and} \quad \text{Max } u = d.$$

Now Method 5.1 implies that the column equations for the tableau described above provide a valid representation of the minimum problem. We shall now show that such a tableau also provides the solution of the minimum problem.

Reading the column equations of the tableau (see Figure 5.4), we have

$$p_j = a_{1j}q_1 + \cdots + a_{mj}q_m - c_j \quad \text{for} \quad j = 1, 2, \ldots, n \tag{3}$$

and

$$w = b_1 q_1 + \cdots + b_m q_m + d. \tag{4}$$

If we now set the nonbasic variables q_i equal to zero for $i = 1, 2, \ldots, m$, then (3) and (4) become

$$p_j = -c_j \quad \text{for} \quad j = 1, 2, \ldots, n \tag{5}$$

and

$$w = d. \tag{6}$$

The basic point thus determined is *feasible* because of (2). That is,

$$p_j \ge 0 \quad \text{for} \quad j = 1, 2, \ldots, n. \tag{7}$$

It remains to be shown that this basic feasible point is also a solution point.

Let us pick any typical feasible point for the minimum problem. For such a point all the q_i's and all the p_j's are nonnegative. Now refer to Equation (4). Each $b_i q_i$ term is nonnegative in view of (1), i.e., we have nonnegative b_i times nonnegative q_i. Thus, we obtain

$$w \geq d \tag{8}$$

for this typical feasible point.

Thus, the basic feasible point determined from $q_i = 0$ for $i = 1$, $2, \ldots, m$ is the solution point of the problem. The basic variables and the minimum w for this point are given by (5) and (6) above; thus,

$$\text{Min } w = d. \tag{9}$$

$\square$

Notes:

a. We see that the solutions for *both* dual programming problems are given by the same tableau; in such a tableau, all the $-b_i$ and c_j entries are nonpositive. In other words, the right-hand column and bottom row entries are nonpositive. The optimal value is the *same* for both objective functions and is given by the corner entry d of this tableau.

b. If we had initially supposed that the minimum problem had a solution, the proof would be essentially the same.

c. One consequence of Theorem 5.4 is the following: If both $\bar{x}^*$ and $\bar{y}^*$ are solutions of the respective dual problems, then $u(\bar{x}^*) = w(\bar{y}^*)$, i.e., Max u = Min w. This statement is the converse of Theorem 5.3.

Theorem 5.5. *If both dual problems are feasible, then both have solutions.*

PROOF OF THEOREM 5.5. If both problems are feasible, then there is at least one $\bar{x}$ that satisfies (2L), (3L) and at least one $\bar{y}$ that satisfies (5L), (6L). Let one such $\bar{y}$ be the constant vector $\bar{y}^\circ$. From Theorem 5.3 we have

$$u(\bar{x}) \leq w(\bar{y}^\circ) \quad \text{for any } \bar{x} \text{ in } S_{c-u}. \tag{1}$$

The quantity $w(\bar{y}^\circ)$ is equal to the finite constant number $b_1 y_1^\circ + \cdots + b_m y_m^\circ$. Thus, (1) indicates that u is bounded from above with respect to the constraint set S_{c-u}. It follows from Note (b) following the Proof of the Corollary of Theorem 5.1 that the maximum problem has a solution (at a b.f.p.). Working with a constant vector $\bar{x}^\circ$ leads to $u(\bar{x}^\circ) \leq w(\bar{y})$; in the same way we conclude that the minimum problem has a solution. $\square$

Theorems 5.4 and 5.5 taken as a unit constitute what is known as the *fundamental theorem of linear programming*. It is also referred to as the *duality theorem*.

As a result of Theorems 5.4 and 5.5 and the nature of the dual tableau, we can obtain the solutions to *both* dual problems from the same tableau. This is particularly advantageous if the primal problem is a minimum one for which the b_i's (the cost terms) are all nonnegative. (This, by the way, is the usual case for a minimum problem.) In this situation the starting tableau corresponds to a b.f.p. for the dual (maximum) problem, since the $-b_i$'s would all be nonpositive. Thus, if we focus on the dual, we can start the Stage Two calculations right away, thereby effectively eliminating the need to consider Stage One. Besides being a lot easier, this often cuts down on the number of pivot steps needed to reach a terminal tableau. Let us summarize the important aspects of this very useful process in Method 5.2.

Method 5.2: Dual Tableau Simplex Algorithm for a Minimum Problem

First, we associate the symbols of (4L), (5L), and (6L) with the minimum problem under consideration. Suppose that the b_i's (cost terms) of the minimum problem are all nonnegative. (Thus, the $-b_i$'s in the starting tableau are all nonpositive.) Line up the minimum problem vertically (i.e., by columns) in a starting dual tableau (Figure 5.3). Fill in the original and slack variables for the dual maximum problem. Focus on the maximum problem, and employ Stage Two calculations until a terminal tableau is reached. If a_{ij} is a typical pivot element, then every time s_j and r_i are interchanged (see Figure 5.4) in the maximum problem, q_i and p_j are to be interchanged in the minimum problem.

The terminal tableau is a solution tableau if all the c_j's are nonpositive. In this case the solution for the minimum problem is determined as follows: Set all the q_i's equal to zero. Then $p_j = -c_j$ for $j = 1, 2, \ldots, n$, and Min $w = d$.

Problem 5.15

Apply Method 5.2 to Problem 1.3.

Solution

First, as in Problem 5.10, we alter the symbols. Thus, w is the cost function and y_1 and y_2 are the original variables. The initial dual tableau is as follows:

	x_1	x_2	x_3	1		
y_1	1	4	6	-200	$= -t_1$	
y_2	2^*	2	2	-300	$= -t_2$	(T.1)
-1	60	120	150	0	$= u$	
	$= v_1$	$= v_2$	$= v_3$	$= -w$		

It is useful to compare this tableau with the data box of Problem 5.10. In the latter, the first main inequality was $y_1 + 2y_2 \geq 60$; after introducing the slack variable v_1, this becomes $y_1 + 2y_2 - 60 = v_1$, which corresponds to the first column of Tableau 1. The other columns are obtained in a similar manner. We read off the standard slack form of the dual maximum problem from the row equations:

$$\text{Maximize} \quad u = 60x_1 + 120x_2 + 150x_3$$

$$\text{subject to} \quad x_1 + 4x_2 + 6x_3 - 200 = -t_1$$

$$2x_1 + 2x_2 + 2x_3 - 300 = -t_2$$

and

$$x_1, x_2, x_3, t_1, t_2 \geq 0.$$

The two t_i equations are equivalent to the main inequalities of Problem 5.10.

Tableau 1 is ready for Stage Two with respect to the maximum problem. Choosing column 1 as the pivot column leads to $a_{21} = 2$ as the pivot element. Pivoting then results in Tableau 2:

	t_2	x_2	x_3	1		
y_1	$-\frac{1}{2}$	3^*	5	-50	$= -t_1$	
v_1	$\frac{1}{2}$	1	1	-150	$= -x_1$	(T.2)
-1	-30	60	90	9000	$= u$	
	$= y_2$	$= v_2$	$= v_3$	$= -w$		

Note that y_2 and v_1 were interchanged simultaneously with x_1 and t_2. Tableau 2 corresponds to the point on the vertical axis of Figure 4.11 where $x = 0$ ($y_1 = 0$) and $r = 0$ ($v_1 = 0$). Choosing column 2 as the pivot column yields $a_{12} = 3$ as the pivot element in T.2. Pivoting then results in Tableau 3:

	t_2	t_1	x_3	1		
v_2	$-\frac{1}{6}$	$\frac{1}{3}$	$\frac{5}{3}$	$-\frac{50}{3}$	$= -x_2$	
v_1	$\frac{2}{3}$	$-\frac{1}{3}$	$-\frac{2}{3}$	$-\frac{400}{3}$	$= -x_1$	(T.3)
-1	-20	-20	-10	$10,000$	$= u$	
	$= y_2$	$= y_1$	$= v_3$	$= -w$		

This tableau is a terminal tableau. We read off the solution of this minimum problem by setting $v_1 = v_2 = 0$. Thus, we have $y_1 = 20$, $y_2 = 20$, $v_3 = 10$, and Min $w = \$10,000$, as in Problem 4.26.

We observe that this method required only three tableaus, while the method employed in Problem 4.26 required four tableaus. It should also be noted that the column part of Tableau 3 here is equivalent to Tableau

4 of Problem 4.26. This can be seen if one recalls the different labeling for symbols in the two problems, as given in the following table:

Variable in Problem 4.26	x	y	r	s	t	u
Variable in Problem 5.15	y_1	y_2	v_1	v_2	v_3	w

For example, column 3 of Tableau 3 here is seen to be equivalent to row 1 of Tableau 4 of Problem 4.26.

The solution to the dual maximum problem is read off from the row system of Tableau 3 in the usual way by setting the nonbasic variables equal to zero. Thus, we have

$$x_1 = \tfrac{400}{3}, \ x_2 = \tfrac{50}{3}, \ x_3 = 0, \ t_1 = t_2 = 0, \quad \text{and} \quad \text{Max } u = \$10,000.$$

Problem 5.16

Rework Problem 1.4 using a dual tableau. Give the solution to both the primal and dual problems.

Solution

The dual problem and data box are given in Problem 5.9, and the starting dual tableau, reproduced below, appears also in Problem 5.12:

	x_1	x_2	x_3	1	
y_1	1	$\tfrac{2}{3}*$	$\tfrac{1}{4}$	-900	$= -t_1$
y_2	0	$\tfrac{1}{3}$	$\tfrac{3}{4}$	-600	$= -t_2$
-1	$\tfrac{1}{4}$	$\tfrac{2}{5}$	$\tfrac{1}{2}$	0	$= u$
	$= v_1$	$= v_2$	$= v_3$	$= -w$	

(T.1)

The dual minimum problem may be read off from the columns of this tableau:

$$\text{Minimize} \quad w = 900 y_1 + 600 y_2$$

$$\text{subject to} \quad y_1 \qquad\qquad -\tfrac{1}{4} = v_1$$

$$\tfrac{2}{3} y_1 + \tfrac{1}{3} y_2 - \tfrac{2}{5} = v_2$$

$$\tfrac{1}{4} y_1 + \tfrac{3}{4} y_2 - \tfrac{1}{2} = v_3$$

and

$$y_1, \ y_2, \ v_1, \ v_2, \ v_3 \geq 0.$$

These constraints are equivalent to the inequality statements of Problem 5.9.

Application of Stage Two of the simplex algorithm to Tableau 1 results in Tableaus 2 and 3, which follow. Tableaus 2 and 3 here are essentially identical to Tableaus 2 and 3 of Problem 4.20, the only difference being that here the dual variables are also interchanged in each pivot step. The reader should again note the relabeling of the original variables to x_1, x_2, and x_3 and of the slack variables to t_1 and t_2 (from r and s):

	x_1	t_1	x_3	1	
v_2	$\frac{3}{2}$	$\frac{3}{2}$	$\frac{3}{8}$	-1350	$= -x_2$
y_2	$-\frac{1}{2}$	$-\frac{1}{2}$	$\frac{5^*}{8}$	-150	$= -t_2$
-1	$-\frac{7}{20}$	$-\frac{3}{5}$	$\frac{7}{20}$	540	$= u$
	$= v_1$	$= y_1$	$= v_3$	$= -w$	

$$(\text{T.2})$$

	x_1	t_1	t_2	1	
v_2	$\frac{9}{5}$	$\frac{9}{5}$	$-\frac{3}{5}$	-1260	$= -x_2$
v_3	$-\frac{4}{5}$	$-\frac{4}{5}$	$\frac{8}{5}$	-240	$= -x_3$
-1	$-\frac{7}{100}$	$-\frac{8}{25}$	$-\frac{14}{25}$	624	$= u$
	$= v_1$	$= y_1$	$= y_2$	$= -w$	

$$(\text{T.3})$$

The solution of the maximum (primal) problem is read off from the equations corresponding to the rows of Tableau 3, after setting the nonbasic variables equal to zero:

$$t_1 = t_2 = 0, \ x_1 = 0, \ x_2 = 1260, \ x_3 = 240, \quad \text{and} \quad \text{Max } u = 624.$$

The solution of the minimum (dual) problem is read off from the equations corresponding to the columns of Tableau 3, again after setting the nonbasic variables equal to zero:

$$v_2 = v_3 = 0, \ v_1 = \tfrac{7}{100}, \ y_1 = \tfrac{8}{25}, \ y_2 = \tfrac{14}{25}, \quad \text{and} \quad \text{Min } w = 624.$$

Problems 5.15 and 5.16 will serve as models for the subsequent work dealing with sensitivity analysis.

The following dual tableau algorithm is fairly useful.

Method 5.3: Dual Tableau Simplex Algorithm for a Maximum Problem

Suppose that we have a typical intermediate dual tableau, as in Figure 5.4, where the maximum problem is lined up by rows and the minimum problem by columns. The method applies when all the c_j's are nonpositive, but at least one of the $-b_i$'s is positive.

Focus on a row for which $-b_i$ is positive. Compute c_j/a_{ij} for all those a_{ij}'s in the row that are *negative*. These ratios will thus be *positive* or zero in value, since $c_j \leq 0$ for all j. Pick the a_{ij} for which c_j/a_{ij} is a minimum. This a_{ij} is then the pivot element. (In case of ties, choose any of the eligible a_{ij}'s.) In the typical pivot operation, s_j and r_i are interchanged in the maximum problem, while q_i and p_j are interchanged in the minimum problem. (It is possible that all the a_{ij}'s in the row under focus are either positive or zero. This indicates that the minimum problem is unbounded.)

In the next tableau, after the pivot transformation has been effected, the c_j's will again be nonpositive, and usually the $-b_i$'s will be less positive in character. A terminal tableau that is a solution tableau is achieved when the $-b_i$'s have all become nonpositive.

Notes:

a. Method 5.3 is equivalent to none other than the Stage Two simplex algorithm applied to the minimum problem.

b. It is a convention of many practitioners in the linear programming field to convert their linear programming problems to the standard maximum form if they are not already in that form (see (1L), (2L), (3L) at the beginning of Section 5 for the standard maximum form). For example, if it is desired to minimize $w = \bar{b}^t \bar{y}$, this is equivalent to maximizing $w' = -w$. Thus, if the maximum of w' occurs at $\bar{y}^\circ$, then the minimum of w also occurs at $\bar{y}^\circ$. The reversal of the sense of the main constraint inequality (5L) (again see the beginning of Section 5) is achieved merely by multiplying through by -1; the resulting inequality (in matrix-vector form) will then have the same appearance as (2L).

   The above-mentioned practitioners refer to the maximum problem as the primal problem, while the related minimum problem (e.g., as determined from Figure 5.2 or 5.3) is called the dual problem. They call Method 5.3 the *dual simplex algorithm*. The reason for this is that the typical Method 5.3 tableau corresponds to *feasibility* with respect to the *dual* (i.e., the minimum) problem. This fact provides the basis for the philosophy behind Method 5.3.

c. It is easy to see that Method 5.3 provides an alternate approach to Stage One (Method 4.3) in situations where all the c_j's are nonpositive.

Problem 5.17

Apply Method 5.3 to the following dual tableau to obtain a terminal tableau:

	s_1	s_2	1	
q_1	$-\frac{2}{3}$	$-\frac{1}{3}*$	12	$= -r_1$
q_2	$\frac{4}{3}$	$\frac{1}{3}$	-60	$= -r_2$
-1	$-\frac{20}{3}$	$-\frac{5}{3}$	200	$= u$
	$= p_1$	$= p_2$	$= -w$	

(T.k)

Solution

The symbols that appear in the margins are in accord with those of Figure 5.4. Clearly, Method 5.3 applies because both c_1 and c_2 are negative, while $-b_1$ is positive. (Note that Tableau k could also be treated by Stage One.) Tableau k corresponds to feasibility with respect to the minimum problem (at the b.f.p., $q_1 = q_2 = 0$, $p_1 = \frac{20}{3}$, $p_2 = \frac{5}{3}$, and $w = 200$). We follow Method 5.3 and compute the relevant c/a ratios for row 1:

$$\frac{c_1}{a_{11}} = \frac{(-\frac{20}{3})}{(-\frac{2}{3})} = +10$$

$$\frac{c_2}{a_{12}} = \frac{(-\frac{5}{3})}{(-\frac{1}{3})} = +5.$$

Thus, a_{12} is the pivot element in Tableau k. Carrying out the pivot transformation leads to Tableau $k + 1$, in which s_2 and r_1 (maximum problem) and q_1 and p_2 (minimum problem), respectively, have been exchanged:

	s_1	r_1	1	
p_2	2	-3	-36	$= -s_2$
q_2	$\frac{2}{3}$	1	-48	$= -r_2$
-1	$-\frac{10}{3}$	-5	140	$= u$
	$= p_1$	$= q_1$	$= -w$	

(T.$k + 1$)

This is a terminal tableau; the solutions of the respective problems are easily read off:

Maximum problem:
$$s_1 = r_1 = 0, \quad s_2 = 36, \quad r_2 = 36, \quad \text{and} \quad \text{Max } u = 140.$$

Minimum problem:
$$p_2 = q_2 = 0, \quad p_1 = \frac{10}{3}, \quad q_1 = 5, \quad \text{and} \quad \text{Min } w = 140.$$

The following two theorems are direct consequences of the *duality theorem* (Theorems 5.4 and 5.5). For this reason they could be (and often are) considered as corollaries of the former.

Theorem 5.6. *If one of the two dual problems is unbounded, then the other is infeasible.*

Theorem 5.7. *If one of the two dual problems is infeasible, then the other is either (a) infeasible or (b) unbounded.*

The reader is referred to Section 2 of this chapter for the meanings of infeasible and unbounded. Note that the expression "feasible but unbounded problem" has been used interchangeably with "unbounded problem."

6. ADDITIONAL TERMINOLOGY

Let us refer to the dual problems as given in standard slack (equality) form (see, e.g., the beginning of Section 5—especially Problem 5.11). Restated, we have the following:

The Maximum Problem

$$\text{Maximize} \quad u = \bar{c}^t \bar{x} \tag{1L}$$

$$\text{subject to} \quad A\bar{x} + \bar{t} = \bar{b} \tag{2L}$$

and

$$\bar{x} \geq \bar{0}; \; \bar{t} \geq \bar{0}. \tag{3L}$$

The Minimum Problem

$$\text{Minimize} \quad w = \bar{b}^t \bar{y} \tag{4L}$$

$$\text{subject to} \quad A^t \bar{y} - \bar{v} = \bar{c} \tag{5L}$$

and

$$\bar{y} \geq \bar{0}; \; \bar{v} \geq \bar{0}. \tag{6L}$$

If the original (primal) problem is of maximum type, then the vector $\bar{x}$ is called the *activity vector*. The vector $\bar{b}$ is called the *capacity-constraint* or *resources* vector; its components give the amounts of the "scarce resources" that can be required by a particular activity vector. The vector $\bar{c}$ is called the *profit* (or *revenue* or *income*) vector. Its entries give, e.g., the unit profit for each component of the activity vector $\bar{x}$. The vector $\bar{y}$ is known as the *imputed value* or *marginal value* or *shadow value* vector (see Notes (a) and (b) below).

If the original program is of minimum type, then the vector $\bar{y}$ is called the *activity vector*. The vector $\bar{c}$ is then referred to as the *requirements vector*; its components give the minimum amounts of each good that must be produced. The vector $\bar{b}$ is called the *cost* vector; its entries give the unit cost of each of the activities. The vector $\bar{x}$ is known as the *imputed cost* or *marginal cost* or *shadow cost* vector (see Notes (a) and (b), which follow).

Notes:
   a. The terms imputed value vector and imputed cost vector will be discussed further in Chapter 6, which deals with sensitivity analysis.
   b. The above terminology is very similar to that found in [8]. In some treatises, both imputed value and imputed cost are called "shadow price."

Another useful bit of terminology is the following: Vectors $\bar{x}$ and $\bar{t}$

satisfying (2L) and (3L) and vectors $\bar{y}$ and $\bar{v}$ satisfying (5L) and (6L) are called *feasible vectors*.

7. COMPLEMENTARY SLACKNESS

The topic complementary slackness is closely related to duality theory (in particular, to various of its important applications). It is relevant also to the study of sensitivity analysis (see Chapter 6). In addition, complementary slackness has played a role in the development of several algorithms relating to linear programming. The Tucker tableau (e.g., see Figures 5.3 and 5.4) and the terminology associated with it greatly facilitate the study of this important subject.

Equations (1L) through (6L) of Section 6 will be relevant here also.

Note: The note that appeared before Theorem 5.2 is relevant here also. Thus, the reader who is not overly concerned with theory should feel free to merely skim the following proofs. However, the author feels that it is important to carefully read the statements of the theorems.

Theorem 5.8. *The following equation, known as Tucker's duality equation, is true for vectors $\bar{t}$, $\bar{x}$ and $\bar{v}$, $\bar{y}$ that satisfy Equations (2L) and (5L), respectively (remember that our vectors are column vectors):*

$$w - u = \bar{t}^t \bar{y} + \bar{v}^t \bar{x}.$$

Notes:

a. If we carry out the indicated vector multiplications, the Tucker duality equation becomes

$$w - u = t_1 y_1 + \cdots + t_i y_i + \cdots + t_m y_m$$
$$+ v_1 x_1 + \cdots + v_j x_j + \cdots + v_n x_n.$$

b. If we introduce the sigma (Σ) notation to denote summation, the previous equation can be made very compact. Briefly, the sigma symbol is defined as follows:

$$\sum_{i=1}^{n} z_i = z_1 + z_2 + \cdots + z_n.$$

The bottom index value ($i = 1$) and top index value ($i = n$) indicate that the summation is on the terms z_1, z_2, etc., up to and including z_n. Thus, we can write the result from Note (a) more compactly as

$$w - u = \sum_{i=1}^{m} t_i y_i + \sum_{j=1}^{n} v_j x_j.$$

PROOF OF THEOREM 5.8. If we subtract Equation (1L) from Equation (4L), we obtain

$$w - u = \bar{b}^t\bar{y} - \bar{c}^t\bar{x}. \tag{1}$$

Suppose that $\bar{t}$, $\bar{x}$ and $\bar{v}$, $\bar{y}$ satisfy Equations (2L) and (5L), respectively. It follows from Equation (2L) that

$$\bar{b}^t = \bar{x}^tA^t + \bar{t}^t \tag{2}$$

and from Equation (5L) that

$$\bar{c}^t = \bar{y}^tA - \bar{v}^t. \tag{3}$$

Note that we have made use of Theorem 2.3 here (recall that vectors are matrices with one row or one column).

Substitution of (2) and (3) into (1) yields

$$w - u = \bar{x}^tA^t\bar{y} + \bar{t}^t\bar{y} - \bar{y}^tA\bar{x} + \bar{v}^t\bar{x}. \tag{4}$$

However, we notice (as in the Proof of Theorem 5.2) that

$$\bar{x}^tA^t\bar{y} = \bar{y}^tA\bar{x}. \tag{5}$$

Thus, Equation (4) becomes

$$w - u = \bar{t}^t\bar{y} + \bar{v}^t\bar{x}, \tag{6}$$

which was to be shown.   □

Problem 5.18

Prove Theorem 5.2 by making use of Theorem 5.8.

Solution

Suppose that vectors $\bar{x}$ and $\bar{y}$ satisfy Inequalities (2L), (3L) and (5L), (6L), respectively, of the inequality form of the dual linear programming problems (see the beginning of Section 5). Noting that $\bar{t}$ equals the difference $\bar{b} - A\bar{x}$ and that $\bar{v}$ equals the difference $A^t\bar{y} - \bar{c}$ leads us to conclude that $\bar{t}$, $\bar{x}$ and $\bar{v}$, $\bar{y}$ satisfy (2L), (3L) and (5L), (6L), respectively, of the equality form (standard slack form) of the dual linear programming problems. (Note that this means, in particular, that $\bar{t} \geq \bar{0}$ and $\bar{v} \geq \bar{0}$). Thus, Theorem 5.8 applies and we have

$$w - u = \sum_{i=1}^{m} t_iy_i + \sum_{j=1}^{n} v_jx_j. \tag{1}$$

However, each of the $m + n$ products on the right-hand side is nonnegative because $\bar{t} \geq \bar{0}$, $\bar{x} \geq \bar{0}$, $\bar{v} \geq \bar{0}$, and $\bar{y} \geq \bar{0}$. (Recall that $\bar{t} \geq \bar{0}$ means $t_1 \geq 0$, $t_2 \geq 0$, . . . , $t_m \geq 0$ and similarly for the other vectors.) Thus, Equation (1) implies that

$$w - u \geq 0 \quad \text{i.e.,} \quad u \leq w, \tag{2}$$

which was to be shown.

The following two theorems are usually known as the *complementary slackness theorems*.

Theorem 5.9. *Suppose that $\bar{x}^*$ and $\bar{y}^*$ are solution vectors for the respective dual linear programming problems. Then:*

    a. Each of the terms $t_i^* y_i^* = 0$ for $i = 1, 2, \ldots, m$. In other words, at least one of t_i^* or y_i^* has to equal zero for each i.
       Each of the terms $v_j^* x_j^* = 0$ for $j = 1, 2, \ldots, n$. In other words, at least one of v_j^* or x_j^* has to equal zero for each j.
    b. For each i value, if $t_i^* \neq 0$, then $y_i^* = 0$.
       For each i value, if $y_i^* \neq 0$, then $t_i^* = 0$.
       For each j value, if $v_j^* \neq 0$, then $x_j^* = 0$.
       For each j value, if $x_j^* \neq 0$, then $v_j^* = 0$.

PROOF OF THEOREM 5.9. Noting that $\bar{t}^* = \bar{b} - A\bar{x}^*$ and $\bar{v}^* = A^t\bar{y}^* - \bar{c}$, we observe that $\bar{t}^*$, $\bar{x}^*$ and $\bar{v}^*$, $\bar{y}^*$ satisfy (2L), (3L) and (5L), (6L), respectively, of the equality form of the dual linear programming problems. Thus, it follows from Theorem 5.8 that

$$w(\bar{y}^*) - u(\bar{x}^*) = \sum_{i=1}^{m} t_i^* y_i^* + \sum_{j=1}^{n} v_j^* x_j^*, \tag{1}$$

where each of the $m + n$ products on the right-hand side is nonnegative. Since $\bar{x}^*$ and $\bar{y}^*$ are solution vectors, it follows from Theorem 5.4 (in particular, see Note (c) after the Proof of Theorem 5.4) that the left-hand side of (1) is zero. Thus, we have

$$\sum_{i=1}^{m} t_i^* y_i^* + \sum_{j=1}^{n} v_j^* x_j^* = 0. \tag{2}$$

Since each of the $m + n$ products is nonnegative, it follows that each of the products must, moreover, *exactly equal zero* (since the right-hand side equals zero). Thus, we have part (a) of Theorem 5.9. Part (b) is easy to prove, and we do so for the first line. From part (a) we have that

$$t_i^* y_i^* = 0 \tag{3}$$

for each i. A product of two real numbers equals zero only if *at least one* of the numbers equals zero. Thus, if $t_i^* \neq 0$ (actually, if $t_i^* > 0$), we have that

$$y_i^* = 0. \tag{4}$$

$\square$

The converse of Theorem 5.9 is also true; it is stated as follows.

Theorem 5.10. *Given two feasible vectors $\bar{x}^\circ$ and $\bar{y}^\circ$ for the dual linear programming problems. If the conditions of complementary slackness* *hold for these vectors, i.e.,*

$$\text{Each of the terms } t_i{}^\circ y_i{}^\circ = 0 \quad \text{for} \quad i = 1, 2, \ldots, m \tag{i}$$

and

$$\text{Each of the terms } v_j{}^\circ x_j{}^\circ = 0 \quad \text{for} \quad j = 1, 2, \ldots, n, \tag{ii}$$

then $\bar{x}^\circ$ and $\bar{y}^\circ$ are solution vectors of the dual linear programming problems.

PROOF OF THEOREM 5.10. Noting that $\bar{t}^\circ = \bar{b} - A\bar{x}^\circ$ and $\bar{v}^\circ = A^t\bar{y}^\circ - \bar{c}$, we observe that $\bar{t}^\circ$, $\bar{x}^\circ$ and $\bar{v}^\circ$, $\bar{y}^\circ$ satisfy (2L), (3L) and (5L), (6L), respectively, of the equality form of the dual linear programming problems. Thus, we have from Theorem 5.8 that

$$w(\bar{y}^\circ) - u(\bar{x}^\circ) = \sum_{i=1}^{m} t_i{}^\circ y_i{}^\circ + \sum_{j=1}^{n} v_j{}^\circ x_j{}^\circ. \tag{1}$$

From the hypothesis of Theorem 5.10, we observe that the right-hand side of Equation (1) will equal zero. This means that

$$u(\bar{x}^\circ) = w(\bar{y}^\circ). \tag{2}$$

It follows from Theorem 5.3 that $\bar{x}^\circ$ and $\bar{y}^\circ$ are solution vectors of the dual linear programming problems. □

Useful discussions on complementary slackness appear in [8, 13, 14].

It is informative to illustrate complementary slackness with respect to one of the sample problems—refer to the tableaus in Problem 5.16 (another form of the "nut problem"). In Tableau 1, which is a special case of Figure 5.3, we can read off the complementary slackness *pairs* from the margins of the columns and rows; e.g., the pairs are x_1 and v_1, x_2 and v_2, etc:

	Column	Row
Original variable	$x_1\ x_2\ x_3$	$y_1\ y_2$
Slack variable	$v_1\ v_2\ v_3$	$t_1\ t_2$

Each pair will always appear in the same *line* (i.e. row or column) in subsequent tableaus; this is a consequence of the nature of the pivot transformation. For example, the pair x_3 and v_3 appear in column 3 of the first two tableaus and in row 2 of the third tableau. The conclusions of Theorem 5.9 have a clear interpretation also with respect to Problem 5.16. To find the primal and dual solutions from the terminal tableau—Tableau 3—we set the nonbasic variables equal to *zero*. This is equivalent to conclusion (a) of Theorem 5.9 because in any tableau one of the members of a complementary slackness pair has to be nonbasic and the other basic. For example, in this problem we have the following correspondence from the terminal tableau, in which the nonbasic variables have been set equal to zero:

$$x_1{}^* = 0 \quad x_2{}^* = 1260 \quad x_3{}^* = 240 \quad y_1{}^* = \tfrac{8}{25} \quad y_2{}^* = \tfrac{14}{25}$$

$$v_1{}^* = \tfrac{7}{100} \quad v_2{}^* = 0 \quad\quad v_3{}^* = 0 \quad\quad t_1{}^* = 0 \quad\quad t_2{}^* = 0.$$

It should be noted that it is possible for both members of a complementary slackness pair to equal zero when the solution is reached; in this case we have *degeneracy* for the basic variable of the pair.

REFERENCES

1. Cooper, L., and Steinberg, D. I. *Introduction to Methods of Optimization.* Philadelphia, Pennsylvania: W. B. Saunders Company, 1970.
2. Dantzig, G. B. "Maximization of a Linear Function of Variables Subject to Linear Inequalities." In *Activity Analysis of Production and Allocation,* edited by T. C. Koopmans. New York: John Wiley and Sons, 1951.
3. Dantzig, G. B. *Linear Programs and Extensions.* Princeton, New Jersey: Princeton University Press, 1963.
4. Gass, S. I. *Linear Programming.* Third edition. New York: McGraw-Hill, 1969.
5. Hadley, G. *Linear Algebra.* Reading, Massachusetts: Addison-Wesley, 1961.
6. Hadley, G. *Linear Programming.* Reading, Massachusetts: Addison-Wesley, 1962.
7. Kemeny, J. G., Mirkil, H., Snell, J. L., and Thompson, G. L. *Finite Mathematics with Business Applications.* Englewood Cliffs, New Jersey: Prentice-Hall, 1962.
8. Kemeny, J. G., Mirkil, H., Snell, J. L., and Thompson, G. L. *Finite Mathematics with Business Applications.* Second edition. Englewood Cliffs, New Jersey: Prentice-Hall, 1972.
9. Kotiah, T. C. T., and Steinberg, D. I. "Occurrence of Cycling and Other Phenomena Arising in a Class of Linear Programming Models." *Commun. ACM,* vol. 20, no. 2 (February 1977), pp. 102–112.
10. Nemhauser, G. L., and Garfinkel, R. *Integer Programming.* New York: John Wiley and Sons, 1972.
11. Owen, G. *Finite Mathematics.* Philadelphia, Pennsylvania: W. B. Saunders Company, 1970.
12. Owen, G. *Game Theory.* Philadelphia, Pennsylvania: W. B. Saunders Company, 1968.
13. Singleton, R. R., and Tyndall, W. F. *Games and Programs.* San Francisco, California: W. H. Freeman and Company, 1974.
14. Strum, J. E. *Introduction to Linear Programming.* San Francisco, California: Holden-Day, 1972.
15. Tucker, A. W. "Combinatorial Algebra of Matrix Games and Linear Programming." In *Applied Combinatorial Mathematics,* edited by E. F. Beckenbach. New York: John Wiley and Sons, 1964.

16. Tucker, A. W., and Balinski, M. L. "Duality Theory of Linear Programs: A Constructive Approach with Applications." *SIAM Review*, vol. 11, 1969.

SUPPLEMENTARY PROBLEMS

Degeneracy

S.P. 5.1: Use the simplex algorithm to solve S.P. 4.13. Follow the approach of Problem 5.1.

S.P. 5.2: Rework S.P. 4.13 using the perturbation approach. In the first tableau, interchange x and s. Display the terminal tableau.

Nonstandard Linear Programming Problems

S.P. 5.3: Solve Problem 3.3 (see also Problem 5.3) by the simplex algorithm.

S.P. 5.4: Work through Problem 5.4 (in text) to completion.

S.P. 5.5: Solve the following problem by using Stage Zero followed by the usual simplex algorithm:

$$\text{Minimize} \quad u = 3x + \ y + 2z$$

$$\text{subject to} \quad x + 2y + 3z \ \geq 24$$

$$2x + 4y + 3z \ = 36$$

and

$$x, \ y, \ z \geq 0.$$

S.P. 5.6: Solve the following linear programming problem in which both x and y are unrestricted. Let $x = a - b$ and $y = c - d$, where $a, b, c, d \geq 0$, and then employ the simplex algorithm:

$$\text{Maximize} \quad u = x + 4y$$

$$\text{subject to} \quad 2x + 3y \leq 24$$

$$2x + \ y \geq 12.$$

S.P. 5.7: Solve S.P. 5.6 by graphical means. Construct some level lines corresponding to the objective function. (Do diagramming in the XY plane.)

Transportation Problems

More efficient methods for solving transportation problems are presented in Chapter 9.

S.P. 5.8: Use the simplex algorithm to solve the balanced 2 by 3 transportation problem in which $a_1 = 60$, $a_2 = 80$, $b_1 = 30$, $b_2 = 50$, and $b_3 = 60$. The total cost function (in dollars) is $u = 2x_{11} + 4x_{12} + 3x_{13} + 5x_{21} + 3x_{22} + 2x_{23}$.

S.P. 5.9: Repeat for the 3 by 3 transportation problem in which $a_1 = 61$, $a_2 = 49$, $a_3 = 90$, $b_1 = 52$, $b_2 = 68$, and $b_3 = 80$. The total cost function (in dollars) is $u = 26x_{11} + 23x_{12} + 10x_{13} + 14x_{21} + 13x_{22} + 21x_{23} + 16x_{31} + 17x_{32} + 29x_{33}$. This problem is a model problem in Chapter 9 (e.g., see Problem 9.1).

Duality Theory

For S.P. 5.10 through S.P. 5.15 give the duals of the linear programming problems listed. Obtain solutions to each pair of problems. (See Problems 5.15 and 5.16 for models.)

S.P. 5.10: The diet problem—Problem 1.12.
S.P. 5.11: S.P. 1.1.
S.P. 5.12: S.P. 1.2.
S.P. 5.13: S.P. 1.3.
S.P. 5.14: S.P. 1.4.
S.P. 5.15: S.P. 1.5.
S.P. 5.16: Apply Method 5.3 to the following dual tableau in order to obtain a terminal tableau (all variables are required to be nonnegative):

	x_1	t_1	t_2	1		
v_2	3	3	-1	-500	$= -x_2$	
v_3	-1	-2	-3	100	$= -x_3$	(T.k)
-1	-8	-20	-50	950	$= u$	
	$= v_1$	$= y_1$	$= y_2$	$= -w$		

Complementary Slackness

Vectors $\bar{x}$, $\bar{t}$ and $\bar{y}$, $\bar{v}$ related by (2L) and (5L), respectively, of Section 6 are said to have the complementary slackness property if $\bar{t}'\bar{y} = 0$ and $\bar{v}'\bar{x} = 0$.

S.P. 5.17: (a) Show that if $\bar{x}$, $\bar{t}$ and $\bar{y}$, $\bar{v}$ satisfy the complementary slackness property, then $w = u$. (b) Is the converse true?
S.P. 5.18: Illustrate Theorem 5.9 with respect to the solutions of the dual problems of Problem 5.15.

ANSWERS TO SUPPLEMENTARY PROBLEMS

S.P. 5.1: Max $u = 48$ when $x = 8$ and $y = 0$. If x and s are interchanged in Tableau 1, then three tableaus are required. If x and r are interchanged in Tableau 1, then Tableau 2 is the terminal tableau.
S.P. 5.2: Start by replacing $-b_1 = 0$ with $-b_1 = -\epsilon$ in Tableau 2. Then obtain the terminal tableau—Tableau 3—given below:

s	r	1	
$-\frac{3}{2}$	$\frac{1}{4}$	$-\epsilon/4$	$= -y$
1	0	-8	$= -x$
$\frac{3}{8}$	$-\frac{1}{4}$	$-6 + \epsilon/4$	$= -t$
-3	$-\frac{1}{2}$	$48 + \epsilon/2$	$= u$

(T.3)

Take the limit as ϵ approaches zero to obtain Max $u = 48$ when $x = 8$ and $y = 0$.

S.P. 5.3: The linear programming problem is unbounded.

S.P. 5.4: Max $u = 39$, $x = 0$, $y = 9$, and $z = 6$.

S.P. 5.5: Min $u = 14$, $x = 0$, $y = 6$, and $z = 4$.

S.P. 5.6: Max $u = 27$, $x = 3$, and $y = 6$.

S.P. 5.7: The constraint set is an unbounded wedge to the right of $(3,6)$, which lies above $2x + y = 12$ and below $2x + 3y = 24$. The level line of maximum u, which cuts the constraint set, is $x + 4y = 27$.

S.P. 5.8: Min $u = \$360$, $x_{11} = 30$, $x_{12} = 30$, $x_{22} = 20$, $x_{23} = 60$, and all other $x_{ij} = 0$. Also, alternate solution where $x_{11} = 30$, $x_{13} = 30$, $x_{22} = 50$, $x_{23} = 30$, and all other $x_{ij} = 0$.

S.P. 5.9: Min $u = \$2877$, $x_{13} = 61$, $x_{22} = 30$, $x_{23} = 19$, $x_{31} = 52$, $x_{32} = 38$, and all other $x_{ij} = 0$.

S.P. 5.10: Maximize $u = 400x_1 + 500x_2 + 300x_3$

subject to $10x_1 + 20x_2 + 15x_3 \leq 0.25$

                     $25x_1 + 10x_2 + 20x_3 \leq 0.40$

and

$$x_1, x_2, x_3 \geq 0.$$

Primal solution: $y_1 = 21.25$ and $y_2 = 7.50$.

Dual solution: $x_1 = .0138$, $x_2 = .00559$, and $x_3 = 0$; common value is 8.3125.

S.P. 5.11: Minimize $w = 12y_1 + 19y_2$

subject to $y_1 + 2y_2 \geq 4$

                     $3y_1 + y_2 \geq 3$

and

$$y_1, y_2 \geq 0.$$

Primal solution: $x_1 = 9$ and $x_2 = 1$.

Dual solution: $y_1 = 0.4$ and $y_2 = 1.8$; common value is 39.

S. P. 5.12: Primal solution: $y_1 = 9$ and $y_2 = 1$.

Dual solution: $x_1 = 1.2$ and $x_2 = 1.4$; common value is 41.

S. P. 5.13: Primal solution: $x_1 = 3$ and $x_2 = 3$.

Dual solution: $y_1 = 2$, $y_4 = 1$, and $y_2 = y_3 = 0$; common value is 27.

S. P. 5.14: Primal solution: $y_1 = \frac{1}{3}$ and $y_2 = \frac{14}{3}$.

Dual solution: $x_1 = \frac{1}{2}$, $x_2 = 1$, and $x_3 = 0$; common value is 21.

S. P. 5.15: Primal solution: $x_1 = 75$, $x_2 = 0$, and $x_3 = 100$.

Dual solution: $y_1 = \frac{1}{8}$ and $y_2 = \frac{9}{20}$; common value is 650.

S.P. 5.16: In Tableau k, $a_{21} = -1$ is the pivot entry. The next tableau, which is given below, is terminal:

	x_3	t_1	t_2	1		
v_2	3	-3	-10	-200	$= -x_2$	
v_1	-1	2	3	-100	$= -x_1$	(T.$k+1$)
-1	-8	-4	-26	150	$= u$	
	$= v_3$	$= y_1$	$= y_2$	$= -w$		

S.P. 5.17: (a) From Theorem 5.8, $w - u = 0 + 0 = 0$, i.e., $w = u$. (b) The converse is false. For example, we could have $\bar{t}^t \bar{y} < 0$, $\bar{v}^t \bar{x} > 0$, and $\bar{t}^t \bar{y} = -\bar{v}^t \bar{x}$.

S.P. 5.18: From the terminal tableau—Tableau 3—of Problem 5.15, $t_1^* = 0$, $y_1^* = 20$; $t_2^* = 0$, $y_2^* = 20$; $v_1^* = 0$, $x_1^* = \frac{400}{3}$; $v_2^* = 0$, $x_2^* = \frac{50}{3}$; and $v_3^* = 10$, $x_3^* = 0$.

Chapter Six

Sensitivity Analysis

1. IMPUTED VALUE AND IMPUTED COST

The vector definitions of imputed value and imputed cost were given in Section 6 of Chapter 5. Let us focus on one of the maximum problems from Chapter 1 (Problem 1.4—the "nut problem"), which was redone in dual tableau format in Problem 5.16. It is possible to give an economic interpretation to the solutions for the dual variables $y_1{}^*$ and $y_2{}^*$ from Tableau 3:

$$y_1{}^* = \tfrac{8}{25} = 0.32 \quad \text{and} \quad y_2{}^* = \tfrac{14}{25} = 0.56.$$

First let us determine the dimensions of these variables. We see from Tucker's duality equation (Theorem 5.8) that all the terms that are summed there must have identical dimensions, i.e., w, u, $t_i y_i$ (for each i), and $v_j x_j$ (for each j) all have the dimension of dollars ($\$$). Using the symbol [ ] for "dimension of," we see that $[y_i] \cdot [t_i] = \$$ and that

$$[y_i] = \frac{\$}{[t_i]}.$$

The quantity t_i has the same dimension as b_i (see, e.g., Section 6 of Chapter 5). Thus,

$$[y_i] = \frac{\$}{[b_i]}.$$

In Problem 5.16 (or Problem 1.4), b_1 is in pounds of peanuts and b_2 is in pounds of walnuts. Thus,

$$[y_1] = \frac{\$}{\text{lb peanuts}} = \frac{\$}{\text{lb } P}$$

and

$$[y_2] = \frac{\$}{\text{lb walnuts}} = \frac{\$}{\text{lb } W}$$

Thus, it appears that $y_1{}^*$ possibly may be interpreted as a *value* for a pound of peanuts, while $y_2{}^*$ may be interpreted as a *value* for a pound of walnuts. Let us now show, through calculations, that this is indeed the case for $y_1{}^*$.

Problem 6.1

Suppose that the amount of peanuts available in Problem 5.16 is increased from 900 pounds to 905 pounds. Analyze the effect of this change in b_1 on Max u and make an interpretation relative to $y_1{}^*$.

Solution

If we make this change in b_1 in Tableau 1 of Problem 5.16 and carry out the same sequence of Stage Two calculations as done there, we obtain the following terminal tableau:

	x_1	t_1	t_2	1		
v_2	$\frac{9}{5}$	$\frac{9}{5}$	$-\frac{3}{5}$	-1269	$= -x_2$	
v_3	$-\frac{4}{5}$	$-\frac{4}{5}$	$\frac{8}{5}$	-236	$= -x_3$	(T.3′)
-1	$-\frac{7}{100}$	$-\frac{8}{25}$	$-\frac{14}{25}$	625.6	$= u$	
	$= v_1$	$= y_1$	$= y_2$	$= -w$		

If we compare Tableau 3 of Problem 5.16 with Tableau 3′, we see that the only locations where changes have occurred are in the right-hand column and the corner entry. In particular, note that the ratio of the change in maximum u value (Δ Max u) to the change in $b_1(\Delta b_1)$ is *exactly equal* to $y_1{}^*$, i.e.,

$$\frac{\Delta \text{ Max } u}{\Delta b_1} = \frac{625.6 - 624}{905 - 900} = \frac{1.6}{5} = 0.32 \frac{\$}{\text{lb } P}.$$

(Note that the change Δ of a quantity q is defined as $\Delta q = q_{\text{new}} - q_{\text{old}}$.) We call $y_1{}^*$ the *imputed value* per pound of additional peanuts available. It should be noted that the imputed value interpretation holds only over a limited range of changes in b_1. The correct statement (to be justified later in this chapter) is that $y_1{}^* = 0.32$ \$/lb P is the imputed value per pound of additional peanuts *provided that the dual solution is not changed* by this change in the quantity of peanuts. A similar statement appears in [1]. We note from Tableau 3 of Problem 5.16 and Tableau 3′ of this problem that the dual solutions for both (excluding Min w) are identical, namely,

$$v_1{}^* = \tfrac{7}{100}, \; v_2{}^* = v_3{}^* = 0, \; y_1{}^* = \tfrac{8}{25}, \quad \text{and} \quad y_2{}^* = \tfrac{14}{25}.$$

We can indicate a general interpretation of the imputed value $y_i{}^*$ by means of the following theorem (to be justified later in this chapter).

Theorem 6.1: Maximum Problem. *If the dual solution (excluding Min w) in the terminal tableau remains unchanged when b_i is changed to $b_i + \Delta b_i$, then the imputed value $y_i{}^*$ associated with b_i is given by*

$$y_i{}^* = \frac{\Delta \; \text{Max} \; u}{\Delta b_i},$$

where Δ Max u is the change in maximum value of the objective function.

Problem 6.2

Suppose that the quantity of walnuts available (b_2) in Problem 5.16 is increased from 600 to 625 pounds. Assume that the solution to the dual problem in the terminal tableau is unaffected by this change. Use Theorem 6.1 to determine the new value of Max u.

Solution

From Tableau 3 of Problem 5.16 we have

$$y_2{}^* = 0.56 \frac{\$}{\text{lb } W}. \tag{1}$$

In addition, $\Delta b_2 = 625 - 600 = 25$ lb W. Thus, from Theorem 6.1,

$$\Delta \; \text{Max} \; u = y_2{}^* \cdot \Delta b_2 = 0.56 \cdot 25 = 14. \tag{2}$$

Therefore, the new maximum value is

$$\text{Max } u = 624 + 14 = \$638. \tag{3}$$

Note: Later in this chapter we shall show that the allowable range (such that the dual solution stays the same) for Δb_2 is $-150 \le \Delta b_2 \le 2100$. The quantity $\Delta b_2 = 25$ is well within this range.

Let us now refer to Problem 5.15 (originally Problem 1.3), which was a minimum problem that dealt with the mining of three types of ore. We will endeavor to show that the solution for a typical dual variable $x_j{}^*$ can be interpreted as a type of cost (known as an imputed cost) associated with the requirements term c_j of the minimum problem. Making a dimensional analysis from Theorem 5.8, we see that $[x_j] = \$/[v_j]$. However, $[v_j] = [c_j]$ from, e.g., Section 6 of Chapter 5. Thus,

$$[x_j] = \frac{\$}{[c_j]}.$$

Thus, x_j^* has the dimension of a cost (in \$) per unit of the requirement term c_j. Consequently, in Problem 5.15 (we abbreviate high-grade ore as *HG*, medium-grade ore as *MG*, and low-grade ore as *LG*) we have the following:

$$x_1^* = \frac{400}{3} = 133.33 \, \frac{\$}{\text{ton } HG},$$

$$x_2^* = \frac{50}{3} = 16.67 \, \frac{\$}{\text{ton } MG},$$

and

$$x_3^* = 0 \, \frac{\$}{\text{ton } LG}.$$

Thus, it seems that each x_j^* may possibly be interpreted as a type of *cost* per unit of c_j. Let us now show, through calculations, that this is indeed the case for x_1^*.

Problem 6.3

Suppose that the tonnage of high-grade ore required (c_1) is increased from 60 to 63 tons in Problem 5.15. Analyze the effect of this change on Min w and make an interpretation relative to x_1^*.

Solution

If we make this change in c_1 in Tableau 1 of Problem 5.15 and carry out the same sequence of pivot calculations (using Method 5.2) as done there, we obtain the following terminal tableau:

	t_2	t_1	x_3	1		
v_2	$-\frac{1}{6}$	$\frac{1}{3}$	$\frac{5}{3}$	$-\frac{50}{3}$	$= -x_2$	
v_1	$\frac{2}{3}$	$-\frac{1}{3}$	$-\frac{2}{3}$	$-\frac{400}{3}$	$= -x_1$	(T.3′)
-1	-22	-19	-8	$10,400$	$= u$	
	$= y_2$	$= y_1$	$= v_3$	$= -w$		

If we compare Tableau 3 of Problem 5.15 with Tableau 3′, we see that the only locations where changes have occurred are in the bottom row and the corner entry. In particular, we note that the ratio of the change in minimum w value (Δ Min w) to the change in c_1 (Δc_1) is *exactly equal* to x_1^*, i.e.,

$$\frac{\Delta \text{ Min } w}{\Delta c_1} = \frac{10{,}400 - 10{,}000}{3} = \frac{400}{3} = 133.33 \frac{\$}{\text{ton } HG}.$$

We call x_1^* the *imputed cost* per ton of additional high-grade ore required (recall that Min w means minimum *cost*). It should be noted that the imputed cost interpretation holds only over a limited range of changes in c_1. The correct statement (to be justified later in this chapter) is that $x_1^* = 133.33$ \$/ton HG is the imputed cost per ton of additional high-grade ore *provided that the dual solution is not changed* by this change in the requirement of high-grade ore. We note from Tableau 3 of Problem 5.15 and Tableau 3' of the current problem that the dual solutions (in the present case, of a maximum problem) for both are identical, namely,

$$t_1^* = t_2^* = 0, \; x_1^* = \tfrac{400}{3}, \; x_2^* = \tfrac{50}{3}, \quad \text{and} \quad x_3^* = 0.$$

(Here we have excluded the change in Max u in comparing dual solutions.)

We can indicate a general interpretation of the imputed cost x_j^* by means of the following theorem (to be justified later in this chapter).

Theorem 6.2: Minimum Problem. *If the dual solution (excluding Max u) in the terminal tableau remains unchanged when c_j is changed to $c_j + \Delta c_j$, then the imputed cost x_j^* associated with c_j is given by*

$$x_j^* = \frac{\Delta \text{ Min } w}{\Delta c_j},$$

where Δ Min w is the change in minimum value of the objective function.

Problem 6.4

Suppose that the tonnage of medium-grade ore required (c_2) in Problem 5.15 is increased from 120 to 126 tons. Assume that the solution to the dual problem in the terminal tableau is unaffected by this change. Use Theorem 6.2 to determine the new value of Min w.

Solution

From Tableau 3 of Problem 5.15 we have

$$x_2^* = \frac{50}{3} \frac{\$}{\text{ton } MG}. \tag{1}$$

In addition, $\Delta c_2 = 126 - 120 = 6$ tons MG. Thus, from Theorem 6.2,

$$\Delta \text{ Min } w = x_2^* \cdot \Delta c_2 = \frac{50}{3} \cdot 6 = 100. \tag{2}$$

Therefore, the new minimum cost is

$$\text{Min } w = \$10,100. \qquad (3)$$

Note: Later in this chapter we shall show that the allowable range (such that the dual solution stays the same) for Δc_2 is $-6 \le \Delta c_2 \le 120$. The quantity $\Delta c_2 = 6$ is within this range.

Problem 6.5

Suppose that the tonnage of low-grade ore required (c_3) in Problem 5.15 is increased from 150 to 154 tons. Assume that the solution to the dual problem in the terminal tableau is unaffected by this change. Use Theorem 6.2 to determine the new value of Min w.

Solution

From Tableau 3 of Problem 5.15, we have

$$x_3{}^* = 0 \frac{\$}{\text{ton } LG}. \qquad (1)$$

In addition, $\Delta c_3 = 154 - 150 = 4$ tons LG. Thus, from Theorem 6.2,

$$\Delta \text{ Min } w = x_3{}^* \cdot \Delta c_3 = 0 \cdot 4 = 0. \qquad (2)$$

Therefore, the new minimum cost is the same as the old minimum cost, namely,

$$\text{Min } w = \$10,000. \qquad (3)$$

Note: Later in this chapter we shall show that the allowable range (such that the dual solution stays the same) for Δc_3 is $\Delta c_3 \le 10$. The quantity $\Delta c_2 = 4$ is within this range.

It is important to note that many practitioners in the field of linear programming use the expression "shadow price" in place of imputed value and imputed cost (see also Section 6 of Chapter 5).

2. SENSITIVITY ANALYSIS

Sensitivity analysis is concerned with varying the numbers that appear in the body of the tableau that represents the initial formulation of the linear programming problem and determining the effect on the final solution and value of the problem. The dual tableau approach provides a very efficient method for doing this. Sensitivity analysis is also known as *postoptimality analysis*. Elementary treatments of this subject appear in [2] and [3].

We begin by analyzing the effects of changing a component (b_i) of the resources vector for a maximum problem; we will be led in a logical and analytical way to the conclusion given in Theorem 6.1. In order to simplify the notation in what follows, we will sometimes replace the symbol for change of b_i, namely, Δb_i, by h_i; analogously, Δc_j will be replaced by k_j.

Problem 6.6

In Problem 5.16 allow the resource term b_1 to change by the amount h_1 (Δb_1). Determine the range of values for h_1 such that Theorem 6.1 applies.

Solution

We refer to Tableau 1 of Problem 5.16. In the first row, we replace $b_1 = 900$ by $900 + h_1$. The equation corresponding to the first row thus becomes

$$x_1 + \tfrac{2}{3}x_2 + x_3 - 900 - h_1 = -t_1.$$

Let us now group the h_1 with t_1. This leads to

$$x_1 + \tfrac{2}{3}x_2 + x_3 - 900 = -(t_1 - h_1).$$

Thus, in Tableau 1 the marginal variable t_1 is replaced by $t_1 - h_1$, while the remainder of the tableau remains unchanged, as can be seen by referring to Tableau 1':

	x_1	x_2	x_3	1	
y_1	1	$\tfrac{2}{3}$	$\tfrac{1}{4}$	-900	$= -(t_1 - h_1)$
y_2	0	$\tfrac{1}{3}$	$\tfrac{3}{4}$	-600	$= -t_2$ (T.1')
-1	$\tfrac{1}{4}$	$\tfrac{2}{5}$	$\tfrac{1}{2}$	0	$= u$
	$= v_1$	$= v_2$	$= v_3$	$= -w$	

We can now conduct the *same* set of pivot steps as in Problem 5.16. The only change is that wherever t_1 appeared previously there now appears $t_1 - h_1$. The terminal tableau of Problem 5.16 is thus replaced by the following, where the *only* change so far is in the top margin (t_1 is replaced by $t_1 - h_1$):

	x_1	$t_1 - h_1$	t_2	1	
v_2	$\tfrac{9}{5}$	$\tfrac{9}{5}$	$-\tfrac{3}{5}$	-1260	$= -x_2$
v_3	$-\tfrac{4}{5}$	$-\tfrac{4}{5}$	$\tfrac{8}{5}$	-240	$= -x_3$ (T.3')
-1	$-\tfrac{7}{100}$	$-\tfrac{8}{25}$	$-\tfrac{14}{25}$	624	$= u$
	$= v_1$	$= y_1$	$= y_2$	$= -w$	

Let us write out the first row equation corresponding to Tableau 3':

$$\tfrac{9}{5}x_1 + \tfrac{9}{5}t_1 - \tfrac{9}{5}h_1 - \tfrac{3}{5}t_2 - 1260 = -x_2.$$

To deduce the effect of the change h_1 it is useful to group the h_1 term with the pure constant term 1260. Thus,

$$\tfrac{9}{5}x_1 + \tfrac{9}{5}t_1 - \tfrac{3}{5}t_2 - 1260 - \tfrac{9}{5}h_1 = -x_2. \tag{1}$$

Proceeding in the same way with the second and bottom rows, we obtain

$$-\tfrac{4}{5}x_1 - \tfrac{4}{5}t_1 + \tfrac{8}{5}t_2 - 240 + \tfrac{4}{5}h_1 = -x_3 \tag{2}$$

$$-\tfrac{7}{100}x_1 - \tfrac{8}{25}t_1 - \tfrac{14}{25}t_2 + 624 + \tfrac{8}{25}h_1 = u. \tag{3}$$

Thus, we may write out the following tableau, which is equivalent to Tableau 3' except that the variable t_1 in the top marginal row is now by itself (in addition, the effects of the change h_1 have been placed into the right-hand column entries and into the corner entry):

	x_1	t_1	t_2	1	
v_2	$\tfrac{9}{5}$	$\tfrac{9}{5}$	$-\tfrac{3}{5}$	$-1260 - \dfrac{9h_1}{5}$	$= -x_2$
v_3	$-\tfrac{4}{5}$	$-\tfrac{4}{5}$	$\tfrac{8}{5}$	$-240 + \dfrac{4h_1}{5}$	$= -x_3$ (T.3″)
-1	$-\tfrac{7}{100}$	$-\tfrac{8}{25}$	$-\tfrac{14}{25}$	$624 + \dfrac{8h_1}{25}$	$= u$
	$= v_1$	$= y_1$	$= y_2$	$= -w$	

Note that it is a mechanical process to go from Tableau 3 of Problem 5.16 to Tableau 3″ (or to Equations (1), (2), and (3)). The coefficients $-\tfrac{9}{5}$, $\tfrac{4}{5}$, and $\tfrac{8}{25}$ all correspond to the entries in the t_1 column of Tableau 3. In particular, the $\tfrac{8}{25}$ is none other than $y_1{}^*$—the imputed value associated with b_1.

Tableau 3″ will be a terminal tableau provided that both the entries in the right-hand column are nonpositive. This leads to

$$-1260 - \tfrac{9}{5}h_1 \le 0 \tag{4}$$

$$-240 + \tfrac{4}{5}h_1 \le 0, \tag{5}$$

and thus

$$-700 \le h_1 \quad \text{(from (4))} \tag{4a}$$

$$h_1 \le 300 \quad \text{(from (5))}. \tag{5a}$$

Thus, if h_1 is in the range

$$-700 \le h_1 \le 300, \tag{6}$$

then Tableau 3″ will be a terminal tableau. Moreover, the change in Max u will then be given by the corner entry:

$$\Delta \text{ Max } u = \frac{8}{25} h_1 \tag{7}$$

or

$$\frac{\Delta \text{ Max } u}{\Delta b_1} = \frac{8}{25} = y_1{}^*. \tag{8}$$

This is none other than the result called for by Theorem 6.1.

Notes:
a. If Δb_1 (h_1) goes outside of the range indicated in (6), then Tableau 3″ will cease to be *terminal*. At least one more pivot would be called for. In this case, Method 5.3 would be most appropriate. We will deal with such a situation in a later problem.
b. We see that the dual solution items in the bottom row have been unaffected by the above manipulations. If Δb_1 stays in the range from -700 to 300, this will be the case. Otherwise, the dual *solution* will no longer be given by $v_2{}^* = v_3{}^* = 0$, $v_1{}^* = \frac{7}{100}$, $y_1{}^* = \frac{8}{25}$, and $y_2{}^* = \frac{14}{25}$. Thus, the dual solution remaining unchanged is intimately related with Δb_1 staying in the acceptable range indicated above.

Problem 6.7

In Problem 5.16 determine the effects on Max u, $x_2{}^*$, and $x_3{}^*$ of changes in b_1 from its initial value $b_1 = 900$ to 905, 895, 1200, and 200.

Solution

All of these changes in b_1 fall within the acceptability range cited in Problem 6.6. Thus, the conclusion of Theorem 6.1 applies. It is easy to indicate the effects of the cited changes by means of a table based on Equations (1), (2), and (3) of Problem 6.6 (alternatively, refer to the right-hand column of Tableau 3″ of that problem and recall that h_1 is Δb_1):

Δb_1	b_1	$x_2{}^*$	$x_3{}^*$	Max u
5	905	1269	236	625.6
-5	895	1251	244	622.4
300	1200	1800	0	720
-700	200	0	800	400

Note that the limiting changes $\Delta b_1 = 300$ and -700 cause degeneracies to appear in the solution of the maximum problem. Changes of b_1 outside of the range indicated in Problem 6.6 (i.e., changes not between -700 and 300 inclusive) will result in Tableau 3″ no longer being terminal.

Problem 6.8

In Problem 5.16 allow b_2 to change by the amount h_2 (Δb_2). Determine the range of values for h_2 such that Theorem 6.1 applies.

Solution

We proceed as in Problem 6.6, only now some of the steps will be abbreviated. Allowing b_2 to change from $b_2 = 600$ to $600 + h_2$ in Tableau 1 and then shifting h_2 will cause $-(t_2 - h_2)$ to appear in the margin of Tableau 1 in place of $-t_2$, with the remainder of the tableau staying the same. We then follow the pivot steps that led to Tableau 3 in Problem 5.16; we obtain a tableau in which the only difference from Tableau 3 of Problem 5.16 is that the top margin variable t_2 is replaced by $t_2 - h_2$. We now write out the row equations corresponding to this new tableau. After simplifying, we obtain the following:

$$\tfrac{8}{5}x_1 + \tfrac{8}{5}t_1 - \tfrac{3}{5}t_2 - 1260 + \tfrac{3}{5}h_2 = -x_2 \tag{1}$$

$$-\tfrac{4}{5}x_1 - \tfrac{4}{5}t_1 + \tfrac{8}{5}t_2 - 240 - \tfrac{8}{5}h_2 = -x_3 \tag{2}$$

$$-\tfrac{7}{100}x_1 - \tfrac{8}{25}t_1 - \tfrac{14}{25}t_2 + 624 + \tfrac{14}{25}h_2 = u. \tag{3}$$

It is useful to compare these equations with their counterparts in Problem 6.6. The basic point corresponding to these equations is obtained by setting $x_1 = t_1 = t_2 = 0$; this leads to

$$-1260 + \tfrac{3}{5}h_2 = -x_2{}^* \tag{4}$$

$$-240 - \tfrac{8}{5}h_2 = -x_3{}^* \tag{5}$$

$$624 + \tfrac{14}{25}h_2 = \text{Max } u \tag{6}$$

Note that the coefficients $\tfrac{3}{5}$, $-\tfrac{8}{5}$, and $\tfrac{14}{25}$ are the entries of the t_2 column of Tableau 3, but with the sign changed.

From (4) and (5), we will have a solution point provided that

$$-1260 + \tfrac{3}{5}h_2 \leq 0 \quad \text{and} \quad -240 - \tfrac{8}{5}h_2 \leq 0.$$

These lead to

$$-150 \leq h_2 \leq 2100. \tag{7}$$

If h_2 is in this range, we have the result called for by Theorem 6.1, namely,

$$\Delta \text{ Max } u = \tfrac{14}{25}h_2 = y_2{}^* \cdot \Delta b_2. \tag{8}$$

This follows from Equation (6).

Problem 6.9

In Problem 5.16 determine the new values for Max u if b_2 is changed from the base value 600 to (a) 650, (b) 550, and (c) 450.

Solution

a. Here $\Delta b_2 = 650 - 600 = 50$, and this is within the acceptable range indicated in Problem 6.8. Thus,

$$\Delta \text{ Max } u = y_2^* \cdot \Delta b_2 = \tfrac{14}{25} \cdot 50 = 28$$

and

$$\text{Max } u = 624 + 28 = 652.$$

b. Here $\Delta b_2 = 550 - 600 = -50$, and this also is within the acceptable range given in Problem 6.8. Thus,

$$\Delta \text{ Max } u = \tfrac{14}{25} \cdot (-50) = -28$$

and

$$\text{Max } u = 624 - 28 = 596.$$

c. Here $\Delta b_2 = 450 - 600 = -150$. This falls on the borderline of the acceptable range; a degeneracy will enter the solution of the maximum problem:

$$\Delta \text{ Max } u = \tfrac{14}{25} \cdot (-150) = -84.$$

and

$$\text{Max } u = 624 - 84 = 540.$$

The above process indicates the ease with which new solutions and Max u values can sometimes be calculated from a terminal tableau of a base problem when we allow for changes in the individual b_i terms.

We will now turn to several problems illustrating the features of Theorem 6.2. Recall that the latter applies for minimum problems. Our problems will involve varying the c_j terms with respect to our model minimum problem—Problem 5.15. For brevity, we shall often use the symbol k_j in place of Δc_j.

Problem 6.10

In Problem 5.15 allow the requirement c_1 to change by the amount $k_1 (\Delta c_1)$. Determine the range of values for k_1 such that Theorem 6.2 applies.

Solution

We refer to Tableau 1 of Problem 5.15. In the first column, we replace $c_1 = 60$ by $c_1 = 60 + k_1$. The equation corresponding to the first column thus becomes

$$y_1 + 2y_2 - 60 - k_1 = v_1.$$

If we group the k_1 with the v_1, we obtain

$$y_1 + 2y_2 - 60 = v_1 + k_1.$$

Thus, in Tableau 1 the marginal variable v_1 is replaced by $v_1 + k_1$, while the remainder of the tableau remains unchanged. This is indicated in Tableau 1', which follows:

	x_1	x_2	x_3	1	
y_1	1	4	6	-200	$= -t_1$
y_2	2	2	2	-300	$= -t_2$
-1	60	120	150	0	$= u$
	$= v_1 + k_1$	$= v_2$	$= v_3$	$= -w$	

(T.1')

We can now conduct the *same* set of pivot steps as in Problem 5.15. The only change is that wherever v_1 appeared previously there now appears $v_1 + k_1$. The terminal tableau of Problem 5.15 is thus replaced by the following, where the *only* change so far is in the left-hand margin (where v_1 is replaced by $v_1 + k_1$):

	t_2	t_1	x_3	1	
v_2	$-\frac{1}{6}$	$\frac{1}{3}$	$\frac{5}{3}$	$-\frac{50}{3}$	$= -x_2$
$v_1 + k_1$	$\frac{2}{3}$	$-\frac{1}{3}$	$-\frac{2}{3}$	$-\frac{400}{3}$	$= -x_1$
-1	-20	-20	-10	$10,000$	$= u$
	$= y_2$	$= y_1$	$= v_3$	$= -w$	

(T.3')

Let us now write out the first column equation corresponding to Tableau 3':

$$-\tfrac{1}{6}v_2 + \tfrac{2}{3}(v_1 + k_1) + 20 = y_2.$$

It is useful to group the k_1 term with the pure constant term 20. Thus,

$$-\tfrac{1}{6}v_2 + \tfrac{2}{3}v_1 + 20 + \tfrac{2}{3}k_1 = y_2. \tag{1}$$

Proceeding in the same way with the other three columns leads to

$$\tfrac{1}{3}v_2 - \tfrac{1}{3}v_1 + 20 - \tfrac{1}{3}k_1 = y_1 \tag{2}$$

$$\tfrac{5}{3}v_2 - \tfrac{2}{3}v_1 + 10 - \tfrac{2}{3}k_1 = v_3 \tag{3}$$

$$-\tfrac{50}{3}v_2 - \tfrac{400}{3}v_1 - (10,000 + \tfrac{400}{3}k_1) = -w. \tag{4}$$

Thus, we can now develop the following tableau, which is equivalent to Tableau 3' except that the marginal variable v_1 is now by itself (in addition, the effects of the change k_1 have been placed into the bottom row entries and into the corner entry):

	t_2	t_1	x_3	1	
v_2	$-\frac{1}{6}$	$\frac{1}{3}$	$\frac{5}{3}$	$-\frac{50}{3}$	$= -x_2$
v_1	$\frac{2}{3}$	$-\frac{1}{3}$	$-\frac{2}{3}$	$-\frac{400}{3}$	$= -x_1$
-1	$-20 - \frac{2}{3}k_1$	$-20 + \frac{1}{3}k_1$	$-10 + \frac{2}{3}k_1$	$10,000 + \frac{400}{3}k_1$	$=\quad u$
	$= y_2$	$= y_1$	$= v_3$	$= -w$	

(T.3″)

Note that going from Tableau 3 of Problem 5.15 to Tableau 3″ (or to Equations (1), (2), (3), and (4)) may be viewed as being a purely mechanical process. The coefficients $-\frac{2}{3}$, $\frac{1}{3}$, $\frac{2}{3}$, and $\frac{400}{3}$ all correspond to the entries in the v_1 row of Tableau 3. In particular, the $\frac{400}{3}$ is none other than $x_1{}^*$—the imputed cost associated with c_1.

Tableau 3″ will be a terminal tableau provided that all the entries in the bottom row are nonpositive. This leads to

$$-20 - \tfrac{2}{3}k_1 \le 0 \tag{5}$$

$$-20 + \frac{k_1}{3} \le 0 \tag{6}$$

$$-10 + \frac{2k_1}{3} \le 0. \tag{7}$$

Inequality (5) leads to $k_1 \ge -30$, while Inequalities (6) and (7) lead to $60 \ge k_1$ and $15 \ge k_1$, respectively. Thus, if k_1 is in the range

$$-30 \le k_1 \le 15, \tag{8}$$

then Tableau 3″ will be a terminal tableau. Moreover, the change in Min w will then be given by the corner entry:

$$\Delta \text{ Min } w = \tfrac{400}{3}k_1 \tag{9}$$

or

$$\frac{\Delta \text{ Min } w}{\Delta c_1} = \tfrac{400}{3} = x_1{}^*. \tag{10}$$

This is none other then the result called for by Theorem 6.2.

Notes:
a. If Δc_1 (k_1) goes outside of the range indicated in (8), then Tableau 3″ will no longer be a *terminal* tableau. At least one more pivot would be needed; the most appropriate rules to then follow are those indicated in Method 5.2.
b. We see that the dual solution items in the right-hand column have been unaffected by the above manipulations. If Δc_1 stays in the range from -30 to 15, this will be the case. Otherwise, the dual *solution* will no longer be given by $t_2{}^* = t_1{}^* = x_3{}^* = 0$, $x_1{}^* = \frac{400}{3}$, and $x_2{}^* = \frac{50}{3}$.

Problem 6.11

In Problem 5.15 determine the effects on Min w, $y_2{}^*$, $y_1{}^*$, and $v_3{}^*$ of changes in c_1 from its initial value $c_1 = 60$ given by $\Delta c_1 = +3$, -6, $+15$, and -30.

Solution

All of these changes in c_1 fall within the acceptability range cited in Problem 6.10. Thus, the conclusion of Theorem 6.2 applies. We indicate the effects of the cited changes by means of a table based on Equations (1), (2), (3), and (4) (or, alternatively, the bottom row of Tableau 3″) of Problem 6.10 (recall that k_1 is Δc_1):

Δc_1	c_1	$y_2{}^*$	$y_1{}^*$	$v_3{}^*$	Min w
+3	63	22	19	8	10,400
−6	54	16	22	14	9,200
+15	75	30	15	0	12,000
−30	30	0	30	30	6,000

Note that the limiting changes $\Delta c_1 = +15$ and -30 cause degeneracies to appear in the solution of the minimum problem. Changes of c_1 outside of the range indicated in Problem 6.10 will result in Tableau 3″ no longer being terminal.

Problem 6.12

In Problem 5.15 allow c_2 to change by the amount k_2 (Δc_2). Determine the range of values for k_2 such that Theorem 6.2 applies.

Solution

We proceed as in Problem 6.11, only now some of the steps will be abbreviated. Allowing c_2 to change from 120 to $120 + k_2$ in Tableau 1 and then shifting k_2 will cause $v_2 + k_2$ to appear in the margin of Tableau 1 in place of v_2, with the remainder of the tableau staying the same. We then follow the pivot steps that led to Tableau 3 in Problem 5.15; we obtain a tableau in which the only difference from Tableau 3 of Problem 5.15 is that the left margin variable v_2 is replaced by $v_2 + k_2$. We now write out the column equations corresponding to this new tableau. After simplifying, we obtain the following:

$$-\tfrac{1}{6}v_2 + \tfrac{2}{3}v_1 + 20 - \tfrac{1}{6}k_2 = y_2 \tag{1}$$

$$\tfrac{1}{3}v_2 - \tfrac{1}{3}v_1 + 20 + \tfrac{1}{3}k_2 = y_1 \tag{2}$$

$$\tfrac{5}{3}v_2 - \tfrac{2}{3}v_1 + 10 + \tfrac{5}{3}k_2 = v_3 \tag{3}$$

$$\tfrac{50}{3}v_2 + \tfrac{400}{3}v_1 + 10{,}000 + \tfrac{50}{3}k_2 = w. \tag{4}$$

To obtain the basic point that corresponds to these equations, we set v_2

$= v_1 = 0$ and obtain

$$20 - \tfrac{1}{6}k_2 = y_2^* \tag{5}$$

$$20 + \tfrac{1}{3}k_3 = y_1^* \tag{6}$$

$$10 + \tfrac{5}{3}k_2 = v_3^* \tag{7}$$

$$10{,}000 + \tfrac{50}{3}k_2 = \text{Min } w. \tag{8}$$

Note that the coefficients $-\tfrac{1}{6}, \tfrac{1}{3}, \tfrac{5}{3}$, and $\tfrac{50}{3}$ correspond to the entries of the v_2 row in Tableau 3.

From (5), (6), and (7) we will have a solution point provided that

$$20 - \tfrac{1}{6}k_2 \geq 0, \quad 20 + \tfrac{1}{3}k_2 \geq 0, \quad \text{and} \quad 10 + \tfrac{5}{3}k_2 \geq 0.$$

These lead to $k_2 \leq 120$, $-60 \leq k_2$, and $-6 \leq k_2$, from which we obtain

$$-6 \leq k_2 \leq 120.$$

If k_2 is in this range, we have the result called for by Theorem 6.2, namely,

$$\Delta \text{ Min } w = \tfrac{50}{3}k_2 = x_2^* \cdot \Delta c_2$$

This follows from Equation (8), since $x_2^* = \tfrac{50}{3}$ and $\Delta c_2 = k_2$.

Problem 6.13

In Problem 5.15, allow c_3 to change by the amount k_3. Determine the range of values for k_3 such that Theorem 6.2 applies.

Solution

Allowing c_3 to change from 150 to $150 + k_3$ in Tableau 1 and then shifting k_3 will cause $v_3 + k_3$ to appear in the margin of Tableau 1 in place of v_3, with the remainder of the tableau staying the same. We then follow the pivot steps that led to Tableau 3 in Problem 5.15; we obtain a tableau in which the only difference from Tableau 3 of Problem 5.15 is that the bottom margin variable v_3 is replaced by $v_3 + k_3$. Notice that $v_3 + k_3$ has *not* been shifted into the left-hand margin as was the case for $v_1 + k_1$ and $v_2 + k_2$ in two previous problems. If we write out the equation corresponding to the third column, we obtain

$$\tfrac{5}{3}v_2 - \tfrac{2}{3}v_1 + 10 = v_3 + k_3$$

or

$$\tfrac{5}{3}v_2 - \tfrac{2}{3}v_1 + 10 - k_3 = v_3. \tag{1}$$

The condition that must be met for Tableau 3 to remain terminal is thus

$$10 - k_3 \geq 0$$

or

$$10 \geq k_3. \tag{2}$$

If (2) holds, then the conclusion of Theorem 6.2 holds, i.e., Min w would stay constant at 10,000 or, in other words,

$$\Delta \text{ Min } w = 0. \tag{3}$$

This is in agreement with Theorem 6.2, since $x_3{}^* = 0$ from Tableau 3 of Problem 5.15.

The next problem deals with how to handle a change in b_i (in a maximum problem) for which Theorem 6.1 does not apply.

Problem 6.14

Suppose that in Problem 5.16 the resource term b_1 is changed by the amount 400 to $b_1 = 1300$. Determine the new solution of the problem and the new value of Max u.

Solution

It is useful to refer to Problem 6.6 at this time. Notice that $\Delta b_1 = 400$ lies outside the range for which Theorem 6.1 holds. If we substitute $h_1 = 400$ into Tableau 3″ of Problem 6.6, we obtain the following tableau:

	x_1	t_1	t_2	1	
v_2	$\frac{9}{5}$	$\frac{9}{5}$	$-\frac{3}{5}$	-1980	$= -x_2$
v_3	$-\frac{4}{5}*$	$-\frac{4}{5}$	$\frac{8}{5}$	$+80$	$= -x_3$
-1	$-\frac{7}{100}$	$-\frac{8}{25}$	$-\frac{14}{25}$	752	$= u$
	$= v_1$	$= y_1$	$= y_2$	$= -w$	

(T.3″)

This tableau is no longer terminal, since $-b_2 = +80$. Thus, the basic feasible point corresponding to the dual (minimum) problem ($v_2 = v_3 = 0$, $v_1 = \frac{7}{100}$, $y_1 = \frac{8}{25}$, and $y_2 = \frac{14}{25}$) is no longer a *solution* point for the dual problem. It is useful at this time to carefully examine the wording of Theorem 6.1; the dual problem b.f.p. *would* be a solution point if the tableau were terminal.

Let us now try to obtain a terminal tableau from Tableau 3″. An algorithm appropriate for our task is Method 5.3 (the Stage One algorithm, i.e., Method 4.3, is also applicable). If we focus on row 2, we see that $a_{21} = -\frac{4}{5}$ is the pivot element. Carrying out the pivoting leads to Tableau 4:

	x_3	t_1	t_2	1	
v_2	$\frac{9}{4}$	0	3	-1800	$= -x_2$
v_1	$-\frac{5}{4}$	1	-2	-100	$= -x_1$
-1	$-\frac{7}{80}$	$-\frac{1}{4}$	$-\frac{7}{10}$	745	$= u$
	$= v_3$	$= y_1$	$= y_2$	$= -w$	

(T.4)

This is clearly a terminal tableau. The solution and the value for Max u of the primal problem are read off as follows:

$$x_3{}^* = t_1{}^* = t_2{}^* = 0, \; x_2{}^* = 1800, \; x_1{}^* = 100, \quad \text{and} \quad \text{Max } u = 745.$$

Note that the dual solution point is a different point from the dual b.f.p. of Tableau 3″.

We now illustrate how to handle a change in c_j (in a minimum problem) for which Theorem 6.2 does not apply.

Problem 6.15

Suppose that in Problem 5.15 the requirement term c_1 is changed by the amount 21 to $c_1 = 81$. Determine the new solution of the problem and the new value of Min w.

Solution

It is useful to refer to Problem 6.10 at this time. Notice that $\Delta c_1 = 21$ lies outside the range for which Theorem 6.2 holds. If we substitute $k_1 = 21$ into Tableau 3″ of Problem 6.10, we obtain the following tableau:

	t_2	t_1	x_3	1		
v_2	$-\frac{1}{6}$	$\frac{1}{3}$	$\frac{5}{3}*$	$-\frac{50}{3}$	$= -x_2$	
v_1	$\frac{2}{3}$	$-\frac{1}{3}$	$-\frac{2}{3}$	$-\frac{400}{3}$	$= -x_1$	(T.3″)
-1	-34	-13	$+4$	$12{,}800$	$= u$	
	$= y_2$	$= y_1$	$= v_3$	$= -w$		

This tableau is no longer terminal, since $c_3 = +4$. Thus, the basic feasible point corresponding to the dual (maximum) problem ($t_2 = t_1 = x_3 = 0$, $x_2 = \frac{50}{3}$, and $x_1 = \frac{400}{3}$) is no longer a *solution* point for the dual program. The reason is that the tableau is not terminal.

To obtain a terminal tableau from Tableau 3″, we apply Method 5.2. This is clearly applicable, since both $-b_i$ terms are nonpositive. Using this method is particularly simple when it is recalled that it is nothing more than Stage Two applied to the dual (maximum) problem. The pivot column is the third one, and $a_{13} = \frac{5}{3}$ is the only candidate for pivot element. Carrying out the pivot operation results in Tableau 4:

	t_2	t_1	x_2	1		
v_3	$-\frac{1}{10}$	$\frac{1}{5}$	$\frac{3}{5}$	-10	$= -x_3$	
v_1	$\frac{3}{5}$	$-\frac{1}{5}$	$\frac{2}{5}$	-140	$= -x_1$	(T.4)
-1	-33.6	-13.8	$-\frac{12}{5}$	$12{,}840$	$= u$	
	$= y_2$	$= y_1$	$= v_2$	$= -w$		

This is clearly a terminal tableau. The solution and value for Min w of the primal problem are read off as follows:

$v_3{}^* = v_1{}^* = 0$, $y_2{}^* = 33.6$, $y_1{}^* = 13.8$, $v_2{}^* = 2.4$, and Min $w = 12{,}840$.

3. A CHANGE OF c_j IN THE MAXIMUM PROBLEM AND A CHANGE OF b_i IN THE MINIMUM PROBLEM

Suppose, as in Problem 5.16, that we have the terminal tableau for a maximum problem and that we wish to ascertain the effect of a change in a profit (or revenue) term c_j. This problem can be very easily resolved by using a dual tableau approach. It is essentially equivalent to the problem of determining the effect of a change in c_j for a minimum problem (the latter was considered in Section 2). In fact, a theorem similar to Theorem 6.2 applies, the only changes being in terminology and emphasis.

Theorem 6.3: Maximum Problem. *If the primal solution (excluding Max u) in the terminal tableau remains unchanged when c_j is changed to $c_j + \Delta c_j$, then the change in Max u is given by Δ Max $u = x_j{}^* \cdot \Delta c_j$.*

We illustrate the effects of changes in c_j by means of some worked-out problems.

Problem 6.16

Refer to the maximum problem given as Problem 5.16 (originally Problem 1.4—the "nut problem"). Suppose that the selling price for the second mixture is changed from $c_2 = \frac{2}{5} = 0.40$ to $0.40 + \Delta c_2$ (both are in \$/lb). Analyze the effects of this change on Tableaus 1 and 3. Determine the range for Δc_2 such that Tableau 3 remains a terminal tableau.

Solution

In Tableau 1 of Problem 5.16 we replace $c_2 = 0.40$ by $c_2 = 0.40 + k_2$ (again we use k_j in place of Δc_j). Then we shift the k_2 so that it appears in combination with v_2, and we obtain the following tableau, which is otherwise identical with Tableau 1:

	x_1	x_2	x_3	1	
y_1	1	$\frac{2}{3}$	$\frac{1}{4}$	-900	$= -t_1$
y_2	0	$\frac{1}{3}$	$\frac{3}{4}$	-600	$= -t_2$
-1	$\frac{1}{4}$	$\frac{2}{5}$	$\frac{1}{2}$	0	$= u$
	$= v_1$	$= v_2 + k_2$	$= v_3$	$= -w$	

(T.1′)

The same pivot steps as in Problem 5.16 are now executed. The only change is that wherever v_2 appeared before there now appears $v_2 + k_2$. The terminal tableau of Problem 5.16 is thus replaced by the following tableau, in which the left margin variable v_2 is replaced by $v_2 + k_2$:

	x_1	t_1	t_2	1		
$v_2 + k_2$	$\frac{9}{5}$	$\frac{9}{5}$	$-\frac{3}{5}$	-1260	$= -x_2$	
v_3	$-\frac{4}{5}$	$-\frac{4}{5}$	$\frac{8}{5}$	-240	$= -x_3$	(T.3')
-1	$-\frac{7}{100}$	$-\frac{8}{25}$	$-\frac{14}{25}$	624	$= u$	
	$= v_1$	$= y_1$	$= y_2$	$= -w$		

So far, the steps parallel those of Problem 6.10. If we now follow what we did there and manipulate the k_2 in the corresponding column equations so that terms containing it appear in combination with the constant terms, we obtain Tableau 3″ below (notice that v_2 again appears alone as a margin variable):

	x_1	t_1	t_2	1		
v_2	$\frac{9}{5}$	$\frac{9}{5}$	$-\frac{3}{5}$	-1260	$= -x_2$	
v_3	$-\frac{4}{5}$	$-\frac{4}{5}$	$\frac{8}{5}$	-240	$= -x_3$	(T.3″)
-1	$-0.07 - \frac{9}{5}k_2$	$-0.32 - \frac{9}{5}k_2$	$-0.56 + \frac{3}{5}k_2$	$624 + 1260k_2$	$= u$	
	$= v_1$	$= y_1$	$= y_2$	$= -w$		

Tableau 3″ will be a terminal tableau provided that all the entries in the bottom row are nonpositive. We obtain

$$-0.07 - \tfrac{9}{5}k_2 \leq 0 \tag{1}$$

$$-0.32 - \tfrac{9}{5}k_2 \leq 0 \tag{2}$$

$$-0.56 + \tfrac{3}{5}k_2 \leq 0. \tag{3}$$

Inequality (1) leads to $k_2 \geq -0.039$, Inequality (2) leads to $k_2 \geq -0.178$, and Inequality (3) leads to $k_2 \leq 0.933$ (or $\frac{14}{15}$). Thus, if k_2 is in the range

$$-0.039 \leq k_2 \leq 0.933, \tag{4}$$

then Tableau 3″ will be a terminal tableau. Moreover, the change in Max u will be given by the corner entry:

$$\Delta \text{ Max } u = 1260 \cdot k_2 = x_2^* \cdot \Delta c_2. \tag{5}$$

This is in agreement with Theorem 6.3. In addition, note that the range of k_2 such that Tableau 3″ is terminal is also the range such that the primal (in the present case, maximum) solution remains unchanged. This also is in accord with the theorem.

Problem 6.17

Suppose that c_2 of Problem 5.16 is changed from 0.4 \$/lb to 1.00 and 0.38 \$/lb, respectively. Determine the effects on the solution and Max u value.

Solution

The changes called for are $\Delta c_2 = 0.6$ and $\Delta c_2 = -0.02$, respectively; both fall within the range cited in Problem 6.16. Thus, the primal (maximum) solution remains unchanged in either case; it is given by

$$x_1^* = t_1^* = t_2^* = 0, x_2^* = 1260, \quad \text{and} \quad x_3^* = 240.$$

The new values for Max u are obtained from Equation (5) of Problem 6.16 and are as follows:

$$\text{Max } u = 624 + 1260 \cdot (0.6) = \$1380 \quad \text{for} \quad \Delta c_2 = 0.6$$

and

$$\text{Max } u = 624 + 1260 \cdot (-.02) = \$598.80 \quad \text{for} \quad \Delta c_2 = -0.02.$$

If the change in c_2 relative to the base value lies outside of the allowable range cited in Problem 6.16 (see Inequality (4)), then Tableau 3″ of that problem will no longer be a terminal tableau. The procedure to follow then is similar to that used in Problem 6.15.

Suppose that we have the terminal tableau for a minimum problem and that we wish to determine the effect of a change in a cost term b_i. The problem is easily resolved using a dual tableau approach. Our task is essentially equivalent to the problem of determining the effect of a change in b_i for a maximum problem (the latter was considered in Section 2). In fact, a theorem similar to Theorem 6.1 applies (note the changes in terminology and emphasis, however).

Theorem 6.4: Minimum Problem. *If the primal solution (excluding Min w) in the terminal tableau remains unchanged when b_i is changed to $b_i + \Delta b_i$, then the change in Min w is given by $\Delta \text{Min } w = y_i^* \cdot \Delta b_i$.*

We illustrate the effects of changes in a b_i term by working out some problems.

Problem 6.18

Refer to the minimum problem given as Problem 5.15 (originally Problem 1.3—the mining problem). Suppose that the cost for working mine B is changed from $b_2 = 300$ to $300 + \Delta b_2$ (both are in \$/day). Analyze the effects of this change on Tableaus 1 and 3. Determine the range for Δb_2 such that Tableau 3 remains a terminal tableau.

Solution

In Tableau 1 of Problem 5.15 we replace $b_2 = 300$ by $b_2 = 300 + h_2$ (again we use h_i in place of Δb_i). Then we shift the h_2 so that it appears in combination with t_2, and we obtain the following tableau, which is otherwise identical with Tableau 1:

	x_1	x_2	x_3	1	
y_1	1	4	6	-200	$= -t_1$
y_2	2	2	2	-300	$= -(t_2 - h_2)$ (T.1')
-1	60	120	150	0	$= u$
	$= v_1$	$= v_2$	$= v_3$	$= -w$	

The same pivot steps as in Problem 5.15 are now executed. The only change is that wherever t_2 appeared before there now appears $t_2 - h_2$. The terminal tableau of Problem 5.15 is thus replaced by the following tableau, in which the top margin variable t_2 is replaced by $t_2 - h_2$:

	$t_2 - h_2$	t_1	x_3	1	
v_2	$-\frac{1}{6}$	$\frac{1}{3}$	$\frac{5}{3}$	$-\frac{50}{3}$	$= -x_2$
v_1	$\frac{2}{3}$	$-\frac{1}{3}$	$-\frac{2}{3}$	$-\frac{400}{3}$	$= -x_1$ (T.3')
-1	-20	-20	-10	$10{,}000$	$= u$
	$= y_2$	$= y_1$	$= v_3$	$= -w$	

So far, the steps parallel those of Problem 6.6. If we now follow what we did there and manipulate the h_2 in the corresponding row equations so that terms containing it appear in combination with the constant terms, we obtain Tableau 3″ below. (notice that t_2 again appears alone as a margin variable):

	t_2	t_1	x_3	1	
v_2	$-\frac{1}{6}$	$\frac{1}{3}$	$\frac{5}{3}$	$-\frac{50}{3} + \frac{1}{6}h_2$	$= -x_2$
v_1	$\frac{2}{3}$	$-\frac{1}{3}$	$-\frac{2}{3}$	$-\frac{400}{3} - \frac{2}{3}h_2$	$= -x_1$ (T.3″)
-1	-20	-20	-10	$10{,}000 + 20h_2$	$= u$
	$= y_2$	$= y_1$	$= v_3$	$= -w$	

Tableau 3″ will be a terminal tableau provided that all the entries in the right-hand column are nonpositive: We obtain

$$-\tfrac{50}{3} + \frac{h_2}{6} \le 0 \tag{1}$$

$$-\tfrac{400}{3} - \frac{2h_2}{3} \le 0 \tag{2}$$

From these we obtain

$$-200 \leq h_2 \leq 100, \tag{3}$$

where the right bound 100 is determined from (1), while the left bound -200 results from (2). If h_2 is in the range given by (3), then Tableau 3″ will be a terminal tableau. Moreover, the change in Min w will be given by the corner entry:

$$\Delta \text{ Min } w = 20h_2 = y_2^* \cdot \Delta b_2. \tag{4}$$

This is in agreement with Theorem 6.4. In addition, note that the range of h_2 such that Tableau 3″ is terminal is also the range such that the primal (in the present case, minimum) solution remains unchanged; this also is in accord with the theorem.

Problem 6.19

Suppose that b_2 of Problem 5.15 is changed from 300 $/day to 350 and 200 $/day, respectively. Determine the effects on the solution and Min w value.

Solution

The changes called for are $\Delta b_2 = +50$ and $\Delta b_2 = -100$, respectively; both fall within the range given in Problem 6.18. Thus, the primal (minimum) solution remains unchanged in either case; it is given by

$$v_1^* = v_2^* = 0, \ y_2^* = 20, \ y_1^* = 20, \quad \text{and} \quad v_3^* = 10.$$

The new values for Min w are obtained from Equation (4) of Problem 6.18 and are as follows:

$$\text{Min } w = 10,000 + 20 \cdot (50) = \$11,000 \quad \text{for} \quad \Delta b_2 = +50$$

and

$$\text{Min } w = 10,000 + 20 \cdot (-100) = \$8,000 \quad \text{for} \quad \Delta b_2 = -100$$

These results are consistent with economic expectations; as the cost for working mine B goes up, so does the minimum cost.

If the change in b_2 relative to the base value lies outside of the allowable range cited in Problem 6.18 (see Inequality (3)), then Tableau 3″ of that problem will no longer be a terminal tableau. The procedure to follow then is similar to that used in Problem 6.14.

4. ADJOINING NEW ROWS AND COLUMNS

We shall consider the effect of adding a new row or a new column to a linear program. The nature of the dual tableau approach enables us to accomplish this without much extra effort. In other words, we will not have to work a problem from scratch; maximum use will be made of the existing set of tableaus for a given problem. To add a practical flavor to matters we shall select Problem 5.15 (Problem 1.3) as our base problem. Restated in standard slack form, using Chapter 5 symbolism, it is as follows:

$$\text{Minimize} \quad w = 200y_1 + 300y_2 \tag{1}$$

$$\text{subject to} \quad y_1 + 2y_2 - 60 = v_1 \text{ (high-grade ore)} \tag{2}$$
$$4y_1 + 2y_2 - 120 = v_2 \text{ (medium-grade ore)} \tag{3}$$

$$6y_1 + 2y_2 - 150 = v_3 \text{ (low-grade ore)} \tag{4}$$

and

$$y_1, y_2, v_1, v_2, v_3 \geq 0.$$

It is useful to recall the different symbolism used in Problem 1.3 (u in place of w; x and y in place of y_1 and y_2).

Problem 6.20

Suppose that the mines of Problem 1.3 are each found to be capable of producing 3 tons of a new grade of ore (call it *premium* grade) per day without increasing the cost of working the mines. Furthermore, suppose that the company requires 60 tons of the premium ore. Determine the effect of this new requirement (constraint) on the optimal solution and Min w value of the problem.

Solution

The new constraint may be written

$$3y_1 + 3y_2 - 60 = v_4 \text{ (premium-grade ore)}, \tag{5}$$

where

$$v_4 \geq 0.$$

Our goal is to add as little work as possible to the work of Problem 5.15. Let us refer to Tableau 3 (the terminal tableau) of that problem. We will append to it a new column representing (5), but with v_4 expressed in terms of v_2 and v_1—the left margin variables of Tableau 3—instead of y_1 and y_2. To do this, we begin by solving for y_1 and y_2 from the first two columns of Tableau 3 in terms of v_2 and v_1:

$$y_1 = \tfrac{1}{3}v_2 - \tfrac{1}{3}v_1 + 20 \tag{6}$$

$$y_2 = -\tfrac{1}{6}v_2 + \tfrac{2}{3}v_1 + 20. \tag{7}$$

If we substitute y_1 and y_2 from Equations (6) and (7) into (5), we obtain

$$v_4 = \frac{v_2}{2} + v_1 + 60. \tag{5'}$$

If we append this equation on Tableau 3 as a new column, we obtain the following:

	t_2	t_1	x_3	x_4	1			
v_2	$-\frac{1}{6}$	$\frac{1}{3}$	$\frac{5}{3}$	$\frac{1}{2}$	$-\frac{50}{3}$	$=$	$-x_2$	
v_1	$\frac{2}{3}$	$-\frac{1}{3}$	$-\frac{2}{3}$	1	$-\frac{400}{3}$	$=$	$-x_1$	(T.3')
-1	-20	-20	-10	-60	$10,000$	$=$	u	
	$= y_2$	$= y_1$	$= v_3$	$= v_4$	$= -w$			

This tableau is a terminal tableau; notice that the $+60$ of Equation (5') is equivalent to $-(-60)$ in the bottom row of Tableau 3'. Thus, the solution for the minimum problem is essentially unchanged from that of Problem 5.15, except that now we have $v_4{}^* = 60$. An important fact is that Min w is still $\$10,000$.

Clearly, events will not always be as simple as in Problem 6.20. Much depends on the coefficients in the new constraint. We illustrate this in the next problem.

Problem 6.21

Suppose that the tonnage of premium ore required in Problem 6.20 is 144 tons instead of 60 tons. Determine the effect of the new constraint on the optimal solution and Min w value of the problem.

Solution

The new constraint is

$$3y_1 + 3y_2 - 144 = v_4 \text{ (premium-grade ore)}.$$

Using Equations (6) and (7) of Problem 6.20, we express v_4 in terms of v_2 and v_1:

$$v_4 = \frac{v_2}{2} + v_1 - 24.$$

If we now incorporate this equation into Tableau 3 of Problem 5.15 as a new column, we obtain the following:

	t_2	t_1	x_3	x_4	1			
v_2	$-\frac{1}{6}$	$\frac{1}{3}$	$\frac{5}{3}$	$\frac{1}{2}{}^*$	$-\frac{50}{3}$	$=$	$-x_2$	
v_1	$\frac{2}{3}$	$-\frac{1}{3}$	$-\frac{2}{3}$	1	$-\frac{400}{3}$	$=$	$-x_1$	(T.3')
-1	-20	-20	-10	$+24$	$10,000$	$=$	u	
	$= y_2$	$= y_1$	$= v_3$	$= v_4$	$= -w$			

This tableau is no longer terminal, since $c_4 = +24$. We now use Method

5.2 to attain a new terminal tableau. The pivot element is $a_{14} = \frac{1}{2}$. Doing the pivot operation leads to Tableau 4:

	t_2	t_1	x_3	x_2	1		
v_4	$-\frac{1}{3}$	$\frac{2}{3}$	$\frac{10}{3}$	2	$-\frac{100}{3}$	$= -x_4$	
v_1	1	-1	-4	-2	-100	$= -x_1$	(T.4)
-1	-12	-36	-90	-48	$10,800$	$= u$	
	$= y_2$	$= y_1$	$= v_3$	$= v_2$	$= -w$		

This tableau is clearly a terminal tableau. We read off the solution and Min w value of the problem as follows:

$$v_4{}^* = v_1{}^* = 0, \; y_2{}^* = 12, \; y_1{}^* = 36,$$

$$v_3{}^* = 90, \; v_2{}^* = 48, \quad \text{and} \quad \text{Min } w = \$10,800.$$

Notice that the requirement on the number of days to keep mines A and B open has changed to 36 and 12 days, respectively.

It is possible to illustrate the effects of the new constraints graphically. This has been done in Figure 6.1, which should be compared to Figure 1.7 (again, note the change in symbolism). The equation parts of the new constraints are given as (4a) and (4b) , respectively, in the graph. The constraint $3y_1 + 3y_2 \geq 60$ of Problem 6.20 (refer to (4a)) does *not alter* the constraint set of Problem 5.15; thus, the solution stayed effectively unchanged. On the other hand, the constraint $3y_1 + 3y_2 \geq 144$ of Problem 6.21 (refer to (4b)) reduces the size of the constraint set, as indicated by the reduced area of shading in Figure 6.1 as compared to Figure 1.7. The old solution point $D(20,20)$ is no longer in the constraint set. The transformation from Tableau 3' to Tableau 4 in Problem 6.21 corresponds to movement from D to the new solution point $H(36,12)$ of Figure 6.1.

The method for handling a new column in a maximum problem (where the maximum problem is lined up by rows) is essentially the same as that illustrated in the previous two problems. This is likewise true with respect to determining the effect of adjoining a new row. The mining problem (Problem 1.3) used in Problems 6.20 and 6.21 also proves to be very useful (from a practical viewpoint) for illustrating the effect of appending a new row.

Suppose the mining company discovers a new mine (call it mine C) that is capable of producing 3 tons per day each of high-, medium-, and low-grade ore. If the other data of Problem 1.3 are unchanged and the cost for working mine C is b \$ per day, the new minimum problem is

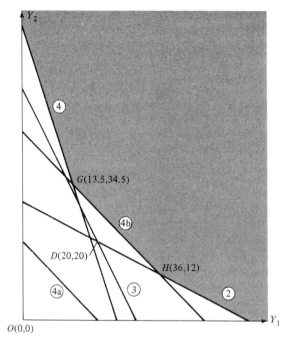

FIGURE 6.1 The constraint set (shown shaded) for Problem 6.21. Encircled numbers indicate equality parts of constraints. Thus, ②represents $y_1 + 2y_2 = 60$, ③represents $4y_1 + 2y_2 = 120$, ④represents $6y_1 + 2y_2 = 150$, ④a represents $3y_1 + 3y_2 = 20$, and ④b represents $3y_1 + 3y_2 = 144$.

stated as follows:

$$\text{Minimize} \quad w = 200y_1 + 300y_2 + by_3 \qquad (1)$$

$$\text{subject to} \quad y_1 + 2y_2 + 3y_3 - 60 = v_1 \qquad (2)$$

$$4y_1 + 2y_2 + 3y_3 - 120 = v_2 \qquad (3)$$

$$6y_1 + 2y_2 + 3y_3 - 150 = v_3 \qquad (4)$$

and the usual nonnegativity constraints on the y_j's and v_i's. Here y_3 denotes the number of days to keep mine C open.

Problem 6.22

Develop the initial tableau corresponding to the modified minimum problem given above. Then using the information available from Problem 5.15, develop the tableau that is the appended version of Tableau 3 of that problem.

Solution

The initial tableau corresponding to the problem involving mine C is given as follows (our minimum problem is lined up vertically):

	x_1	x_2	x_3	1	
y_1	1	4	6	-200	$= -t_1$
y_2	2	2	2	-300	$= -t_2$
y_3	3	3	3	$-b$	$= -t_3$
-1	60	120	150	0	$= u$
	$= v_1$	$= v_2$	$= v_3$	$= -w$	

(T.1′)

The only change from Tableau 1 of Problem 5.15 is in the adjoining of the new third row. Note that the new variable t_3 is the other member of the complementary slackness pair (see Section 7 of Chapter 5) that contains y_3. We now proceed in a fashion similar to that of Problem 6.20. First we write out the equation corresponding to the appended row:

$$-t_3 = 3x_1 + 3x_2 + 3x_3 - b. \tag{1}$$

Note that this equation corresponds to a new main constraint of the dual (maximum) problem. We now refer to Tableau 3 of Problem 5.15. Our goal is to express $-t_3$ in terms of the (dual) nonbasic variables of Tableau 3, namely, t_2, t_1, and x_3. From rows 1 and 2 of Tableau 3 we have

$$-x_2 = -\tfrac{1}{6}t_2 + \tfrac{1}{3}t_1 + \tfrac{5}{3}x_3 - \tfrac{50}{3} \tag{2}$$

$$-x_1 = \tfrac{2}{3}t_2 - \tfrac{1}{3}t_1 - \tfrac{2}{3}x_3 - \tfrac{400}{3}. \tag{3}$$

If we substitute the expressions for x_2 and x_1 from (2) and (3) into (1), we obtain

$$-t_3 = -\tfrac{3}{2}t_2 + 0{\cdot}t_1 - 3x_3 + (450 - b). \tag{4}$$

If we now introduce Equation (4) into Tableau 3 as the third row, we obtain the following appended tableau:

	t_2	t_1	x_3	1	
v_2	$-\tfrac{1}{6}$	$\tfrac{1}{3}$	$\tfrac{5}{3}$	$-\tfrac{50}{3}$	$= -x_2$
v_1	$\tfrac{2}{3}$	$-\tfrac{1}{3}$	$-\tfrac{2}{3}$	$-\tfrac{400}{3}$	$= -x_1$
y_3	$-\tfrac{3}{2}$	0	-3	$-(b - 450)$	$= -t_3$
-1	-20	-20	-10	$10{,}000$	$= u$
	$= y_2$	$= y_1$	$= v_3$	$= -w$	

(T.3′)

Problem 6.23

Suppose that the b value in Problem 6.22 is $b = 500$ \$/day. Determine the solution and Min w value of the modified minimum problem. Comment on the range of b values which will yield a similar solution.

Solution

If we substitute $b = 500$ into Tableau 3′ of Problem 6.22, the third entry in the right-hand column becomes -50, i.e., the tableau is still terminal. Thus, the solution and Min w are essentially unchanged from those of Problem 5.15. The only addition involves $y_3{}^*$. Thus, the solution of the modified problem is

$$v_2{}^* = v_1{}^* = y_3{}^* = 0, \ y_2{}^* = y_1{}^* = 20,$$

$$v_3{}^* = 10, \quad \text{and} \quad \text{Min } w = 10{,}000.$$

It is readily apparent that the above primal solution and Min w value hold for *all* b values such that Tableau 3′ remains terminal. The latter will be true provided that

$$-(b - 450) \leq 0,$$

i.e., if $b \geq 450$ \$/day. On the other hand, if

$$b < 450 \, \frac{\$}{\text{day}},$$

then Tableau 3′ will no longer be a terminal tableau. The following problem illustrates how to handle such a situation.

Problem 6.24

Suppose that the b value in Problem 6.22 is $b = 429$ \$/day. Determine the solution and Min w value of the modified minimum problem.

Solution

Substitution of $b = 429$ into Tableau 3′ of Problem 6.22 yields the following tableau:

	t_2	t_1	x_3	1			
v_2	$-\frac{1}{6}$	$\frac{1}{3}$	$\frac{5}{3}$	$-\frac{50}{3}$	$= -x_2$		
v_1	$\frac{2}{3}$	$-\frac{1}{3}$	$-\frac{2}{3}$	$-\frac{400}{3}$	$= -x_1$	(T.3′)	
y_3	$-\frac{3}{2}$	0	-3^*	$+21$	$= -t_3$		
-1	-20	-20	-10	$10{,}000$	$= u$		
	$= y_2$	$= y_1$	$= v_3$	$= -w$			

This tableau is no longer terminal, since $-b_3 = +21$. Method 5.3 is applicable here, and the pivot entry candidates are a_{31} and a_{33}. The pivot entry turns out to be $a_{33} = -3$, and the pivoting process leads to Tableau 4:

	t_2	t_1	t_3	1		
v_2	-1	$\frac{1}{3}$	$\frac{5}{9}$	-5	$= -x_2$	
v_1	1	$-\frac{1}{3}$	$-\frac{2}{9}$	-138	$= -x_1$	(T.4)
v_3	$\frac{1}{2}$	0	$-\frac{1}{3}$	-7	$= -x_3$	
-1	-15	-20	$-\frac{10}{3}$	9930	$= u$	
	$= y_2$	$= y_1$	$= y_3$	$= -w$		

This tableau is terminal. The solution and Min w value of the minimum problem are read off as follows:

$$v_1{}^* = v_2{}^* = v_3{}^* = 0, \quad y_1{}^* = 20, \quad y_2{}^* = 20,$$

$$y_3{}^* = \tfrac{10}{3}, \quad \text{and} \quad \text{Min } w = \$9{,}930.$$

Notice that the new mine now has a strong effect on the solution of the problem and that the minimum cost has been reduced from its prior value of \$10,000. This makes sense from an intuitive point of view. It stands to reason that it will become economically feasible to work mine C if the cost per day is sufficiently low; in the previous problem, the cost to work mine C (500 \$/day) was too high, while the current cost (429 \$/day) is low enough.

5. CONCLUDING REMARKS

So far we have considered the effects of various changes on the known solutions of linear programming problems. When given a complete set of tableaus leading to the terminal (i.e., solution) tableau of some base problem, the goal has been to introduce as few extra calculations as possible in determining the solution of a problem which deviates slightly from the base problem. The changes considered have been in a b_i or a c_j value (for both maximum and minimum problems) or in the adjoining of a new row or column to the dual tableau.

There are efficient methods for handling changes in other coefficients. The reader is referred to Strum [3] for an approach on how to determine the effect of a change in an a_{ij} coefficient.

REFERENCES

1. Kemeny, J. G., Mirkil, H., Snell, J. L., and Thompson, G. L. *Finite Mathematics with Business Applications.* Second edition. Englewood Cliffs, New Jersey: Prentice-Hall, 1972.

2. Singleton, R. R., and Tyndall, W. F. *Games and Programs*. San Francisco, California: W. H. Freeman and Company, 1974.

3. Strum, J. E. *Introduction to Linear Programming*. San Francisco, California: Holden-Day, 1972.

SUPPLEMENTARY PROBLEMS

Imputed Value and Imputed Cost

S.P. 6.1: Refer to the terminal tableau of Problem 5.16. Determine the new value of Max u if the amount of peanuts available is changed to (a) 1000 pounds and (b) 700 pounds. Assume that Theorem 6.1 is applicable.

S.P. 6.2: Refer to the terminal tableau of Problem 5.15. Determine the new value of Min w if the tonnage of medium-grade ore required is changed to (a) 150 tons and (b) 117 tons. Assume that Theorem 6.2 is applicable.

S.P. 6.3: Determine the terminal dual tableau for S.P. 1.16. List the solutions to the primal (maximum) and dual problems.

The next two problems pertain to S.P. 6.3. Assume that Theorem 6.1 is applicable.

S.P. 6.4: Determine the new value of Max u (here u is total profit) if the available man-hours per week in shop 1 (b_1) is changed to (a) 180 hr and (b) 130 hr.

S.P. 6.5: Determine the new value of Max u if the available man-hours per week in shop 2 (b_2) is changed to (a) 145 hr and (b) 110 hr.

S.P. 6.6: Determine the terminal dual tableau for S.P. 1.12(a). List the solutions to the primal (minimum) and dual problems.

The next three problems pertain to S.P. 6.6. Assume that Theorem 6.2 is applicable.

S.P. 6.7: Determine the new value of Min w (here w is total cost) if the amount of high-grade oil required is changed to (a) 15,000 barrels and (b) 9,000 barrels.

S.P. 6.8: Determine the new value of Min w if the amount of low-grade oil required is changed to (a) 18,000 barrels and (b) 12,000 barrels.

S.P. 6.9: Determine the new value of Min w if the amount of medium-grade oil required is changed to (a) 23,000 barrels and (b) 17,000 barrels.

Sensitivity Analysis

S.P. 6.10: In S.P. 1.16 allow the resource term b_1 to change by the amount h_1. Determine the range of values for h_1 such that Theorem 6.1 applies. (*Suggestion:* Make use of the terminal dual tableau of S.P. 6.3. Use the method of Problem 6.6.)

S.P. 6.11: Rework S.P. 6.10 for $h_2 = \Delta b_2$.

S.P. 6.12: In S.P. 1.12(a) allow the requirement term c_1 to change by the amount k_1. Determine the range of values for k_1 such that Theorem 6.2 applies. (*Suggestion:* Make use of the terminal dual tableau of S.P. 6.6. Use the method of Problem 6.10.)

S.P. 6.13: Rework S.P. 6.12 for $k_2 = \Delta c_2$.

S.P. 6.14: Rework S.P. 6.12 for $k_3 = \Delta c_3$.

S.P. 6.15: Suppose that in S.P. 1.16 the resource term b_1 is changed to 250 man-hours ($\Delta b_1 = 100$). Determine the new solution and value of Max u. (Refer to Problem 6.14 for a model and to S.P. 6.10 for a useful tableau.)

S.P. 6.16: Suppose that in S.P. 1.12(a) the requirement term c_1 is changed to 6000 barrels of high-grade oil ($\Delta c_1 = -6000$). Determine the new solution and value of Min w. (Refer to Problem 6.15 for a model and to S.P. 6.12 for a useful tableau.)

A Change of c_j in the Maximum Problem and a Change of b_i in the Minimum Problem

S.P. 6.17: In S.P. 1.16 allow the profit term c_1 to change by the amount Δc_1. Determine the range of values for Δc_1 such that Theorem 6.3 applies. (*Suggestion:* Make use of the terminal dual tableau of S.P. 6.3. Use the method of Problem 6.16.)

S.P. 6.18: Refer to S.P. 6.17. Determine the new value of Max u if the unit profit for a standard car is changed to (a) $460 and (b) $350.

S.P. 6.19: Refer to S.P. 6.17. Determine the new value of Max u if the unit profit for a standard car is changed to $550. (*Suggestion:* Refer to Problem 6.15 for a problem where a Δc_j value lies outside of the "acceptable range.")

S.P. 6.20: In S.P. 1.12(a) allow the cost term b_1 (daily cost for refinery A) to change by the amount Δb_1. Determine the range of values for Δb_1 such that Theorem 6.4 applies. (*Suggestion:* Make use of the terminal dual tableau of S.P. 6.6. Use the method of Problem 6.18.)

S.P. 6.21: Refer to S.P. 6.20. Determine the new value of Min w if the daily cost for refinery A is changed to (a) $500 and (b) $250.

S.P. 6.22: Refer to S.P. 6.20. Determine the new value of Min w if the daily cost for refinery A is changed to $100. (*Suggestion:* Refer to Problem 6.14 for a problem where a Δb_i value lies outside of the "acceptable range.")

Adjoining New Rows and Columns

S.P. 6.23: Refer to S.P. 1.16. Suppose that the factory contains a third shop (shop 3) and that the work requirements for shop 3 are 3 man-hours for each standard car and 2 man-hours for each compact car. In addition, 100 man-hours per week are available in shop 3. Determine the effect of this new constraint on the optimal solution of S.P. 1.16. (*Hint:* The problem can be resolved by adjoining a new row to the terminal tableau associated with S.P. 1.16 (see S.P. 6.3 for the terminal tableau).)

S.P. 6.24: Rework S.P. 6.23 for the case where 90 man-hours per week are available in shop 3. The rest of the data are unchanged.

S.P. 6.25: Refer to S.P. 1.16. Suppose that the manufacturer wants to produce luxury cards in addition to standard and compact cars (let x_3 denote the number of luxury cars produced per week). Suppose that the work requirements in shops 1 and 2 are 6 and 4 man-hours per luxury car, respectively. The profit is $450 on each luxury car. Determine the effect of this new information on the optimal solution of S.P. 1.16. (*Hint:* The problem can be resolved by adjoining a new column to the terminal tableau associated with S.P. 1.16. Think in terms of the dual problem.)

S.P. 6.26: Rework S.P. 6.25 for the case where the profit per luxury car is $550. The rest of the data are unchanged.

ANSWERS TO SUPPLEMENTARY PROBLEMS

S.P. 6.1: (a) $656; (b) $560.

S.P. 6.2: (a) $10,500; (b) $9,950.

S.P. 6.3:

	t_1	t_2	1	
v_1	$\frac{1}{2}$	$-\frac{1}{2}$	-15	$=-x_1$
v_2	$-\frac{1}{2}$	$\frac{5}{6}$	-25	$=-x_2$
-1	-50	-50	$13{,}500$	$=u$
	$=y_1$	$=y_2$	$=-w$	

(T.3)

Maximum problem solution: $x_1{}^* = 15$, $x_2{}^* = 25$, $t_1{}^* = t_2{}^* = 0$, and Max $u = \$13{,}500$.

Dual solution: $y_1{}^* = 50$, $y_2{}^* = 50$, and $v_1{}^* = v_2{}^* = 0$.

S.P. 6.4: (a) $15,000; (b) $12,500.

S.P. 6.5: (a) $14,750; (b) $13,000.

S.P. 6.6:

	x_2	t_1	t_2	1	
v_3	$\frac{2}{3}$	$\frac{1}{300}$	$-\frac{1}{300}$	$-\frac{5}{3}$	$=-x_3$
v_1	$\frac{5}{3}$	$-\frac{1}{300}$	$\frac{2}{300}$	$-\frac{2}{3}$	$=-x_1$
$-10{,}000$	-60	-30	$33{,}000$		$=u$
	$=v_2$	$=y_1$	$=y_2$	$=-w$	

(T.4)

Minimum problem solution: $y_1{}^* = 60$, $y_2{}^* = 30$, $v_1{}^* = v_3{}^* = 0$, $v_2{}^* = 10{,}000$, and Min $w = 33{,}000$.

Dual solution: $t_1{}^* = t_2{}^* = x_2{}^* = 0$, $x_1{}^* = \frac{2}{3}$, and $x_3{}^* = \frac{5}{3}$.

S.P. 6.7: (a) $35,000; (b) $31,000.

S.P. 6.8: (a) $38,000; (b) $28,000.

S.P. 6.9:   (a) \$33,000; (b) \$33,000. In other words, no change in Min w, since $x_2^* = 0$.

S.P. 6.10: $-30 \le h_1 \le 50$.

S.P. 6.11: $-30 \le h_2 \le 30$.

S.P. 6.12: $-4,500 \le k_1 \le 18,000$.

S.P. 6.13: $k_2 \le 10,000$. No lower limit.

S.P. 6.14: $-9,000 \le k_3 \le 9,000$.

S.P. 6.15: $x_1^* = 40$, $x_2^* = 0$, $t_1^* = 50$, $t_2^* = 0$, and Max $u = \$16,000$.

S.P. 6.16: $y_1^* = 75$, $y_2^* = 0$, $v_1^* = 1,500$, $v_2^* = 2,500$, $v_3^* = 0$, and Min $w = \$30,000$.

S.P. 6.17: $-100 \le \Delta c_1 \le 100$.

S.P. 6.18:   (a) \$14,400; (b) \$12,750.

S.P. 6.19:   Max $u = \$16,500$; new solution is $x_1^* = 30$, $x_2^* = 0$, $t_1^* = 0$, and $t_2^* = 30$.

S.P. 6.20: $-250 \le \Delta b_1 \le 200$.

S.P. 6.21:   (a) \$39,000; (b) \$24,000.

S.P. 6.22:   Min $w = \$12,000$; new solution is $y_1^* = 120$ days, $y_2^* = 0$, $v_1^* = 0$, $v_2^* = 16,000$, and $v_3^* = 9,000$.

S.P. 6.23:   Optimal solution stays the same. Max $u = \$13,500$, $x_1^* = 15$, and $x_2^* = 25$.

S.P. 6.24:   New optimal solution and Max u value: Max $u = \$13,000$, $x_1^* = 10$, and $x_2^* = 30$.

S.P. 6.25:   Optimal solution stays the same. Max $u = \$13,500$, $x_1^* = 15$, and $x_2^* = 25$ ($x_3^* = 0$).

S.P. 6.26:   New optimal solution and Max u value: Max $u = \$14,250$, $x_1^* = 0$, $x_2^* = 20$, and $x_3^* = 15$.

Chapter Seven

Further Calculational Aspects

In the previous chapters, several calculational aspects used in the linear programming field were either glossed over or omitted completely. One purpose of the current chapter is to fill in some of the gaps. In particular, several new techniques that are now being used by practitioners in the linear programming field will be covered. It should be noted, however, that the material covered up to now was intended by the author to constitute a streamlined foundation for understanding the field of mathematical programming and other fields related to linear programming.

1. CHOOSING A PIVOT COLUMN IN STAGE TWO

The symbolism we shall use is consistent with that of Chapter 4. The student is referred back to Method 4.2, i.e., Stage Two of the simplex algorithm. In the algorithm for the maximum case, the pivot column selected was *any* column for which c_j was positive. We can do better than this, however; i.e., it is possible that some columns are more desirable to use than others. In [3, pp. 70, 71] (also, e.g., in [1]), it is indicated that the number of iterations necessary to reach a terminal tableau can be greatly reduced by selecting the column that causes the greatest immediate change in the corner entry (equivalently, in the objective function). It is useful here to introduce some new symbolism and terminology. Suppose that we have a tableau for which Stage Two is applicable. An *eligible* column is said to be one for which c_j is positive in the maximum case and negative in the minimum case.

Definition 7.1. Given an eligible column j. The pivot ratio for the column (formerly referred to as b_k/a_{kj} in Method 4.2), denoted by θ_j, is the *smallest* b_i/a_{ij} ratio for those a_{ij}'s in the column that are positive; i.e.,

$$\theta_j = \text{Min } b_i/a_{ij} \text{ (for those } a_{ij} > 0).$$

Part (b) of Problem 4.25 indicates that the change in the corner entry from the current to the next tableau is given by $\theta_j \cdot c_j$ if eligible column j is selected as the pivot column. In other words,

$$d' - d = \theta_j \cdot c_j.$$

This corresponds to an increase in a maximum problem and a decrease in a minimum one (neglecting degeneracies).

Thus, the eligible column that yields the greatest change in the corner entry from the current to the next tableau is the one for which $\theta_j \cdot c_j$ is greatest in absolute value. In the case of ties, a random selection technique is recommended for making a definite choice.

Method 7.1: Greatest Change in Corner Entry Criterion

Given a tableau for which the Stage Two algorithm applies. Do the following.

Part (a), Maximum Case

Determine θ_j and $\theta_j c_j$ for all columns for which c_j is positive. Select the pivot column to be the one for which $\theta_j c_j$ is most positive.

Part (b), Minimum Case

Determine θ_j and $\theta_j c_j$ for all columns for which c_j is negative. Select the pivot column to be the one for which $\theta_j c_j$ is most negative.

If there are many columns that are eligible, Method 7.1 becomes rather complicated to apply. A criterion that has been found to be quite satisfactory in practice is to choose as the pivot column that eligible column possessing the most positive (maximum case) or most negative (minimum case) c_j. This approach has been employed by many computation centers (see [3, p. 71]). It has been found to be quite efficient at reducing the number of steps required in going from the first Stage Two tableau to the terminal tableau. Its key advantage is in its simplicity: A preliminary calculation of θ_j values for eligible columns is no longer required.

Method 7.2

Given a tableau for which the Stage Two algorithm applies. Do the following.

Part (a), Maximum Case

For those columns for which $c_j > 0$, select the pivot column to be the one for which c_j is most positive.

Part (b), Minimum Case

For those columns for which $c_j < 0$, select the pivot column to be the one for which c_j is most negative.

In the case of ties, a random selection technique is recommended. We illustrate the application of Methods 7.1 and 7.2 by working out several sample problems.

Problem 7.1

Consider the simplex algorithm solution of the "nut problem" (Problem 1.4) as given by Problem 4.20. Refer to Tableau 1 of Problem 4.20. Determine the pivot columns recommended by the application of Methods 7.1 and 7.2.

Solution

We begin by noting that Stage Two is applicable to Tableau 1. Columns 1, 2, and 3 are all eligible, since c_1, c_2, and c_3 are all positive. The θ_j values, together with other relevant data, are given in the following table (recall that the only b_i/a_{ij} ratios that need to be considered in the computation of a θ_j value are those for which $a_{ij} > 0$):

Column	c_j	θ_j	$c_j\theta_j$
1	$\frac{1}{4}$	900	225
2	$\frac{2}{5}$	1350	540
3	$\frac{1}{2}$	800	400

In column 1, $\theta_1 = \frac{900}{1} = 900$, since $a_{11} = 1$ is the only positive a_{i1} entry in the column. In column 2, $b_1/a_{12} = 1350$, while $b_2/a_{22} = 1800$, and θ_2 is the smaller of the two, i.e., $\theta_2 = 1350$.

According to Method 7.1, column 2 should be selected as the pivot column (with a_{12} as the corresponding pivot entry), since 540 is the most positive $c_j\theta_j$ value. In fact, this was done in Problem 4.20. Note that $c_2\theta_2 = 540$ is equal to the change in the corner entry from Tableau 1 to Tableau 2 in Problem 4.20.

Method 7.2 would lead us to choose column 3 as the pivot column (with $a_{23} = \frac{3}{4}$ as the corresponding pivot entry), since $\frac{1}{2}$ is more positive than $\frac{1}{4}$ and $\frac{2}{5}$. If that had been done in Problem 4.20, then Tableau 2 would correspond to point C in Figure 4.8, where u is equal to 400. This is in accord with the equation $d' - d = c_3\theta_3 = 400$, since $d = 0$ for Tableau 1.

Problem 7.2

Given the following Stage Two tableau for a minimum problem. Determine the pivot columns recommended by the application of Methods 7.1 and 7.2:

x	y	s	1	
1	1**	2	-10	$= -r$
16*	5	-4	-80	$= -z$
-12	-9	6	600	$= u$

(T.k)

Solution

The symbols here are in accord with those of Chapter 4, e.g., u (not w) is the symbol for the objective function to be minimized.

Columns 1 and 2 are the eligible columns, since c_1 and c_2 are both negative. We have $\theta_1 = 5$ because $b_2/a_{21} = 5$ is smaller than $b_1/a_{11} = 10$. The key data are summarized in the following table:

Column	c_j	θ_j	$c_j\theta_j$
1	−12	5	−60
2	−9	10	−90

According to Method 7.1, we should choose column 2 as the pivot column, since −90 is more negative than −60. If we did, then the corner entry of the next tableau would be $600 - 90 = 510$.

Method 7.2 would lead us to choose column 1 as the pivot column $(-12 < -9)$. This, in turn, would lead to a new corner entry value of $600 - 60 = 540$, which is not as much an improvement as the value associated with column 2.

The appropriate pivot entries for these columns are marked with asterisks in Tableau k (* denotes pivot entry corresponding to the application of Method 7.2, while ** denotes that corresponding to the application of Method 7.1).

2. ALTERNATE SOLUTION POINTS

On occasion, the solution to a linear programming problem will occur at more than one basic feasible point (b.f.p.). This situation is sometimes referred to as one in which alternate optima exist (e.g., see [4]). In other words, the objective function is optimized at more than one point.

Alternate solution points are easily revealed using the scanning technique of Chapter 1. However, as we know, that approach is recommended only for fairly small problems. Fortunately, the simplex algorithm approach also reveals the existence of alternate solution points in a very simple way.

Suppose that we have arrived at a terminal tableau that is a solution tableau. We shall be using the terminology and symbolism of Chapter 4 here (thus, refer to Figure 4.5). We know that in such a terminal tableau all the $-b_i$'s are nonpositive and all the c_j's nonpositive (maximum problem) or nonnegative (minimum problem). The optimum value of the objective function is the corner entry of the tableau. However, the existence of an alternate solution b.f.p. is revealed by the presence of at least one c_j that is *zero* in the terminal tableau. To determine an alternate solution b.f.p. we merely apply the Stage Two rules to such a column.

Method 7.3: Alternate Solution Points

Suppose that in a terminal solution tableau there exists a column j for which $c_j = 0$. To determine an alternate solution b.f.p., apply the Stage Two rules to that column. The b.f.p. corresponding to the next tableau is an *alternate* solution b.f.p.

We illustrate Method 7.3 with some solved problems that relate to several of our frequently considered model problems.

Problem 7.3

Refer to the mining problem as given by Problems 1.3 and 4.26. Suppose that the objective function is changed to

$$u = 200x + 400y,$$

i.e., c_2 is increased by 100. Determine the point(s) where u is minimized.

Solution

The problem is easily solved by the scanning technique of Chapter 1. It is found that Min $u = \$12{,}000$ at point D, where $x = y = 20$, *and* at point C, where $x = 60$ and $y = 0$. It is useful here to refer to Figures 1.7 and 4.11. Let us draw some level lines for different values of u (see Chapter 3 for a discussion of level lines and hyperplanes); this is done in Figure 7.1. We see that the level line $200x + 400y = 12{,}000$ corresponding to the minimum u passes through points C and D. It also contains all points on the line segment connecting C and D. Thus, u is also minimized at all such points. A formal demonstration pertaining to this will be given in a later problem.

If we apply the simplex algorithm as in Problem 4.26, we obtain the following terminal tableau (which is similar to Tableau 4 of Problem 4.26):

r	s	1			
$\frac{2}{3}$	$-\frac{5}{3}$	-10	$= -t$		
$-\frac{2}{3}$	$\frac{1}{6}^*$	-20	$= -y$	$D(20,20)$	(T.4')
$\frac{1}{3}$	$-\frac{1}{3}$	-20	$= -x$		
200	0	12,000	$= u$		

It should be noted that the calculational effort needed to obtain Tableau 4' can be reduced by using some of the methods of Chapter 6.

Tableau 4' is a terminal tableau, and the b.f.p. D clearly corresponds to it. The presence of an alternate solution point is revealed by the fact that $c_2 = 0$ in Tableau 4'. If we apply Method 7.3 to Tableau 4', it is easily established that a_{22}—the only candidate—is the pivot entry. Carrying out the pivot operation leads to Tableau 5:

r	y	1		
−6	10	−210	= −t	
−4	6	−120	= −s	C(60,0)
−1	2	−60	= −x	
200	0	12,000	= u	

(T.5)

This alternate terminal tableau clearly corresponds to b.f.p. C, where x = 60 and y = 0, and Min u is again \$12,000. Application of Method 7.3 to Tableau 5 would lead back to Tableau 4′.

Problem 7.4

Suppose that the extreme value (i.e., maximum or minimum value) of u is achieved at basic feasible points $\bar{A}$ and $\bar{B}$. Show that this same value of u is achieved at any point on the line segment connecting $\bar{A}$ and $\bar{B}$.

Solution

Note, first of all, that we are using vector notation (bar above a letter) to represent points. The formal definition of a line segment is given in Definition 2.13. Any point $\bar{x}$ on the line segment connecting $\bar{A}$ and $\bar{B}$ (and

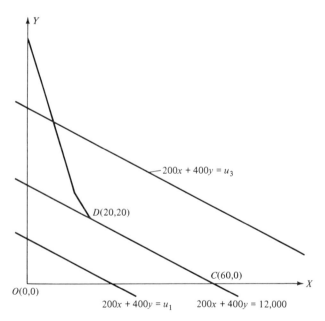

FIGURE 7.1 Alternate solution points for a minimum problem (Problem 7.3). *Note:* For the three level lines shown $(200x + 400y = u_i)$, $u_1 < 12,000 < u_3$.

strictly between them) is given by

$$\bar{x} = t\bar{A} + (1 - t)\bar{B}, \text{ where } 0 < t < 1. \tag{1}$$

Thus,

$$u(\bar{x}) = u[t\bar{A} + (1 - t)\bar{B}]. \tag{2}$$

The objective function u is a linear function and thus possesses the linearity property (Theorem 2.8). Thus, Equation (2) can be written as

$$u(\bar{x}) = tu(\bar{A}) + (1 - t)u(\bar{B}). \tag{3}$$

However, $u(\bar{A})$ and $u(\bar{B})$ are equal, and thus Equation (3) reduces to

$$u(\bar{x}) = u(\bar{A}), \tag{4}$$

which was to be shown.

Notes:
a. The above proof is easily extended to the case where the extreme value of u is achieved at three or more b.f.p.s (e.g., $\bar{A}$, $\bar{B}$, $\bar{C}$, etc.). Thus, if $\bar{x}$ is any *convex combination* of such b.f.p.s (see Definition 3.1), then u achieves the same extreme value at $\bar{x}$. This result is a generalization of the result of Problem 7.4, since any point on a line segment connecting two points is also a convex combination of the two points (see Problem 3.2).
b. The existence of multiple solution points is important from a practical point of view. Knowledge of this may allow the decision-maker to satisfy other criteria. For example, suppose that in Problem 7.3 one wanted a solution point for which all of x, y, s, and t were nonzero. This can be achieved, e.g., if we select $\bar{x} = \frac{1}{2}\bar{C} + \frac{1}{2}\bar{D}$ as a solution point. For this point, which is halfway between $\bar{C}$ and $\bar{D}$ in Figure 7.1, we have $x = 40$, $y = 10$, $s = 60$, and $t = 110$.

Problem 7.5

Refer to the "nut problem" as given by Problems 1.4 and 4.20. Suppose that the value for c_1 (cost of the all-peanuts mixture) is changed to 32¢ per pound, thus causing the objective function to become

$$u = 0.32x + 0.4y + 0.5z.$$

Determine the point(s) where u is maximized.

Solution

If we apply the simplex algorithm as in Problem 4.20, we obtain the following terminal tableau (which is similar to Tableau 3 of Problem 4.20):

x	r	s	1	
$\frac{9}{5}*$	$\frac{8}{5}$	$-\frac{3}{5}$	-1260	$= -y$
$-\frac{4}{5}$	$-\frac{4}{5}$	$\frac{8}{5}$	-240	$= -z$ $A(0,1260,240)$ (T.3′)
0	$-\frac{8}{25}$	$-\frac{14}{25}$	624	$= u$

The only change is that the c_1 entry is equal to zero in this tableau, which indicates the existence of an alternate b.f.p. It should be noted that the calculational effort needed to derive Tableau 3′ can be reduced by using some of the ideas from Chapter 6.

The solution b.f.p. A (see Figure 1.8 or 4.8) corresponds to Tableau 3′. Applying Method 7.3 to Tableau 3′ leads to Tableau 4, in which x and y have been interchanged:

y	r	s	1	
$\frac{5}{9}$	1	$-\frac{1}{3}$	-700	$= -x$
$\frac{4}{9}$	0	$\frac{4}{3}$	-800	$= -z$ $B(700,0,800)$ (T.4)
0	$-\frac{8}{25}$	$-\frac{14}{25}$	624	$= u$

This alternate terminal tableau corresponds to b.f.p. B, which is an alternate solution point (e.g., refer to Figure 4.8). From Problem 7.4 it follows that any point on the line segment connecting A and B is also a solution point. Such points would be of practical value if one desired a solution for which x, y, and z were all positive. For example, the solution point $\frac{1}{2}\bar{A} + \frac{1}{2}\bar{B}$, halfway between $\bar{A}$ and $\bar{B}$, yields $x = 350$, $y = 630$, and $z = 520$ (note that we returned to vector notation here).

A geometrical interpretation is possible. If we set $u = 624$ in the objective function equation, we obtain $0.32x + 0.4y + 0.5z = 624$. This is the equation of the level plane (or level hyperplane—see Chapter 3) that passes through points A and B, and hence through every point on the line segment connecting them.

It should be apparent that the presence of alternate solution points in a primal problem implies that a degeneracy exists in the solution point of the dual problem. This can be made quite clear by employing the dual tableau format of Chapter 5. For example, the presence of the zero in the bottom row of the terminal tableaus of Problem 7.3 or 7.5 means that a basic variable *equals zero* in the solutions of the respective dual problems.

3. THE EXTENDED TABLEAU

Up to now, we have employed the condensed (or Tucker) tableau in most of our calculational procedures. As indicated in Section 1 of Chapter 4, many authors use a different tableau configuration, which shall be referred to as the extended (or Dantzig) tableau. The approach for this section will be similar to that of Strum [4].

First of all, let us consider a standard maximum problem in which m slack variables have been introduced (see Section 2 of Chapter 4). In addition, suppose that all the b_i's are nonnegative. Since it is illuminating to work with a definite problem, let us again turn to the "nut problem" (Problems 1.4 and 4.20). Restated, it is as follows:

$$\text{Maximize} \quad u = \tfrac{1}{4}x + \tfrac{2}{5}y + \tfrac{1}{2}z \tag{1}$$

$$\text{subject to} \quad x + \tfrac{2}{3}y + \tfrac{1}{4}z + r = 900 \tag{2}$$

$$\tfrac{1}{3}y + \tfrac{3}{4}z + s = 600 \tag{3}$$

and

$$x, y, z, r, s \geq 0.$$

Since both b_i's are positive, an easy-to-obtain b.f.p. is that for which the nonbasic variables x, y, and z are set equal to zero. In this case the basic variable values are as follows: $r = 900$ and $s = 600$. This also will be the case, in general, for a standard maximum problem with all b_i's nonnegative, i.e., the initial nonbasic variables are the original variables, the initial basic variables are the slack variables, and the initial b.f.p. is obtained by setting the initial nonbasic variables equal to zero.

The extended tableau equivalent of Equations (1), (2), and (3) is as follows:

	x	y	z	r	s		
r	1	$\tfrac{2}{3}$*	$\tfrac{1}{4}$	1	0	900	
s	0	$\tfrac{1}{3}$	$\tfrac{3}{4}$	0	1	600	(T.1)
	$\tfrac{1}{4}$	$\tfrac{2}{5}$	$\tfrac{1}{2}$	0	0	u	

The first two rows correspond to Equations (2) and (3), and the *bottom* row corresponds to Equation (1). The double vertical lines symbolize the "equals" signs in the equations. The current basic variables (in the present case, r and s) are listed in the left margin. The column of a basic variable contains all zeros, except for a one in the row of the basic variable. (These columns correspond to unit column vectors; this is relevant to a vector interpretation of linear programming.) The other labeled columns correspond to the nonbasic variables.

The relevant pivot transformation is essentially the same as that discussed in Chapter 4 (recall that the purpose of pivoting is to interchange a basic and nonbasic variable). The algorithm that applies to the extended tableau is quite similar to that outlined in Method 4.1. The pivot entry p is the item in the row and column, respectively, of the basic and nonbasic variables that are to be interchanged. An asterisk is placed adjacent to it. In the next tableau, the current basic variable listed in the left margin is replaced by the new basic variable. The variables listed in the top margin remain in their same respective locations, as do the other basic variables in the left margin. In the following, the pivot row and column refer to the row and column, respectively, of the pivot entry p.

Method 7.4: Pivot Transformation for the Extended Tableau

1. The pivot row is transformed by dividing each of its entries by the pivot entry (in particular, p is transformed to 1).
2. The *other* entries in the pivot column (besides p) are transformed to zeros.
3. The transformation for an entry s that is neither in the pivot row nor in the pivot column is identical to that of Method 4.1; i.e., s is transformed to $s - qr/p$, where q is in the column of s and row of p and r is in the row of s and column of p.

Notes: As a result of applying this algorithm, the column of the new basic variable will contain all zeros, except for a 1 in the row of the new basic variable. The columns for other basic variables are the same before and after applying the algorithm.

Problem 7.6

To illustrate Method 7.4, interchange the variables y and r in Tableau 1 and develop the next extended tableau.

Solution

We begin by placing an asterisk next to the pivot entry $\frac{2}{3}$, which is in the row of r and column of y. Following the rules of Method 7.4, we obtain the following tableau:

	x	y	z	r	z			
y	$\frac{3}{2}$	1	$\frac{3}{8}$	$\frac{3}{2}$	0		1350	
s	$-\frac{1}{2}$	0	$\frac{5}{8}*$	$-\frac{1}{2}$	1		150	(T.2)
	$-\frac{7}{20}$	0	$\frac{7}{20}$	$-\frac{3}{5}$	0		$u - 540$	

It is instructive to compare this tableau with Tableau 2 of Problem 4.13 (or 4.20), which is the analogous condensed tableau. Also note what happened in the corner entry: A minus zero (e.g., in $u - 0 = u$) was transformed to $-0 - 900 \cdot (\frac{2}{3})/(\frac{2}{3}) = -540$. To determine the b.f.p. corresponding to Tableau 2, we set $x = z = r = 0$ and obtain $y = 1350$ and $s = 150$. The value of u at this point is $u = 540$, since $u - 540 = 0$.

In linear programming there usually is a purpose for doing a pivot transformation besides merely interchanging a basic and nonbasic variable. For example, in Stage Two the goal was to develop a sequence of tableaus, all corresponding to b.f.p.'s, for which the u value kept improving. Ultimately, under normal circumstances, a tableau is reached such that the signs of the bottom row entries indicate that the optimal u has been achieved. The Stage Two process for the extended tableau is very similar to that for the condensed tableau (Method 4.2). The following discussion and the rules of Method 7.5 pertain to a maximum problem. (It should be noted that certain authors refer to Stage Two as Phase Two.) The normal terminal tableau of Stage Two is one in which the bottom row entries in the nonbasic columns are all nonpositive. The maximum value of u then appears in the corner entry. The pathological terminal situation (unbounded problem) occurs when there is a column for which the bottom row entry is positive, but all other entries in the column are nonpositive (compare with Problem 4.24).

Method 7.5: Stage Two for the Extended Tableau—Maximum Case

1. The current tableau is presumed to correspond to a b.f.p.; all entries to the right of the double vertical line (except for the corner entry) are nonnegative. If all the bottom row entries in the nonbasic columns are nonpositive, the existing tableau is a normal terminal tableau from which the solution point and Max u value may be determined. If there is a positive bottom row entry in a nonbasic column, go to step (2).

2. Let c_j be any positive entry in the bottom row. Thus, the column of this entry is an acceptable pivot column. For each positive a_{ij} in this column calculate the ratio b_i/a_{ij} by dividing the entry in the right column (to the right of double vertical line) by a_{ij}. Such b_i/a_{ij} ratios will all be nonnegative. Suppose that b_k/a_{kj} is the smallest of the ratios. Then row k is the pivot row and a_{kj} is the *pivot entry*. (If there is a tie for the smallest ratio, the pivot entry may be chosen to be *any* of those corresponding to the tie.) After employing the pivot transformation (Method 7.4) to obtain the next tableau, go to step (1).

Note: Some practitioners employ the analog of Method 7.1 (or 7.2) together with Method 7.5. For example, Strum [4] incorporates Method 7.2 into the Stage Two algorithm; the pivot column is thus the one for which c_j is most positive.

Problem 7.7

Use Method 7.5 to develop the terminal tableau for Problem 4.20. Read off the values of the variables at the solution point and also the value for Max u from the terminal tableau.

Solution

The very first extended tableau is Tableau 1, which is given at the beginning of the current section. Each of the x, y, and z columns is an eligible column. (We would choose the z column as the pivot column if we were incorporating Method 7.2 into the Stage Two process.) Suppose that we choose the y column as the pivot column. The pivot entry turns out to be $\frac{2}{3}$, since its b_i/a_{ij} ratio (1350) is smaller than that associated with $\frac{1}{3}$ (namely, 1800). Carrying out the pivot operation leads to Tableau 2 of Problem 7.6, in which the new basic variable pair consists of y and s. The only eligible pivot column here is the z column, and $\frac{5}{8}$ turns out to be the pivot entry. Thus, in the next tableau z replaces s as a basic variable. The pivot transformation is employed, giving rise to Tableau 3, which follows:

	x	y	z	r	s			
y	$\frac{9}{5}$	1	0	$\frac{9}{5}$	$-\frac{3}{5}$		1260	
z	$-\frac{4}{5}$	0	1	$-\frac{4}{5}$	$\frac{8}{5}$		240	(T.3)
	$-\frac{7}{100}$	0	0	$-\frac{8}{25}$	$-\frac{14}{25}$		$u - 624$	

According to step (1) of Method 7.5, this tableau is a terminal tableau. The solution point values are obtained by setting the nonbasic variables x, r, and s equal to zero. We thus obtain

$$x = r = s = 0, \quad y = 1260, \quad z = 240, \quad \text{and} \quad \text{Max } u = 624.$$

This corresponds to point A in Figure 4.8. It is useful to compare the extended Tableaus 1, 2, and 3 with their condensed counterparts of Problem 4.20. The essential equivalence then becomes apparent.

Note: At this point it appears that the chief advantage of the condensed tableau format is that it takes up less space. In other respects, the two approaches are essentially the same. However, it is this author's belief that the key advantage of the condensed tableau format is that it leads to a much simpler interpretation of duality theory and its computational aspects.

At this juncture, we shall end our discussion of the extended tableau approach. The interested reader is referred to Strum [4] for an elementary but thorough discussion of this approach. It should be noted that various alternatives to Stage One (and Stage Zero) are traditionally used in conjunction with the extended tableau approach. Two such alternatives are the "Big M" method and the "Phase One" method. These are also covered in [4].

REFERENCES

1. Dantzig, G. B. "Maximization of a Linear Function of Variables Subject to Linear Inequalities." In *Activity Analysis of Production and Allocation*, edited by T. C. Koopmans. New York: John Wiley and Sons, 1951.

2. Dantzig, G. B. *Linear Programs and Extensions*. Princeton, New Jersey: Princeton University Press, 1963.
3. Gass, S. I. *Linear Programming*. Third edition. New York: McGraw-Hill, 1969.
4. Strum, J. E. *Introduction to Linear Programming*. San Francisco, California: Holden-Day, 1972.

SUPPLEMENTARY PROBLEMS

Choosing a Pivot Column in Stage Two

S.P. 7.1: Given the following Stage Two tableau for a maximum problem, determine the pivot columns and entries recommended by the application of Methods 7.1 and 7.2:

x	y	z	1	
3	6	4	-20	$= -r$
2	7	10	-25	$= -s$
6	-5	8	400	$= u$

S.P. 7.2: Given the following Stage Two tableau for a minimum problem, determine the pivot columns and entries recommended by the application of Methods 7.1 and 7.2:

x	y	s	1	
5	7	4	-10	$= -r$
10	-8	5	-15	$= -t$
8	5	6	-20	$= -z$
-10	9	-8	500	$= u$

Alternate Solution Points

S.P. 7.3: Given the following terminal tableau for a maximum problem, determine all the b.f.p.s at which u is maximized:

x	r	s	1	
2	4	6	-10	$= -y$
-3	2	4	-8	$= -z$
-5	0	-7	500	$= u$

S.P. 7.4:   Given the following terminal tableau for a minimum problem, determine all the b.f.p.s at which w is minimized:

x	r	s	1	
5	10	2	-5	$= -y$
6	6	4	-4	$= -z$
8	0	0	400	$= w$

S.P. 7.5:   To what condition does the existence of alternate solution basic feasible points in the dual linear programming problem correspond?

The Extended Tableau

S.P. 7.6:   Use Method 7.5 (Stage Two for the Extended Tableau—Maximum Case) to solve S.P. 1.1.
S.P. 7.7:   Use Method 7.5 to solve S.P. 1.3.
S.P. 7.8:   Use Method 7.5 to solve S.P. 1.5.
S.P. 7.9:   Use Method 7.5 to solve S.P. 1.11.
S.P. 7.10:   Use Method 7.5 to solve S.P. 1.14.
S.P. 7.11:   Use Method 7.5 to solve S.P. 1.15(a).
S.P. 7.12:   Use Method 7.5 to solve S.P. 1.16.

ANSWERS TO SUPPLEMENTARY PROBLEMS

S.P. 7.1:   Method 7.1: pivot entry is $a_{11} = 3$; Method 7.2: pivot entry is $a_{23} = 10$.
S.P. 7.2:   Method 7.1: pivot entry is $a_{13} = 4$; Method 7.2: pivot entry is $a_{21} = 10$.
S.P. 7.3: $x = r = s = 0$, $y = 10$, and $z = 8$; $x = y = s = 0$, $r = \frac{5}{3}$, and $z = 3$.
S.P. 7.4:   There are 4 b.f.p.s: $x = r = s = 0$, $y = 5$, and $z = 4$; $x = r = z = 0$, $y = 3$, and $s = 1$; $x = y = s = 0$, $r = \frac{1}{2}$, and $z = 1$; $x = y = z = 0$, $r = \frac{3}{4}$, and $s = \frac{5}{11}$.
S.P. 7.5:   A degenerate solution basic feasible point in the primal problem.
S.P. 7.6:   Max $u = 39$, $x = 9$, and $y = 1$.
S.P. 7.7:   Max $u = 27$, $x = 3$, and $y = 3$.
S.P. 7.8:   Max $u = 605$, $x = 75$, $y = 0$, and $z = 100$.
S.P. 7.9:   Max $u = \$635.71$, $x = 5.71$, and $y = 3.57$.
S.P. 7.10:   Max $u = \$283.33$, $x = 5.10$, $y = 4.83$, and $z = 3.67$.
S.P. 7.11:   Maximum profit equals $\$42.00$, $x_1 = 12$, and $x_2 = 6$.
S.P. 7.12:   Maximum profit equals $\$13,500$ per week, $x_1 = 15$, and $x_2 = 25$.

Chapter Eight

Real–World and Computer Applications

The first section of this chapter will be concerned with real-world problems, which can be posed as linear programming problems. The latter part will deal with a simple and easy-to-use computer package that was designed for solving linear programming problems. It was developed by Professor Jerome D. Herniter of Boston University and is known as the EASYLP package.

1. SELECTED REAL-WORLD APPLICATIONS

The real-world applications in this section will, for the most part, be miniature versions of actual large-scale applications. (This will also be true in Chapter 9, which deals with applications relating to the classical transportation and assignment problems.) However, the major aspects of how to formulate the various linear programming models will be emphasized, even though the problems studied here may be "small" in scope. The references listed at the end of this chapter contain treatments of most of the various applications of linear programming techniques to problems occurring in industry, business, economics, and engineering. In particular, a bibliography at the end of [5], entitled "Bibliography of Linear Programming Applications," contains a list of books and articles—categorized by topic—dealing with a myriad of linear programming applications.

The sample problems in this section were all solved by using a computer package (see, e.g., Section 2). Many of them can also be solved without much difficulty by using the computational techniques of Chapters 1, 4, 5, and 7.

Problem 8.1: A Production Problem—Television Sets

A certain company manufactures both black-and-white and color television sets (one model of each). The black-and-white sets are put together on assembly line A, which is limited to a daily production of 90 sets. The color sets are assembled on assembly line B, which is limited to at most 50 sets per day. A total of 980 electronic components is supplied daily; each black-and-white set requires 7 components, while each color set needs 14 components. The labor supply per day is 1080 man-hours. Each black-and-white set requires 9 man-hours of labor, while each color set requires 12 man-hours. Each black-and-white set can be sold at a profit of \$60, while the profit on each color set is \$100. How many television sets of each type should be manufactured each day so as to maximize profits? (Assume that the company is able to sell all the sets that it manufactures.)

Formulate this problem as a linear programming problem. Determine its optimal solution.

Solution

Let x_1 be the number of black-and-white television sets to be produced per day, and let x_2 be the number of color television sets to be produced per day. Let u symbolize the total profit (in dollars) per day. Thus, the problem is as follows:

$$\text{Maximize} \quad u = 60x_1 + 100x_2$$

$$\text{subject to} \quad x_1 \qquad\qquad \leq \quad 90 \text{ (assembly line } A)$$

$$x_2 \leq \quad 50 \text{ (assembly line } B)$$

$$7x_1 + 14x_2 \leq \quad 980 \text{ (electronic components)}$$

$$9x_1 + 12x_2 \leq 1080 \text{ (labor available)}$$

and

$$x_1, x_2 \geq 0.$$

The optimal solution is as follows:

$$x_1{}^* = 80 \text{ black-and-white sets per day}$$

$$x_2{}^* = 30 \text{ color sets per day}$$

$$\text{Max } u = \$7800 \text{ per day}.$$

Four tableaus were required in a computer solution.

Strictly speaking, Problem 8.1 is an integer programming problem, i.e., there is an implied requirement that x_1 and x_2 be integers. In the solution obtained, both $x_1{}^*$ and $x_2{}^*$ are integers, so no further analysis is necessary. If it had turned out that $x_1{}^*$ and $x_2{}^*$ were not integers, the recommended procedure would be to choose rounded-off integer values for x_1 and x_2 which have the property of yielding both a feasible solution as well as a value for u that is close to Max u. (By feasible solution, we mean that all the constraints are satisfied.) This statement is not strictly correct; the best approach to use is one in which *integer programming* methods are used. However, this subject is beyond the scope of the current book and thus will not be pursued further here.

Problem 8.2: A Vehicle-Production Problem

A vehicle manufacturer produces compact cars, regular-size cars, and small trucks. The factory consists of an assembly shop that can provide at most 200 man-hours per week and a finishing shop that can provide at most 150 man-hours per week. Both these capacities are limited by the naturally occurring interactions between people and machines. A compact car requires 3 man-hours of assembly and 4 man-hours of finishing. For a regular-size car, the figures are 4 and 5 man-hours, respectively. A small truck requires 5 man-hours of assembly and 3 man-hours of finishing.

The profit figures are \$150 for each compact car sold, and \$250 for each regular-size car sold, and \$200 for each truck sold. How many vehicles of each type should be produced per week so as to maximize profits? (We again assume that the company can sell as many vehicles as it manufactures.)

Formulate this problem as a linear programming problem. Determine and discuss its optimal solution.

Solution

Our notation is as follows:

x_1 equals the number of compact cars to be produced per week.

x_2 equals the number of regular size cars to be produced per week.

x_3 equals the number of trucks to be produced per week.

The total profit in dollars per week, u, is expressed as

$$u = 150x_1 + 250x_2 + 200x_3 \quad \text{(to be maximized).} \quad (1)$$

The two main constraints are as follows:

$$3x_1 + 4x_2 + 5x_3 \le 200 \quad \text{(assembly)} \quad (2.1)$$

$$4x_1 + 5x_2 + 3x_3 \le 150 \quad \text{(finishing).} \quad (2.2)$$

In addition, x_1, x_2, and x_3 are required to be nonnegative.

The optimal solution, as determined by computer, is

$$x_1^* = 0, \ x_2^* = 11.538, \ x_3^* = 30.769, \quad \text{and} \quad \text{Max } u = \$9038.30.$$

This indicates that no compact cars, 11.54 regular-size cars, and 30.77 trucks should be produced per week so as to yield the maximum weekly profit. Of course, this solution does contain a flaw; in reality, the values for x_1, x_2, and x_3 are required to be whole numbers. Proceeding according to the discussion following Problem 8.1, we attempt to find a suitable "rounded-off" solution.

If we use $x_2 = 11$ and $x_3 = 31$, we find that $u = \$8950$. This constitutes a feasible solution, since the left-hand sides of (2.1) and (2.2) become 199 and 148, respectively.

Employing $x_2 = 12$ and $x_3 = 30$ yields the better value $u = \$9000$. Here, too, we have a feasible solution; the left-hand side of (2.1) becomes 198, while the left-hand side of (2.2) becomes 150.

Thus, it appears that a satisfactory realistic solution is to produce 12 regular cars and 30 small trucks per week at the vehicle factory. Selling all of these vehicles will yield a profit of $9000.

The two product mix problems just discussed are typical of a whole class of standard maximum problems. Two other problems in this class are Problem 1.1, which dealt with furniture production, and Problem 1.4, which dealt with producing mixtures of nuts (the so-called "nut problem").

Problem 8.3: A Gasoline-Blending Problem

A small petroleum refinery has three types of raw gasoline available, and it wishes to blend them in order to produce regular and premium gasoline. The performance rating (e.g., a type of octane rating), the daily output (capacity) in barrels per day, and the cost per barrel of the raw gasolines are given in the following table:

Gasoline type	Performance rating	Capacity (barrels/day)	Cost/barrel
1	70	5000	$3.00
2	90	3000	$4.00
3	100	4000	$6.00

The performance rating for regular gasoline is required to be at least 87, and the selling price is $11.00 per barrel. The performance rating of premium gasoline is required to be at least 97, and its selling price is $13.00 per barrel.

Develop a linear programming formulation to determine what type of blend to use in order to maximize the daily profit. Determine and discuss the optimal solution of the problem.

Solution

We introduce the following six variables (the dimensions are in barrels per day):

x_1 is the amount of raw gasoline 1 to be used in the regular gasoline

x_2 is the amount of raw gasoline 2 to be used in the regular gasoline

x_3 is the amount of raw gasoline 3 to be used in the regular gasoline
x_4 is the amount of raw gasoline 1 to be used in the premium gasoline
x_5 is the amount of raw gasoline 2 to be used in the premium gasoline
x_6 is the amount of raw gasoline 3 to be used in the premium gasoline.

The term p_i (for $i = 1, 2, 3$) denotes the profit per barrel when raw gasoline i is used to make regular gasoline. The term p_{i+3} (for $i = 1, 2, 3$, i.e., p_4 etc.) denotes the profit per barrel when raw gasoline i is used to make premium gasoline.

We assume that the densities (mass per volume) of all five liquids are approximately equal. Another assumption we make is that the performance rating of regular (or premium) gasoline is a weighted average by volume of the performance ratings of the component raw gasolines. Therefore,

$$p_1 = 11 - 3 = 8, \quad p_2 = 11 - 4 = 7, \quad p_3 = 11 - 6 = 5,$$

$$p_4 = 13 - 3 = 10, \quad p_5 = 13 - 4 = 9, \quad p_6 = 13 - 6 = 7.$$

Thus, the quantity u, denoting profit per day, is given by

$$u = 8x_1 + 7x_2 + 5x_3 + 10x_4 + 9x_5 + 7x_6. \tag{1}$$

The capacity constraints are

$$x_1 + x_4 \le 5000 \tag{2.1}$$

$$x_2 + x_5 \le 3000 \tag{2.2}$$

$$x_3 + x_6 \le 4000. \tag{2.3}$$

Let R_r and R_p denote the respective performance ratings of the regular and premium gasolines. Thus, we have the following requirements:

$$R_r \ge 87 \tag{a}$$

$$R_p \ge 97. \tag{b}$$

According to our assumption with respect to performance rating,

$$R_r = \frac{70x_1 + 90x_2 + 100x_3}{x_1 + x_2 + x_3} \tag{c}$$

$$R_p = \frac{70x_4 + 90x_5 + 100x_6}{x_4 + x_5 + x_6}. \tag{d}$$

Substitution of (c) into (a) and of (d) into (b) leads to the following:

$$70x_1 + 90x_2 + 100x_3 \ge 87x_1 + 87x_2 + 87x_3$$

$$70x_4 + 90x_5 + 100x_6 \ge 97x_4 + 97x_5 + 97x_6.$$

Simplifying these leads to the two final main constraints:

$$-17x_1 + 3x_2 + 13x_3 \geq 0 \tag{2.4}$$

$$-27x_4 - 7x_5 + 3x_6 \geq 0. \tag{2.5}$$

In addition, x_1 through x_6 are required to be nonnegative.

The optimal solution of the problem, obtained by computer, (four tableaus), is as follows:

$$x_1^* = x_2^* = x_3^* = 0, \ x_4^* = 5000, \ x_5^* = 3000,$$

$$x_6^* = 4000, \quad \text{and} \quad \text{Max } u = 105{,}000 \ \$/\text{day}.$$

In other words, the company should blend all of its raw gasolines into premium gasoline in order to maximize profit. The total capacities of the raw gasolines are used in this solution, i.e., constraints (2.1), (2.2), and (2.3) are satisfied as equations.

Problem 8.4: A Whiskey-Blending Problem

A whiskey-blending company has three types of spirits available in the following amounts:

Type of spirits	Amount (in gallons)
A	3,000
B	5,000
C	10,000

From these three types of spirits the company will blend three grades of whiskey, the compositions of which are indicated in the following table:

| Whiskey grade | Percentage of spirits | | |
	A	B	C
1	90	10	0
2	30	50	20
3	10	30	60

The profits per gallon (in dollars) from selling grades 1, 2, and 3 are $12, $6, and $4, respectively. Develop a linear programming formulation to determine how best to blend spirits A, B, and C in order to derive a maximum profit. Determine the optimal solution of the problem.

Solution

It is assumed that all quantities of whiskey produced from the blending can be sold. We also assume that the six relevant liquids have equal densities (mass per volume).

Let x_i (for $i = 1, 2, 3$) denote the quantity (in gallons) of whiskey grade

i to be produced. Thus, the total profit u (in dollars) is given by

$$u = 12x_1 + 6x_2 + 4x_3 \quad \text{(to be maximized).} \tag{1}$$

By making use of the information in the above tables, we obtain the following main constraints:

$$0.9x_1 + 0.3x_2 + 0.1x_3 \leq 3,000 \tag{2.1}$$

$$0.1x_1 + 0.5x_2 + 0.3x_3 \leq 5,000 \tag{2.2}$$

$$0.2x_2 + 0.6x_3 \leq 10,000. \tag{2.3}$$

As an example, consider the terms that appear in (2.1). The term $0.9x_1$ indicates that 90% of x_1 gallons of grade 1 whiskey consists of spirits type A. Similarly, 30% of x_2 gallons of grade 2 whiskey and 10% of x_3 gallons of grade 3 whiskey consists of spirits type A. The right-hand side of (2.1) indicates that at most 3,000 gallons of spirits type A are available. The other two constraints pertain to the other types of spirits. Of course, x_1, x_2, and x_3 have to be nonnegative.

The optimal solution, obtained by computer, occurred in the third tableau. The relevant values are as follows:

$x_1{}^* = 1,538.5$ gallons, $x_2{}^* = 0$, $x_3{}^* = 16,153.8$ gallons,

and   Max $u = \$83,077.20$.

Thus, it is best for the whiskey-blending company to produce whiskey grades 1 and 3 in the above amounts and not to produce any grade 2 whiskey. The computer solution also reveals that the first two constraints are satisfied as equations (the available spirits types A and B are used completely), but that 307.7 gallons of spirits type C are left over. These results can also be checked by substituting the above numerical values into the three main constraints.

The next two problems illustrate how linear programming techniques may be applied in the investment area.

Problem 8.5

An investment company wishes to invest a large amount of money. There are five different investment choices: common stocks, real estate, savings certificates, municipal bonds, and mutual funds. The current annual yields on investment are known for each of these investments and are given in the following table:

Investment choice	Current annual yield (%)
1. Common stocks	12
2. Real estate	9
3. Savings certificates	7
4. Municipal bonds	6
5. Mutual funds	10

The company experts theorize that these current yields will stay roughly the same in the immediate future. They wish to diversify investments so as to obtain maximum returns. Based on detailed analyses of the investment market, and with a view toward reducing risk factors, the company experts have decided to use the criteria set forth in the following statements: The sum of the investments in common stock and savings certificates should not exceed the sum of the investments in municipal bonds and mutual funds; the total investment in real estate and mutual funds combined should be at least as great as the investment in savings certificates; finally, the investment in real estate should not exceed the investment in mutual funds.

Develop a linear programming formulation to determine the distribution of investment funds among the five choices that will yield the maximum total returns. Determine and discuss the optimal solution of the problem.

Solution

Let x_1, x_2, x_3, x_4, and x_5 represent the fractions of the total investment to be allocated to common stocks, real estate, savings certificates, municipal bonds, and mutual funds, respectively. Let the objective function u denote the total annual returns from investment in the five categories. Thus,

$$u = 0.12x_1 + 0.09x_2 + 0.07x_3 + 0.06x_4$$

$$+ 0.10x_5 \quad \text{(to be maximized).} \quad (1)$$

We shall now restate the restrictions cited above (in the problem statement) as inequalities:

$$x_1 + x_3 \leq x_4 + x_5 \quad\quad\quad (2.1)$$

$$x_2 + x_5 \geq x_3 \quad\quad\quad (2.2)$$

$$x_2 \leq x_5. \quad\quad\quad (2.3)$$

In addition, the sum of the fractions must equal one:

$$x_1 + x_2 + x_3 + x_4 + x_5 = 1. \quad\quad\quad (2.4)$$

Finally, each of the x_i's is required to be nonnegative. The optimal solution values for this maximum problem, obtained by computer, are as follows:

$$x_1{}^* = 0.5, \; x_2{}^* = x_3{}^* = x_4{}^* = 0, \; x_5{}^* = 0.5, \quad \text{and} \quad \text{Max } u = 0.110.$$

In other words, the investment company should allocate 50% of its total investment to common stocks (choice 1) and 50% to mutual funds (choice 5). This will cause a maximum annual return of 11%. In other words, if the company has $2,000,000 to invest, it should invest $1,000,000 in both common stocks and mutual funds. This will yield an annual return of $220,000.

Problem 8.6

Three stocks have been analyzed in terms of growth in market value during the next year (short-term growth), growth in market value over the next 12 years (long-term growth), and anticipated dividend rate. The key data appear in the table presented below (the data are in terms of the percentage of the amount originally invested):

	Stock 1	Stock 2	Stock 3
Anticipated growth rate during the next year	3.2%	3.0%	2.5%
Anticipated growth rate in the next 12 years	100%	140%	70%
Anticipated dividend rate	3.0%	3.6%	3.9%

The investor has the following investment goals:

  a. Appreciation of at least $200 over the next year
  b. Appreciation of at least $3000 over the next 12 years
  c. Income of at least $150 each year

Develop a linear programming formulation to determine the minimum amount that can be invested in stocks 1, 2, and 3 while still meeting goals (a), (b), and (c). Present the numerical results for the optimal solution of the problem.

Solution

Let x_1, x_2, and x_3 denote the amount (in dollars) invested in stocks 1, 2, and 3, respectively. Thus, the desire is to minimize

$$w = x_1 + x_2 + x_3. \tag{1}$$

We now express goals (a), (b), and (c) as mathematical inequalities by referring back to the data in the table:

$$0.032x_1 + 0.030x_2 + 0.025x_3 \geq 200 \tag{2.1}$$

$$1.0x_1 + 1.4x_2 + 0.7x_3 \geq 3000 \tag{2.2}$$

$$0.030x_1 + 0.036x_2 + 0.039x_3 \geq 150. \tag{2.3}$$

In addition, x_1, x_2, and x_3 are all required to be nonnegative.

The computer solution required five tableaus. The relevant optimal solution values are

$$x_1^* = \$6250, \ x_2^* = x_3^* = 0, \quad \text{and} \quad \text{Min } w = \$6250.$$

Thus, the least expensive (minimum cost) way to satisfy the investment goals is to invest $6250 in stock 1 and nothing at all in the other stocks.

Goal (a) will be satisfied exactly, i.e., (2.1) is satisfied as an equation. An anticipated appreciation of $6250 over the next 12 years results from the above investment scheme, exceeding the lower limit cited in goal (b) by \$3250. Likewise, the minimum dividend income in goal (c) will be exceeded by \$37.50 as a result of the above investment scheme.

Problem 8.7: A Scheduling Problem

At a certain hospital, the day is divided into six periods of 4 hours each. The minimum number of doctors required during the different periods (labeled 1 to 6, inclusive) are given in the following table:

Label	Time period	Minimum number of doctors
1.	Midnight to 4 AM	4
2.	4 AM to 8 AM	7
3.	8 AM to noon	9
4.	Noon to 4 PM	12
5.	4 PM to 8 PM	8
6.	8 PM to midnight	6

Doctors report for work at midnight, 4 AM, 8 AM, Noon, 4 PM, and 8 PM; each works an 8-hour shift. Set up the linear programming problem to *minimize* the total number of doctors required during a typical day's operation. (If the doctors are all paid at the same rate, then the problem is equivalent to minimizing the total salary paid to all doctors during a typical day.) Determine and discuss the optimal solution of the problem.

Solution

It should be noted that any particular doctor will work two periods, e.g., a doctor reporting to work at midnight will work both the midnight to 4 AM and 4 AM to 8 AM periods. Let x_i (for $i = 1, 2, 3, 4, 5, 6$) denote the number of doctors who report to work at the beginning of period i. Thus, the linear programming problem is to minimize

$$w = x_1 + x_2 + x_3 + x_4 + x_5 + x_6 \quad \text{(total doctors)} \qquad (1)$$

subject to

$$x_1 + x_2 \geq 7 \quad \text{(4 AM to 8 AM)} \qquad (2.1)$$

$$x_2 + x_3 \geq 9 \quad \text{(8 AM to noon)} \qquad (2.2)$$

$$x_3 + x_4 \geq 12 \quad \text{(Noon to 4 PM)} \qquad (2.3)$$

$$x_4 + x_5 \geq 8 \quad \text{(4 PM to 8 PM)} \qquad (2.4)$$

$$x_5 + x_6 \geq 6 \quad \text{(8 PM to midnight)} \qquad (2.5)$$

$$x_6 + x_1 \geq 4 \quad \text{(Midnight to 4 AM)}. \qquad (2.6)$$

Each x_i is required to be nonnegative. Moreover, note that each x_i is required to be an integer if the problem is to make any sense.

A brief explanation of how Constraints (2.1) through (2.6) were derived is in order. As an example, consider Constraint (2.1); the number of doctors working the 4 AM to 8 AM period consists of these who start work at midnight (x_1) plus those who start at 4 AM (x_2). The total number $(x_1 + x_2)$ is required to be at least 7. In the midnight to 4 AM period, the doctors working are those who start at 8 PM (x_6) plus those who start at midnight (x_1), and the total number must be at least 4; thus, we obtain Constraint (2.6).

The optimal solution was obtained by computer; a total of seven tableaus was required. The optimal values for the x_i's are as follows:

$$x_1^* = 4, \ x_2^* = 3, \ x_3^* = 10, \ x_4^* = 2, \ x_5^* = 6, \quad \text{and} \quad x_6^* = 0.$$

Also, Min $w = 25$.

It can be seen that all the main constraints are satisfied as equations except for (2.2), in which the left-hand side exceeds the right-hand side by 4.

Problem 8.8: Production Scheduling

A manufacturer of a particular product is planning the production schedule for the months of June, July, and August. The demands for the product are for exactly 70 units in June, 90 units in July, and 120 units in August. The manufacturer has facilities capable of producing at most 100 units in each of the three months.

The unit production cost is $15 per unit in June and $16 in both July and August. The inventory storage cost per item produced is zero during the month of production and $3 per item in any following month. The storage cost is figured by multiplying the number of items in storage on the first day of the month by $3. The company does not have any items in storage at the beginning of the three-month period, and it does not wish to have any items in storage at the end of the three-month period.

Set up the linear programming problem to schedule monthly production and delivery in order to minimize the total production and storage costs for the three-month period. Determine and discuss the optimal solution of the problem.

Solution

First of all, we label the months June, July, and August as 1, 2, and 3, respectively. The symbol x_{ij} denotes the number of items produced in month i that are to be delivered in month j. Clearly, j cannot be lower than i, i.e., an item cannot be delivered before it has been produced. Thus, x_{11}, x_{12}, and x_{13} refer to items produced in month 1 (June) to be delivered in months 1, 2, and 3, respectively. The total cost (for production and storage) is given by

$$w = 15x_{11} + 18x_{12} + 21x_{13} + 16x_{22} + 19x_{23} + 16x_{33}. \tag{1}$$

The goal is to minimize w. The 18 preceding x_{12} represents the $15 unit production cost in June plus a \$3 one-month storage charge. The monthly production capacity conditions are stated as the following inequalities:

$$x_{11} + x_{12} + x_{13} \leq 100 \tag{2.1}$$

$$x_{22} + x_{23} \leq 100 \tag{2.2}$$

$$x_{33} \leq 100. \tag{2.3}$$

The product demands for the three months are indicated by the following three *equations*:

$$x_{11} = 70 \tag{2.4}$$

$$x_{12} + x_{22} = 90 \tag{2.5}$$

$$x_{13} + x_{23} + x_{33} = 120. \tag{2.6}$$

Each x_{ij} is required to be nonnegative and to have an integer value.

The optimal solution was obtained by computer (six tableaus were required). The relevant optimal solution values are as follows:

$$x_{11}^* = 70, \; x_{12}^* = 10, \; x_{13}^* = 0, \; x_{22}^* = 80,$$

$$x_{23}^* = 20, \; x_{33}^* = 100, \quad \text{and} \quad \text{Min } w = \$4490.$$

Constraints (2.2) and (2.3) are satisfied as equations, while the slack in (2.1) is 20 units.

This same problem is treated in Chapter 9 (Problem 9.36) using a specialized technique employed for transportation problems (the Stepping Stone Algorithm).

Problem 8.9: A Cargo-Loading Problem

A cargo airplane has one front compartment and one rear compartment. The front compartment has a weight capacity limit of 20,000 pounds and a volume capacity limit of 10,000 cubic feet. The rear compartment has a weight capacity limit of 25,000 pounds and a volume capacity limit of 15,000 cubic feet. Two commodities (referred to as commodities 1 and 2) are available for shipment. Commodity 1 totals 28,000 pounds, and the volume per mass is 0.6 cubic feet per pound. The revenue rate for shipping commodity 1 is \$1.20 per pound. Commodity 2 totals 22,000 pounds, and its volume per mass is 1.2 cubic feet per pound. The revenue rate for shipping commodity 2 is \$1.50 per pound. The cargo plane may accept all or any part of either commodity.

Set up the linear programming problem to determine how much of each commodity should be accepted so as to maximize the total revenue. Determine the optimal solution of the problem.

Solution

We label the front and rear compartments as compartments 1 and 2, respectively. By x_{ij} we denote the amount of commodity i (in pounds) to be put into compartment j. Thus, for example, x_{21} denotes the poundage of commodity 2 to be placed into the front compartment.

The total revenue in dollars, u, is to be maximized:

$$u = 1.20x_{11} + 1.20x_{12} + 1.50x_{21} + 1.50x_{22} . \tag{1}$$

The weight restrictions for the two compartments are expressed as follows:

$$x_{11} + x_{21} \leq 20{,}000 \quad \text{(front)} \tag{2.1}$$

$$x_{12} + x_{22} \leq 25{,}000 \quad \text{(rear).} \tag{2.2}$$

Let us now consider the volume restrictions for the front and rear compartments. The volume taken up by the x_{11} pounds of commodity 1 that goes in the front compartment is x_{11} pounds times 0.6 cubic feet per pound, i.e., $0.6x_{11}$ cubic feet. Likewise, $1.2x_{21}$ is the volume occupied by commodity 2 in the front compartment. The volume restrictions for the two compartments are

$$0.6x_{11} + 1.2x_{21} \leq 10{,}000 \quad \text{(front)} \tag{2.3}$$

$$0.6x_{12} + 1.2x_{22} \leq 15{,}000 \quad \text{(rear).} \tag{2.4}$$

The supply limitations for the two commodities are expressed by

$$x_{11} + x_{12} \leq 28{,}000 \quad \text{(commodity 1)} \tag{2.5}$$

$$x_{21} + x_{22} \leq 22{,}000 \quad \text{(commodity 2).} \tag{2.6}$$

In addition, each x_{ij} must be nonnegative.

The computer solution required six tableaus. The relevant optimal solution values are

$$x_{11}^* = 3{,}000, \ x_{12}^* = 25{,}000, \ x_{21}^* = 6{,}833.33,$$

$$x_{22}^* = 0, \quad \text{and} \quad \text{Max } u = \$43{,}850.$$

Notice that all of the available weight capacity for the rear compartment is used up. In addition, the volume capacities of both the front and rear compartments are completely used up.

The next problem is an example of a "cutting stock" or "trim loss" problem. Illustrations of this important type of problem appear in Strum [18], Kim [11], and Taha [19]. Two of the original papers dealing with the linear programming approach to the cutting stock problem are those of Gilmore and Gomory [7, 8].

Problem 8.10

Paper rolls are produced by a paper mill in standard rolls of width 16 feet. (By width we mean the dimension perpendicular to the circular base when the roll is in its unrolled form.) The mill receives orders from its customers for 200 rolls of width 3 feet, 50 rolls of width 4 feet, and 140 rolls of width 5 feet. It is assumed that all four types of roll have the same unrolled length.

To satisfy the orders, the mill will have to cut a certain number of standard rolls into smaller rolls. Various efficient (or plausible) cutting patterns are possible, e.g., a standard roll can be cut into three smaller rolls of width 3 feet and one smaller roll of width 5 feet, resulting in a trim loss of 2 feet $(16 - 3\cdot3 - 1\cdot5)$. Another cutting pattern consists of two smaller rolls of width 3 feet, one smaller roll of width 4 feet, and one smaller roll of width 5 feet; the trim loss here is 1 foot $(16 - 2\cdot3 - 1\cdot4 - 1\cdot5)$.

Determine all the efficient cutting patterns. (An efficient cutting pattern is one for which the trim loss is less than the smallest required width—3 feet in the present case.)

Then, set up the linear programming problem to determine the minimum number of standard rolls to be cut so as to satisfy the customer orders. Determine and discuss the optimal solution of the problem.

Solution

It is possible to determine all efficient cutting patterns for this problem by using trial and error. Cutting pattern i is indicated by the symbol P_i. The cutting patterns are given by the columns in the following table:

Efficient Cutting Pattern

Width	↓	P_1	P_2	P_3	P_4	P_5	P_6	P_7	P_8	P_9	P_{10}
3 feet		5	0	0	4	3	2	2	1	1	0
4 feet		0	4	0	1	0	1	0	2	3	1
5 feet		0	0	3	0	1	1	2	1	0	2
Trim loss		1	0	1	0	2	1	0	0	1	2

Some illustrations would be helpful. Cutting patterns P_5 and P_6 were cited in the statement of the problem. In cutting pattern P_1, the standard roll is cut into five smaller rolls of width 3 feet, resulting in a trim loss of 1 foot. In cutting pattern P_8, the standard roll is cut into one smaller roll of width 3 feet, two smaller rolls of width 4 feet, and one smaller roll of width 5 feet. The trim loss is zero, since $1\cdot3 + 2\cdot4 + 1\cdot5 = 16$.

In the notation for the linear programming setup, we let x_i represent the number of rolls to be cut up using cutting pattern P_i. For example, suppose $x_6 = 10$. This means that 10 rolls are to be cut up according to pattern P_6; this results in 20 rolls of width 3 feet, 10 rolls of width 4 feet, and 10 rolls of width 5 feet. The trim loss is 10 rolls of width 1 foot.

The total number of rolls to be cut up, denoted by w, is given by

$$w = x_1 + x_2 + x_3 + x_4 + x_5 + x_6 + x_7 + x_8 + x_9 + x_{10}. \qquad (1)$$

The goal is to minimize w. The constraints pertain to the number of orders for rolls of width 3, 4, and 5 feet. They are given as follows:

$$5x_1 + 4x_4 + 3x_5 + 2x_6 + 2x_7 + x_8 + x_9 \geq 200 \qquad (2.1)$$

$$4x_2 + x_4 + x_6 + 2x_8 + 3x_9 + x_{10} \geq 50 \qquad (2.2)$$

$$3x_3 + x_5 + x_6 + 2x_7 + x_8 + 2x_{10} \geq 140. \qquad (2.3)$$

In addition, $x_i \geq 0$ for $i = 1, 2, \ldots, 10$.

As an example, consider constraint (2.1), which deals with the order for 200 rolls of width 3 feet. The term $5x_1$ indicates that x_1 rolls in pattern P_1 will yield x_1 times $5 = 5x_1$ rolls of width 3 feet. The term $4x_4$ indicates that x_4 rolls in pattern P_4 will yield x_4 times $4 = 4x_4$ rolls of width 3 feet. Likewise, $3x_5$ is the number of rolls of width 3 feet that result when x_5 rolls are cut up in pattern P_5. The other terms on the left-hand side of constraint (2.1) are found in a similar fashion from the cutting-pattern table. Thus, (2.1) indicates that the total number of resulting rolls of width 3 feet (left-hand side) should be at least 200.

It should be noted that in determining an optimal solution it is necessary to consider only cutting patterns that are efficient. All those in the table have trim losses less than 3 feet and hence are efficient. Another requirement is that all the x_i's have to be integer in value.

The relevant optimal solution values, obtained by computer, are as follows:

$x_2{}^* = 8.75$, $x_4{}^* = 15$, $x_7{}^* = 70$, and all other $x_i{}^*$'s $= 0$; Min $w = 93.75$.

We observe that constraints (2.1), (2.2), and (2.3) are all satisfied as equations. The solution value $x_2{}^* = 8.75$ is unacceptable, since it is not an integer. We follow the type of procedure discussed after Problem 8.1, and round off $x_2{}^*$ to 9. This value, together with $x_4{}^* = 15$ and $x_7{}^* = 70$, causes constraint (2.2) to be satisfied as a strict inequality, while Min w becomes 94.

Thus, to satisfy the requirements in the minimum-cost fashion, one should cut up 9 rolls in pattern P_2, 15 rolls in pattern P_4, and 70 rolls in pattern P_7. All other patterns can be ignored.

2. A COMPUTER PACKAGE

In recent years, organizations and individuals have produced large-scale computer programs and have made them available for general use. Such programs are called computer "packages." They have been developed for many areas of application, including statistics, mathematical programming, and numerical analysis. If a specific package is available at a company or school computer center, it is often quite easy to use. All the user needs to know are the required control cards (or lines) and the particular required format for input data. The user does not have to know much about the structure of the computer program.

The package is stored in a storage unit of the computer, e.g., magnetic disks. It remains there, ready to be used by any party interested in its special capabilities.

At present, there exist several computer packages that are useful for handling linear programming problems. Among these are the MPSX system developed by the IBM Corporation [9] and the SOUPAC package developed at the University of Illinois.

Recently, Professor Jerome D. Herniter of Boston University developed a computer package for solving linear programming problems. The package is entitled EASYLP, and it involves the conversational programming concept in which the user communicates with the computer. The device for the communication is usually a time-sharing terminal, and the communication consists of the user answering questions posed by the computer. The questions are determined by the computer package. It is very easy to use the EASYLP package; essentially, no prior knowledge of computer functioning is necessary. All one has to do is answer simple questions posed by the program (typical answers are ''YES'' and ''NO'').

Printouts of runs of several model linear programming problems will be presented later in this section to illustrate the functioning of the package.*

It should be noted that the program is capable of handling up to 30 main constraints and 60 nonnegative variables. The program has the capability of producing and solving the dual problem of the linear programming problem under consideration and also of performing sensitivity analyses. Both features will be illustrated in the sample problems that are presented in this section.

The directions pertaining to the format for the constraints indicate that the symbol $<$ is to be interpreted as $\leq$, while $>$ is to be interpreted as $\geq$. The reason for this is related to the scarcity of symbols on older time-sharing terminal keyboards. Another unconventional aspect of the package is that it refers to all constraints as ''equations.'' Outside of these items, most of the remaining format rules are consistent with the usual linear programming symbolism and conventions.

Another point concerning the format of variables is that each variable name must have not more than four characters (alphabetic or numerical), where the first character is alphabetic. Subscripts are not used in variable names. Each variable that appears is assumed to be nonnegative unless a constraint indicates otherwise (e.g., $X \leq 0$).

---

* Permission to display the EASYLP printouts was kindly granted by Jerome D. Herniter, Professor of Business Administration, School of Management, Boston University. Inquiries concerning how to acquire copies of the EASYLP computer programming package should be addressed to Professor Herniter.

A single constant must appear to the right of the symbols $>$, $<$, or $=$ in the constraints. Actually, the rules are really not too restrictive (this will be made clear below). It is noted that the key rules are repeated for the user's benefit each time the EASYLP package is run. Other rules are available in a pamphlet from Professor Herniter.

Let us illustrate the use of the package by means of some runs of familiar linear programming problems. The first run will involve the model problem known as the "nut problem," which was previously considered in Problems 1.4, 4.20, 5.16, 6.1, and elsewhere. The symbolism will be consistent with that used in Problem 5.16.

In the run, the first main constraint $x_1 + \frac{2}{3}x_2 + \frac{1}{4}x_3 \le 900$ will appear as X1 + .66667X2 + .25X3 < 900. The goal in the "nut problem" was to maximize $u = \frac{1}{4}x_1 + \frac{2}{5}x_2 + \frac{1}{2}x_3$; an equivalent statement will be MAX .25X1 + .4X2 + .5X3. Note that a symbol for u does not appear.

The Sigma 7 computer at Queens College in New York City was used for the runs discussed in this section.

Problem 8.11: Simple Run of the "Nut Problem"

Use the EASYLP computer package to obtain the solution of the "nut problem." (Here we shall not be concerned with duality or sensitivity studies, although we shall be later).

Solution

The first part of the printout on the time-sharing terminal follows:

```
!SET   F:1/RRTEM;OUTIN

!SET   F:9 ME

!SET   F:6 ME

!RUN   JSLPBO.ACCT
       LINKING   JSLPBO
'P1'   ASSOCIATED.
       LINKING   SYSTEM LIB
```

CONVERSATIONAL LINEAR PROGRAMMING

EASYLP
COPYRIGHT 1974
JEROME D. HERNITER
BOSTON UNIVERSITY

The first few lines are control statements that are peculiar to the computer being used. They are typed by the *user* (in this case, the author) on his terminal and are instructions to the computer to activate the EASYLP package. The last five lines indicate that the package has been activated and is ready to operate.

HOW MANY CONSTRAINT EQUATIONS DO YOU HAVE?
?2
WRITE THE TITLE OF THIS PROBLEM.
? NUTS
HAVE YOU INCLUDED YOUR EQUATIONS IN A /DATA INPUT FILE?
?NO
WRITE YOUR EQUATIONS. LESS THAN OR EQUAL TO, WRITE <. GREATER THAN
OR EQUAL TO WRITE >. EQUAL TO WRITE =.

EQUATION 1
? X1 + .66667X2 + .25X3 < 900

EQUATION 2
?   .33333X2 + .75X3 < 600
WRITE YOUR OBJECTIVE FUNCTION IN THE FORM MIN A+B OR MAX A+B. IF
YOUR FUNCTION IS LONGER THAN 80 CHARACTERS, YOU MAY CONTINUE THE
LINE BY PLACING AN * AT THE END.
?MAX .25X1 + .4X2 + .5X3

The computer (under the control of the EASYLP package) has asked
questions concerning the constraints, title ("NUTS"), and objective
function. The answers supplied by the *user* appear after the ? symbols.
These answers were 2, NUTS, NO, etc.

WOULD YOU LIKE TO PRINT YOUR EQUATIONS?
?YES

		X1	X2	X3		
EQ	1	1.000	.667	.250	<	900.000
EQ	2	0.000	.333	.750	<	600.000
OBJ	MAX	.250	.400	.500	=	0.000

DO YOU WANT TO CHANGE ANYTHING?
?NO
DO YOU WANT ONLY THE VALUES OF THE BASIC VARIABLES AND THE
OBJECTIVE FUNCTION?
?YES

PROBLEM   NUTS

BASIC VARIABLE	VALUE
X2	1259.9897
X3	240.0103
OBJ	624.0010

OPTIMAL SOLUTION FOUND

DO YOU WANT TO PERFORM SENSITIVITY ANALYSIS?
?NO
DO YOU WANT TO CONSTRUCT THE DUAL TABLEAU?
?NO

The computer first provided the printout of the constraints *and* objective function. Then the solution to the problem was presented. (Note that the results agree with those obtained in Problem 5.16 and elsewhere: Max u = 624, $x_2^* = 1260$, $x_3^* = 240$, and $x_1^* = 0$.) The user then indicated NO to questions about conducting sensitivity and duality studies.

We shall now use the EASYLP package to run another model linear programming problem that has been considered in this book—the so-called "mining problem," which was considered in Problems 1.3, 4.26, 5.15, 6.3, and elsewhere. The symbolism will be consistent with that used in Problem 5.15.

The printout in Problem 8.12 continues from the last printout line of Problem 8.11 (after the question about the dual tableau).

Problem 8.12: Simple Run of the "Mining Problem"

Use the EASYLP computer package to obtain the solution of the "mining problem."

Solution

DO YOU WANT ANOTHER RUN?
?YES
DO YOU WANT TO USE THE SAME DATA?
?NO
HOW MANY CONSTRAINT EQUATIONS DO YOU HAVE?
?3
WRITE THE TITLE OF THIS PROBLEM.
?MINING
HAVE YOU INCLUDED YOUR EQUATIONS IN A /DATA INPUT FILE?
?NO
WRITE YOUR EQUATIONS. LESS THAN OR EQUAL TO, WRITE <. GREATER THAN OR EQUAL TO WRITE >. EQUAL TO WRITE =

EQUATION 1
?   Y1 + 2Y2 >   60

EQUATION 2
? 4Y1 + 2Y2 > 120

EQUATION 3
? 6Y1 + 2Y2 > 150
WRITE YOUR OBJECTIVE FUNCTION IN THE FORM MIN A+B OR MAX A+B IF YOUR FUNCTION IS LONGER THAN 80 CHARACTERS, YOU MAY CONTINUE THE LINE BY PLACING AN * AT THE END.

?MIN 200Y1 + 300Y2
 WOULD YOU LIKE TO PRINT YOUR EQUATIONS?
?YES

		Y1	Y2		
EQ	1	1.000	2.000	>	60.000
EQ	2	4.000	2.000	>	120.000
EQ	3	6.000	2.000	>	150.000
OBJ	MIN	200.000	300.000	=	0.000

DO YOU WANT TO CHANGE ANYTHING?
?NO
 DO YOU WANT ONLY THE VALUES OF THE BASIC VARIABLES AND THE
 OBJECTIVE FUNCTION?
?YES

PROBLEM   MINING

BASIC VARIABLE	VALUE
Y2	20.0000
SL 3	10.0000
Y1	20.0000
OBJ	10000.0039

OPTIMAL SOLUTION FOUND

Note that the results obtained agree with those obtained in Problem 5.15 and elsewhere : Min $w = 10,000$, $y_1^* = 20$, $y_2^* = 20$, and $y_3^* = 0$. The indication SL3 = 10. refers to the value of the slack variable v_3. The indication that Y2, SL3, and Y1 are the final basic variables agrees with Tableau 3 of Problem 5.15.

It is interesting to use the EASYLP package to study duality and sensitivity analysis. We shall first do this for the "nut problem." The reader is referred to Problem 5.16 for a duality analysis and to Problems 6.1, 6.2, 6.6, etc., for various sensitivity analysis calculations. The only change called for in running the EASYLP package is for the user to answer YES to the questions on performing a sensitivity analysis and constructing a dual tableau.

Problem 8.13

Use the EASYLP package to perform a sensitivity analysis and to construct a dual tableau for the "nut problem."

Solution

The following printout is merely a continuation of the printout from Problem 8.11, except that the sensitivity question was answered YES.

```
DO YOU WANT TO PERFORM SENSITIVITY ANALYSIS?
?YES
 DO YOU KNOW THE SELECTION CODES?
?NO
        1-CONSTRAINT COEFFICIENTS
        2-OBJECTIVE FUNCTION
        3-RIGHT HAND CONSTANTS
 WRITE THE DESIRED CODE.
?3
 SENSITIVITY ANALYSIS FOR CONSTRAINT CONSTANTS
    EQUATION   1
      VALUE OF AN ADDITIONAL UNIT  =  .320
      LOWER LIMIT       =     200.000
      UPPER LIMIT       =    1200.018
    EQUATION   2
      VALUE OF AN ADDITIONAL UNIT  =  .560
      LOWER LIMIT       =     449.993
      UPPER LIMIT       =    2700.000
```

The sensitivity analysis was performed for b_1 and b_2—the constraint (right-hand) constants. We shall now make a partial interpretation of the printout. Note that $\Delta \, \text{Max} \, u / \Delta b_1 \, = \, 0.32 \, = \, y_1{}^*$, in agreement with Problem 6.1. The related lower and upper limits for $b_1 (200 \le b_1 \le 1200)$ agree with the calculations in Problem 6.6. In that problem, we had $-700 \le h_1 \le 300$, which is equivalent to $200 \le b_1 \le 1200$ if we recall that $h_1 = \Delta b_1$ and $b_1 = 900 + \Delta b_1 = 900 + h_1$. Thus, the lower and upper limits for b_1 are 200 and 1200.

```
 WOULD YOU LIKE MORE SENSITIVITY ANALYSIS?
?YES   ·
 WRITE THE DESIRED CODE.
?2

 SENSITIVITY ANALYSIS FOR OBJECTIVE FUNCTION COEFFICIENTS

    VARIABLE X1
      NO LOWER LIMIT.
      UPPER LIMIT       =    .320

    VARIABLE X2
      LOWER LIMIT       =    .361
      UPPER LIMIT       =   1.333

    VARIABLE X3
      LOWER LIMIT       =    .150
      UPPER LIMIT       =    .588
```

Next, the sensitivity analysis was performed for c_1, c_2, and c_3—the objective function coefficients. We shall now make a partial interpretation of the printout. The limits for c_2 (namely, $0.361 \leq c_2 \leq 1.333$) are consistent with the calculations in Problem 6.16. In that problem, we had $-0.039 \leq k_2 \leq 0.933$, which is equivalent to the limits obtained here, since $k_2 = \Delta c_2$ and $c_2 = 0.4 + \Delta c_2 = 0.4 + k_2$.

```
WOULD YOU LIKE MORE SENSITIVITY ANALYSIS?
?NO
DO YOU WANT TO CONSTRUCT THE DUAL TABLEAU?
?YES

WOULD YOU LIKE TO PRINT YOUR EQUATIONS?
?YES


                  U1              U2
EQ     1        1.000           0.000  >   .250
EQ     2         .667            .333  >   .400
EQ     3         .250            .750  >   .500


OBJ    MIN     900.000         600.000  =  0.000

DO YOU WANT TO CHANGE ANYTHING?
?NO
DO YOU WANT ONLY THE VALUES OF THE BASIC VARIABLES AND THE
OBJECTIVE FUNCTION?
?YES

PROBLEM   NUTS

        BASIC
        VARIABLE           VALUE
        ---------------    ---------------
          U1                .3200
          SL  1             .0700
          U2                .5600

          OBJ             624.0010

OPTIMAL SOLUTION FOUND
```

Next, the dual tableau was constructed. It is useful to compare this printout with the results of Problem 5.16. The dual problem listed above is the same as that given in Problem 5.16 except that the dual variables are U1 and U2 here. The solution results listed above are identical to those obtained from Tableau 3 of Problem 5.16.

Problem 8.14

Use the EASYLP package to perform a sensitivity analysis and to construct a dual tableau for the "mining problem."

Solution

The following printout continues from Problem 8.12:

```
OPTIMAL SOLUTION FOUND

DO YOU WANT TO PERFORM SENSITIVITY ANALYSIS?
?YES
 DO YOU KNOW THE SELECTION CODES?
?NO
        1-CONSTRAINT COEFFICIENTS
        2-OBJECTIVE FUNCTION
        3-RIGHT HAND CONSTANTS
 WRITE THE DESIRED CODE.
?3

 SENSITIVITY ANALYSIS FOR CONSTRAINT CONSTANTS
   EQUATION  1
     VALUE OF AN ADDITIONAL UNIT  =   133.333
     LOWER LIMIT       =    30.000
     UPPER LIMIT       =    75.000
   EQUATION  2
     VALUE OF AN ADDITIONAL UNIT  =    16.668
     LOWER LIMIT       =   114.000
     UPPER LIMIT       =   240.000
   EQUATION  3
     VALUE OF AN ADDITIONAL UNIT  =      .000
     NO LOWER LIMIT.
     UPPER LIMIT       =   160.000

 WOULD YOU LIKE MORE SENSITIVITY ANALYSIS?
?NO
```

The sensitivity analysis was performed for c_1, c_2, and c_3. We shall now make a partial interpretation of the printout. Note that Δ Min $w/\Delta c_1 = 133.33 = x_1^*$, in agreement with Problem 6.3. The related upper and lower limits for c_1 ($30 \le c_1 \le 75$) agree with the calculations of Problem 6.10. In that problem, we had $-30 \le k_1 \le 15$, which is equivalent to $30 \le c_1 \le 75$ if we recall that $k_1 = \Delta c_1$ and $c_1 = 60 + \Delta c_1 = 60 + k_1$.

```
DO YOU WANT TO CONSTRUCT THE DUAL TABLEAU?
?YES

 WOULD YOU LIKE TO PRINT YOUR EQUATIONS?
?YES

                 U1            U2            U3
  EQ    1       1.000         4.000         6.000  <   200.000
  EQ    2       2.000         2.000         2.000  <   300.000

  OBJ   MAX    60.000       120.000       150.000  >     0.000
```

```
 DO YOU WANT TO CHANGE ANYTHING?
?NO
 DO YOU WANT ONLY THE VALUES OF THE BASIC VARIABLES AND THE
 OBJECTIVE FUNCTION?
?YES

 PROBLEM   MINING

        BASIC
        VARIABLE           VALUE
     ------------      ------------
        U2                  16.6667
        U1                 133.3334

        OBJ               9999.9922

 OPTIMAL SOLUTION FOUND
```

Next, the dual tableau was constructed. It is useful to compare this printout with the results of Problems 5.10 and 5.15. The dual program listed above is the same as that given in Problem 5.10 except that the dual variables are U1, U2, and U3 here. The solution results listed above are identical to those obtained from Tableau 3 of Problem 5.15.

The EASYLP computer package has been shown to be highly flexible and very easy to use. If a problem is unbounded or infeasible, the package merely indicates this is the case by means of a printed message ("UN-BOUNDED SOLUTION" or "NO FEASIBLE SOLUTION").

We shall now use the package to solve S.P. 5.9—a transportation problem that happens to be a model problem for Chapter 9 (see, e.g., Problems 9.1 and 9.6).

Problem 8.15

Use the EASYLP package to obtain the solution of S.P. 5.9. (Note that all the main constraints are equation constraints.)

Solution

The relevant printout is as follows:

```
 HOW MANY CONSTRAINT EQUATIONS DO YOU HAVE?
?5
 WRITE THE TITLE OF THIS PROBLEM.
?TRANSP
 HAVE YOU INCLUDED YOUR EQUATIONS IN A /DATA INPUT FILE?
?NO

 EQUATION  1
?  X11  +  X12  +  X13  =  61
```

```
EQUATION  2
? X21  +  X22  +  X23  =  49

EQUATION  3
? X31  +  X32  +  X33  =  90

EQUATION  4
? X11  +  X21  +  X31  =  52

EQUATION  5
?  X12  +  X22  +  X32  =  68
```

WRITE YOUR OBJECTIVE FUNCTION IN THE FORM MIN A+B OR MAX A+B. IF
YOUR FUNCTION IS LONGER THAN 80 CHARACTERS, YOU MAY CONTINUE THE
LINE BY PLACING AN * AT THE END.

?MIN 26X11 + 23X12 + 10X13 + 14X21 + 13X22 + 21X23 + 16X31 + 17X32 +
29X33

WOULD YOU LIKE TO PRINT YOUR EQUATIONS?
?NO

DO YOU WANT TO CHANGE ANYTHING?
?NO
DO YOU WANT ONLY THE VALUES OF THE BASIC VARIABLES AND THE
OBJECTIVE FUNCTION?
?YES

PROBLEM   TRANSP

BASIC VARIABLE	VALUE
X13	61.0000
X22	30.0000
X23	19.0000
X31	52.0000
X32	38.0000
OBJ	2877.0000

OPTIMAL SOLUTION FOUND

REFERENCES

1. Converse, A. O. *Optimization.* New York: Holt, Rinehart and Winston, Incorporated, 1970.
2. Cooper, L., and Steinberg, D. I. *Introduction to Methods of Optimization.* Philadelphia, Pennsylvania: W. B. Saunders Company, 1970.
3. Cooper, L., and Steinberg, D. I. *Methods and Applications of Linear Programming.* Philadelphia, Pennsylvania: W. B. Saunders Company, 1974.

4. Dantzig, G. *Linear Programming and Extensions.* Princeton, New Jersey: Princeton University Press, 1963.

5. Gass, S. *Linear Programming.* Third edition. New York: McGraw-Hill, 1969.

6. Gaver, D. P., and Thompson, G. L. *Programming and Probability Models in Operations Research.* Monterey, California: Brooks/Cole Publishing Company, 1973.

7. Gilmore, P. C., and Gomory, R. E. "A Linear Programming Approach to the Cutting Stock Problem. Part I." *Operations Research,* vol. 9, 1961.

8. Gilmore, P. C., and Gomory, R. E. "A Linear Programming Approach to the Cutting Stock Problem. Part II." *Operations Research,* vol. 11, 1963.

9. International Business Machines Corporation, *Introduction to Mathematical Programming System—Extended (MPSX).* Fourth edition. GH20-0849-3, 1973.

10. Kemeny, J. G., Mirkil, H., Snell, J. L., and Thompson, G. L. *Finite Mathematics with Business Applications.* Second edition. Englewood Cliffs, New Jersey: Prentice-Hall, 1972.

11. Kim, C. *Introduction to Linear Programming.* New York: Holt, Rinehart and Winston, Incorporated, 1971.

12. Kwak, N. K. *Mathematical Programming with Business Applications.* New York: McGraw-Hill, 1973.

13. McCracken, D. D. *A Guide to FORTRAN IV Programming.* Second edition. New York: John Wiley and Sons, 1972.

14. Owen, G. *Finite Mathematics.* Philadelphia, Pennsylvania: W. B. Saunders Company, 1970.

15. Shim, J. K., and Zecher, I. *The Computerized Business and Economic Analysis.* Department of Accounting, Queens College, Flushing, New York, 11367, 1974.

16. Singleton, R. R., and Tyndall, W. F. *Games and Programs.* San Francisco, California: W. H. Freeman and Company, 1974.

17. Spivey, W. A., and Thrall, R. M. *Linear Optimization.* New York: Holt, Rinehart and Winston, Incorporated, 1970.

18. Strum, J. E. *Introduction to Linear Programming.* San Francisco, California: Holden-Day, 1972.

19. Taha, H. A. *Operations Research—An Introduction.* New York: Macmillan Publishing Company, 1971.

SUPPLEMENTARY PROBLEMS

Selected Real World Applications

For the following, the reader should attempt to set up the linear programs representing the problems. Complete numerical solutions are given in the answers. In some cases, the linear program is also presented. The numerical solutions, sometimes difficult to obtain by hand, were obtained by using a computer package.

S.P. 8.1:   An electronics company manufactures three types of radios: types I, II, and III. Type I radios are assembled on an assembly line that is limited to no more than 90 sets per day, while not more than 100 type

II radios and 120 type III radios can be assembled per day on their respective assembly lines. In addition, 1000 electronic components and 1500 man-hours of labor are available each day at the company. The number of components, man-hours required, and profits for the three types of radios are given in the following table:

Radio type	Number of components/radio	Man-hours/radio	Profit ($/radio)
I	12	5	8
II	8	4	7
III	7	4	6

How many radios of each type should be manufactured each day so as to maximize total profits?

S.P. 8.2: An automobile manufacturer makes compact, regular, and luxury cars. The factory is divided into shops A, B, and C. In shop A, 150 man-hours are available per week. Each compact car requires 3 man-hours in shop A; the corresponding figures for regular and luxury cars are 4 and 5 man-hours per car. The following table summarizes the shop requirements and profits per car, as well as the weekly capacities of the three shops:

Car type	Shop A	Shop B	Shop C	Profit ($/car)
Compact	3 man-hours	4 man-hours	4 man-hours	150
Regular	4 "	5 "	3 "	180
Luxury	5 "	3 "	6 "	200
Weekly capacity (man-hours)	150	120	100	

How many cars of each type should be produced in order to maximize profits?

S.P. 8.3: A manufacturer makes small, medium, and large boxes from large sheets of cardboard. The small boxes require 3 sq ft per box, the medium 4 sq ft, and the large 5 sq ft per box. It is required that at least four large boxes be made. Furthermore, the number of small boxes should be at least as great as the total number of medium and large boxes. If 200 sq ft of cardboard are in stock, and if the profits per box for small, medium, and large boxes are $1.00, $1.50, and $2.00, respectively, how many of each box should be made so as to maximize total profits?

S.P. 8.4: A petroleum refinery has three types of raw gasoline available, and it wishes to blend them into regular and premium gasoline. The performance ratings, daily outputs, and costs per barrel are given in the following table:

Gasoline type	Performance rating	Output (barrels/day)	Cost ($/barrel)
1	90	6000	4.00
2	100	5000	5.00
3	105	4000	7.00

The performance ratings for the regular and premium gasolines are required to be at least 95 and 102, respectively. The selling prices are $10 and $12 per barrel, respectively. Determine what type of blend of gasolines 1, 2, and 3 to use such that daily profits are maximized.

S.P. 8.5: A whiskey blender has three types of spirits available: 8,000 gal of type A, 12,000 gal of type B, and 30,000 gal of type C. From these, the company will blend three grades of whiskey, with compositions and profits as follows:

Whiskey grade	Percentage of spirits			Profit ($/gal)
	A	B	C	
1	80	20	0	8
2	30	40	30	5
3	0	30	70	2

Determine the best way to blend the three types of spirits so as to maximize profits.

S.P. 8.6: An investor gets advice from a stock-brokerage firm on various investments in terms of expected annual yield, risk factor, and average term of investment in years. The data are given in the following table:

Investment	Expected annual yield (%)	Risk factor (fraction)	Average term of investment (years)
Bonds (1)	6.00	0.06	10
Savings and loan deposits (2)	5.25	0.02	4
Preferred stock (3)	6.50	0.13	5
Growth stock (4)	14.00	0.50	2

Let x_i denote the fraction invested in investment type i. The investor wishes to distribute his investments so that the weighted average risk factor is at most 0.25 and the weighted average investment period is at least 5 years. How should he distribute his investments in order to maximize the overall expected annual yield?

S.P. 8.7: At a certain hospital, the day is divided into four periods of 6 hr each. The minimum number of doctors required during each period (labeled 1 to 4) is as follows:

Time period	Minimum number of doctors
1. Midnight to 6 AM	9
2. 6 AM to Noon	11
3. Noon to 6 PM	8
4. 6 PM to midnight	6

Doctors report for work at midnight, 6 AM, noon, and 6 PM; each works a 12-hr shift. (a) How should the doctors be scheduled to work so as to minimize the total number of doctors required in a day? (b) Suppose that doctors reporting to work at 6 PM and at midnight get $15 per hour; those starting at 6 AM, $10 per hour; and those starting at noon, $12 per hour. How should the doctors be scheduled so as to minimize the total salary paid per day?

S.P. 8.8: A manufacturer of a product is planning the production for October, November, and December. The demands for the product are exactly 80 units in October, 100 in November, and 110 in December. The production facilities are capable of producing at most 100 units in any one month. The unit production cost is $15 in October, $17 in November, and $18 in December. The storage cost is zero during the production month and $3 per item in any following month. The company has no items in storage on October 1 and wishes to have no items in storage on December 31. Determine the least expensive way to schedule production and delivery.

S.P. 8.9: A shipper is asked to ship three kinds of packaged goods in the company truck and is supplied with 30 packages of type 1 goods, 35 packages of type 2 goods, and 35 packages of type 3 goods. The following table gives the weights, volumes, and shipping rates per package:

Type of goods	Weight/package	Volume/package	Rate/package
1	150 lb	60 cu ft	$70
2	200 lb	40 cu ft	$60
3	220 lb	80 cu ft	$80

The truck cannot carry more than 12,000 lb or 2,500 cu ft of goods. How many packages of each type should the shipper put on the truck so as to maximize the total shipping rate?

S.P. 8.10: Paper rolls are produced by a paper mill in standard rolls of width 13 ft. The mill has orders for 150 rolls of width 3 ft, 200 rolls of width 4 ft, and 100 rolls of width 5 ft. Determine all *efficient* cutting patterns (see Problem 8.10) and then determine the minimum number of standard rolls to be cut so as to satisfy the orders.

S.P. 8.11:   In a certain country, the Department of Labor has made the following preliminary estimates: Total labor force = 50 million, employed = 47 million, and unemployed = 5 million. Since the first number should equal the sum of the next two, these estimates are in error. Set up a linear programming problem to determine new estimates of the above three numbers (T, E, and U) such that $T = E + U$ and such that the maximum of the absolute values of the differences between new and old estimates is minimized. (Recall that the "absolute value" of x, denoted by $|x|$, refers to the value of a number x without regard to sign; thus, $|x| = x$ if $x \geq 0$ and $|x| = -x$ if $x < 0$.) *Hints:* Let M be the maximum of $|T - 50|$, $|E - 47|$, and $|U - 5|$. Thus, for T we have $|T - 50| \leq M$, which is equivalent to $T - 50 \leq M$ and $-M \leq T - 50$.

ANSWERS TO SUPPLEMENTARY PROBLEMS

S.P. 8.1:   See Problem 8.1; Max u = $871.43, $x_1 = 0$, $x_2 = 100$, and $x_3 = 28.57$.

S.P. 8.2:   See Problem 8.2; Max u = $4933.33, $x_1 = 0$, $x_2 = 20$, and $x_3 = 6.67$.

S.P. 8.3:   Let x_1 denote small boxes and x_3 large boxes: Max u = $75, $x_1 = 25$, $x_2 = 0$, and $x_3 = 25$.

S.P. 8.4:   See Problem 8.3; let R_i denote the amount of raw gasoline i to use in regular gasoline and P_i be the corresponding amount for premium gasoline: Max u = $83,000, $R_1 = 5000$, $R_2 = 5000$, $R_3 = 0$, $P_1 = 1000$, $P_2 = 0$, and $P_3 = 4000$.

S.P. 8.5:   See Problem 8.4; Max u = $146,666.67, $x_1 = 10,000$, $x_2 = 0$, and $x_3 = 33,333.33$.

S.P. 8.6:   See Problem 8.3 for the weighted average concept. The constraints are $x_1 + x_2 + x_3 + x_4 = 1$, $0.06x_1 + 0.02x_2 + 0.13x_3 + 0.5x_4 \leq 0.25$, and $10x_1 + 4x_2 + 5x_3 + 2x_4 \geq 5$. The goal is to maximize overall expected annual yield, $u = 0.06x_1 + 0.0525x_2 + 0.065x_3 + 0.14x_4$; Max u = 0.0945 (9.45%), when $x_1 = 0.568$, $x_4 = 0.432$, and $x_2 = x_3 = 0$.

S.P. 8.7:   See Problem 8.7; (a) Min w = 17: $y_1 = 9$, $y_2 = 2$, $y_3 = 6$, and $y_4 = 0$ *or* $y_1 = 3$, $y_2 = 8$, $y_3 = 0$, and $y_4 = 6$; (b) Min w' = $2580 (cost per day); $y_1 = 3$, $y_2 = 8$, $y_3 = 0$, and $y_4 = 6$.

S.P. 8.8:   See Problem 8.8. Let i = 1, 2, 3 refer to October, November, and December, respectively. Thus, x_{ij} refers to items produced in month i and delivered in month j: Min w = $4910: $x_{11} = 80$, $x_{12} = 0$, $x_{13} = 10$, $x_{22} = 100$, $x_{23} = 0$, and $x_{33} = 100$ *or* $x_{11} = 80$, $x_{12} = 10$, $x_{13} = 0$, $x_{22} = 90$, $x_{23} = 10$, and $x_{33} = 100$.

S.P. 8.9:   Let x_i denote number of packages of goods i. Our goal is to maximize $u = 70x_1 + 60x_2 + 80x_3$ subject to $x_1 \leq 30$, $x_2 \leq 35$, $x_3 \leq 35$; $150x_1 + 200x_2 + 220x_3 \leq 12,000$, $60x_1 + 40x_2 + 80x_3 \leq 2,500$; and x_1, x_2, $x_3 \geq 0$. The answer: Max u = $3,383.33, $x_1 = 18.33$, $x_2 = 35$, and $x_3 = 0$.

S.P. 8.10:   See Problem 8.10; the eight efficient cutting patterns are as follows:

Width	P_1	P_2	P_3	P_4	P_5	P_6	P_7	P_8
3 ft	4	3	2	1	1	1	0	0
4 ft	0	1	0	2	0	1	2	3
5 ft	0	0	1	0	2	1	1	0
Trim loss (ft)	1	0	2	2	0	1	0	1

Let y_i represent the number of rolls cut up using cutting pattern P_i. The answer: Min $w = 134.62$, $y_2 = 46.15$, $y_5 = 11.54$, and $y_7 = 76.92$.

S.P. 8.11: Let $M = \text{Max}\{|T - 50|, |E - 47|, |U - 5|\}$. Thus, $|T - 50| \le M$, which is equivalent to $T - 50 \le M$ and $-M \le T - 50$. Likewise, $|E - 47| \le M$ and $|U - 5| \le M$, with appropriate equivalent linear statements. The goal is to minimize M subject to the six linear constraints indicated above and the constraint $T = E + U$. The answer: Min $M = 0.667$, $T = 50.667$, $E = 46.333$, and $U = 4.333$.

Chapter Nine

Transportation and Assignment Problems

The balanced transportation problem has already been considered in Section 4 of Chapter 5, where the general problem involving m' warehouses and n' markets was stated (Problem 5.7). In addition, a simple and specific transportation problem was solved (Problem 5.6). The very laborious solution of the problem by the traditional simplex approach involved Stages Zero, One, and Two.

In this chapter, we shall develop more efficient computational procedures for handling transportation problems and the related assignment problems. In addition, some typical applications will be considered.

1. THE BALANCED TRANSPORTATION PROBLEM

First of all, it is appropriate to redefine some of the terms and symbols previously encountered. The symbols for the number of warehouses and markets will be changed from m' and n' to m and n, respectively. The following table provides a comparison between the former terminology and the new terminology that will be used throughout much of this chapter:

Former terminology	New terminology
Feasible point	Feasible solution
Basic point	Basic solution
Basic feasible point	Basic feasible solution
(Optimal) solution point	Optimal solution

The new terminology is widely used in discussions of transportation problems. In addition, it is fairly common in discussions of traditional linear programming situations (e.g., the simplex algorithm). The main distinction involves the use of the word solution in place of point.

The balanced transportation problem is restated as follows (the symbols have already been defined in Section 4 of Chapter 5; here we shall use w to denote the total shipping cost):

Minimize $\quad w = c_{11}x_{11} + c_{12}x_{12} + \cdots + c_{mn}x_{mn}$ $\hspace{3cm}$ (1)

subject to $\quad x_{i1} + x_{i2} + \cdots + x_{in} = a_i \quad$ for $\quad i = 1, 2, \ldots, m$ $\hspace{1cm}$ (2)

$\hspace{2.3cm} x_{1j} + x_{2j} + \cdots + x_{mj} = b_j \quad$ for $\quad j = 1, 2, \ldots, n$ $\hspace{1cm}$ (3)

and

$$x_{ij} \geq 0 \quad \text{for all} \quad i \text{ and } j. \hspace{3cm} (4)$$

In addition, the balance of supplies (a_i) and demands (b_j) is given by

$$a_1 + a_2 + \cdots + a_m = b_1 + b_2 + \cdots + b_n. \hspace{2cm} (5)$$

The table in Problem 5.7 clearly indicates the relationships between the various symbols.

We now introduce a model problem, so that the later calculational approaches will have more meaning. This problem appeared previously as Supplementary Problem 5.9.

Problem 9.1

The Johnson Typewriter Company has warehouses in Cleveland, Pittsburgh, and Baltimore and markets in New York, Detroit, and Chicago. At a particular time, the company has 61 typewriters in Cleveland, 49 in Pittsburgh, and 90 in Baltimore. The company plans to ship 52 typewriters to New York, 68 to Detroit, and 80 to Chicago. The transportation costs per unit (e.g., in dollars) as well as the above data are given in the following table, in which the warehouse and market locations have been relabeled with numbers:

	M_1	M_2	M_3	
Cleveland—W_1	26	23	10	61
Pittsburgh—W_2	14	13	21	49 Supplies
Baltimore—W_3	16	17	29	90 (typewriters)
	52	68	80	

Demands (typewriters)

M_1 is New York, M_2 is Detroit, and M_3 is Chicago

Thus, from this table we see, e.g., that the cost per typewriter when shipping from Cleveland (W_1) to New York (M_1) is \$26.

Interpret the various symbols that appear in (2) through (5) with respect to the current data and verify that the current problem is balanced.

Solution

From the data table given in the problem statement, it is easy to see that $m = 3$, $a_1 = 61$, $a_2 = 49$, and $a_3 = 90$. In addition, we have $n = 3$, $b_1 = 52$, $b_2 = 68$, and $b_3 = 80$. The problem is balanced, since $a_1 + a_2 + a_3 = b_1 + b_2 + b_3 = 200$. Typical cost symbols are as follows: $c_{11} = 26$, $c_{12} = 23$, $c_{13} = 10$, . . . , $c_{32} = 17$, $c_{33} = 29$.

We shall now develop methods for solving the Problem 9.1 (for the optimal x_{ij}'s). In the first stage, using the *Minimum Entry Method* [3], we shall attempt to locate a basic feasible solution. This process is the counterpart of Stage One (possibly preceded by Stage Zero) of the simplex algorithm. Once a basic feasible solution (abbreviated b.f.s.) is determined, a second stage—analogous to Stage Two of the simplex algorithm—is employed to determine an optimal solution. The overall algorithm is known as the *Stepping Stone Method* (Algorithm), and it occurs as Method 9.4 in this chapter.

Any set of $m \cdot n$ variables x_{ij} that satisfies (2), (3), and (4) is said to be a *feasible solution* of the transportation problem. A *basic feasible solution* (b.f.s.) is a feasible solution that has at most $m + n - 1$ positive x_{ij}'s. (This corresponds to the result of Problem 5.8—recall the different symbolism, however.) Equivalently, the balanced transportation problem has $m + n - 1$ basic variables. The cells pertaining to a transportation problem will be denoted by (i, j), where i refers to the row (warehouse) and j to the column (market). Thus, cell $(2,3)$ refers to the cell in row 2 and column 3.

We shall now state the Minimum Entry Method for finding an initial b.f.s. In this approach, a positive shipment (i.e., a positive x_{ij}) will be indicated by encircling the c_{ij} and putting the amount for x_{ij} above and to the right of the circle.

Method 9.1: Minimum Entry Method

1. Find the cell (p,q) such that c_{pq} is the smallest cost in the unchecked part of the data table.
2. Ship as much as possible by cell (p,q). Thus, let x_{pq} be the smaller of a_p and b_q. Then adjust a_p and b_q to account for the amount x_{pq} to be shipped. Expressing this by means of equations, we have

    Let $x_{pq} = $ minimum of a_p and b_q.

    Replace a_p by $a_p - x_{pq}$.

    Replace b_q by $b_q - x_{pq}$.

    Encircle the cost c_{pq} in cell (p,q). At the end of this step, either a_p or b_q (or both) will equal zero.
3. (a) If $a_p = 0$ and $b_q > 0$ after step 2, check row p on the left. This means that warehouse p has been emptied, but the demand at market q has not yet been satisfied.

(b) If $a_p > 0$ and $b_q = 0$ after step 2, check column q on the top. This means that market q has been satisfied, but warehouse p has not yet been emptied.

(c) If both $a_p = 0$ and $b_q = 0$ after step 2 (degenerate case), check row p *unless* it is the only unchecked row remaining, in which case check column q.

4. (a) If there is a total of two or more unchecked rows *and* columns, return to step 1.

(b) If there is one unchecked row and no unchecked column, halt. The basic feasible solution can be determined from the encircled cells.

The following theorem is proved in [3].

Theorem 9.1. *The Minimum Entry Method produces a basic feasible solution for any transportation problem.*

Let us illustrate Method 9.1 by applying it to Problem 9.1.

Problem 9.2

Use Method 9.1 to find an initial b.f.s. (i.e. basic feasible solution) for Problem 9.1.

Solution

First, refer to step 1 of Method 9.1. We notice that initially the unchecked part of the data table comprises the *entire* data table.

The smallest cost of the data table of Problem 9.1 is the 10 in cell (1,3). Since $a_1 = 61$ and $b_3 = 80$, we set $x_{13} = 61$ according to step 2. Carrying out the rest of step 2, we see that the *new* $a_1 = 0$. Thus, in accord with step 3(a), we check row 1 with $\sqrt{1}$. (The numbers following checks indicate the sequence in which rows and columns have been checked—this is very useful when retracing the steps in a previously solved problem.) The resulting partial solution is given in Table 1 (T.1), which follows:

$$
\begin{array}{c|ccc|cl}
\sqrt{1} & 26 & 23 & \circled{10}\ ^{61} & \cancel{61}\ 0 & \\
         & 14 & 13 & 21 & 49 & \text{(T.1)} \\
         & 16 & 17 & 29 & 90 & \\
\hline
         & 52 & 68 & \cancel{80} & & \\
         &   &   & 19 & &
\end{array}
$$

Step 4 indicates a return to step 1, where we consider the unchecked portion of the data table, namely rows 2 and 3 (remember, row 1 has been checked). The smallest entry in this part is the 13 in cell (2,2). Since $a_2 = 49$ and $b_2 = 68$, we set $x_{22} = 49$; thus, row 2 is checked, and the new $a_2 = 0$, while the new $b_2 = 19$ (from $68 - 49$). The resulting table (Table 2) is as follows:

√1	26	23	⑩ ^{61}	0	
√2	14	⑬ ^{49}	21	4̶9̶ 0	(T.2)
	16	17	29	90	
	52	6̶8̶	19		
		19			

A return to step 1 is again indicated. The remainder of the table consists only of row 3, where $c_{31} = 16$ is the smallest cost. We set $x_{31} = 52$, and we have $a_3 = 90 - 52 = 38$ and $b_1 = 0$; thus, column 1 is checked. Table 3 results:

	√3				
√1	26	23	⑩ ^{61}	0	
√2	14	⑬ ^{49}	21	0	(T.3)
	⑯ ^{52}	17	29	9̶0̶ 38	
	5̶2̶	19	19		
	0				

The remaining unchecked part of the table consists of just cells (3,2) and (3,3). Thus, cell (3,2) is assigned the shipment, since it has the smaller cost. Thus, $x_{32} = 19$, $a_3 = 38 - 19 = 19$, $b_2 = 0$, and column 2 is checked. The next table is as follows:

	√3	√4			
√1	26	23	⑩ ^{61}	0	
√2	14	⑬ ^{49}	21	0	(T.4)
	⑯ ^{52}	⑰ ^{19}	29	3̶8̶ 19	
	0	1̶9̶	19		
		0			

The remaining unchecked part of the table consists only of cell (3,3). Since $a_3 = b_3 = 19$ (these *have to be* equal if no prior errors were committed), we set $x_{33} = 19$ and then set $a_3 = b_3 = 0$. Following step

3(c) (degenerate case), we check column 3 and arrive at Table 5:

$$
\begin{array}{c|ccc|c}
 & \sqrt{3} & \sqrt{4} & \sqrt{5} & \\
\hline
\sqrt{1} & 26 & 23 & \circled{10}\,^{61} & 0 \\
\sqrt{2} & 14 & \circled{13}\,^{49} & 21 & 0 \\
 & \circled{16}\,^{52} & \circled{17}\,^{19} & \circled{29}\,^{19} & \cancel{19}\ 0 \\
\hline
 & 0 & 0 & \cancel{19} & \\
\end{array}
$$

(T.5)

$$0$$

Step 4(b) of the algorithm indicates we should halt, since all the columns and all but one row have been checked.

We read off the initial b.f.s. from Table 5 as follows:

$x_{13} = 61$, $x_{22} = 49$, $x_{31} = 52$, $x_{32} = 19$, $x_{33} = 19$, and all other $x_{ij} = 0$.

Observe that constraints (2) and (3) are satisfied for all rows and columns. Since there are exactly five ($m + n - 1 = 5$) positive x_{ij}'s, the feasible solution above is also a b.f.s. To find the cost of the shipping pattern corresponding to this b.f.s., we multiply each shipment by the corresponding unit cost. Thus,

$$\text{Cost} = w = 10\cdot 61 + 13\cdot 49 + 16\cdot 52 + 17\cdot 19 + 29\cdot 19 = \$2953.$$

This is not the minimum shipping cost; that will be determined later by the Stepping Stone Method.

Another method that produces an initial b.f.s. from the starting data table (or matrix) is the *Northwest Corner Rule,* which is discussed, e.g., in [2,3,4,6]. This algorithm is given as follows:

Method 9.2: Northwest Corner Rule

1. Locate the cell (p,q) that is in the northwest (i.e., upper left hand) corner of the unchecked part of the data matrix. The first time around, this would be cell $(1,1)$.

Steps 2, 3, and 4 are identical to those of Method 9.1.

Problem 9.3

Use Method 9.2 to find an initial b.f.s. for Problem 9.1.

Solution

It is possible to trace through all the operations involved in using Method 9.2 (or Method 9.1, for that matter) in one table diagram. The relevant table is given below; the numbered checks and the successive a_i and b_j values make the task of determining the correct sequence of operations fairly straightforward:

	$\sqrt{1}$	$\sqrt{4}$	$\sqrt{5}$	
$\sqrt{2}$	㉖ 52	㉓ 9	10	6̸1̸ 9̸ 0
$\sqrt{3}$	14	⑬ 49	21	4̸9̸ 0
	16	⑰ 10	㉙ 80	9̸0̸ 8̸0̸ 0
	5̸2̸	6̸8̸	8̸0̸	
	0	5̸9̸	0	
		1̸0̸		
		0		

Since $52 < 61$, we first ship 52 by cell $(1,1)$; we check column 1, since b_1 has been reduced to 0. In addition, a_1 is reduced to 9. The remaining (unchecked) matrix consists of columns 2 and 3, for which cell $(1,2)$ is the northwest corner. Since $9 < 68$, we ship 9 by cell $(1,2)$. Thus, a_1 is reduced to zero, row 1 is checked, and b_2 is reduced to 59. The unchecked matrix now consists of those cells that are neither in row 1 nor column 1. The northwest corner cell is $(2,2)$, by which 49 is shipped; row 2 is checked ($\sqrt{3}$), a_2 is reduced to zero, and b_2 is reduced to 10. The unchecked matrix now contains cells $(3,2)$ and $(3,3)$, of which the former is the northwest corner. Thus, 10 is shipped via cell $(3,2)$, b_2 is reduced to zero, column 2 is checked, and a_2 is reduced to 80. Finally, the remaining 80 items are shipped by cell $(3,3)$; following step 3(c), we check column 3. Since there remains only unchecked row 3, we halt according to step 4(b).

We read off the initial b.f.s. from the circled entries in the above table:

$x_{11} = 52$, $x_{12} = 9$, $x_{22} = 49$, $x_{32} = 10$, $x_{33} = 80$, and all other $x_{ij} = 0$.

The cost associated with this b.f.s. is computed as follows:

$$w = 26{\cdot}52 + 23{\cdot}9 + \cdots + 29{\cdot}80 = \$4686.$$

Notes:
a. The expression to ship K units by cell (p,q) signifies the shipment of K units from warehouse p to market q. In other words, $x_{pq} = K$.
b. The Northwest Corner Rule is (usually) a faster way to obtain a starting b.f.s. than the Minimum Entry Method. However, the latter method usually yields a b.f.s. with a lower shipping cost; this was the case here, as a comparison of the w values in Problems 9.2 and 9.3 indicates. The advantage of a lower shipping cost is that fewer computations are then required (usually) to reach the optimal solution by the Stepping Stone Algorithm.

c. Another method for obtaining an initial b.f.s. is the *Vogel Advanced Start Method (VAM)*; this is discussed in [2]. In the current book, Method 9.1 will be used most often.

The set B of $m + n - 1$ cells corresponding to the possible nonnegative shipments $(x_{ij}$'s$)$ of any b.f.s. will be called the *basis* corresponding to the b.f.s. Both Method 9.1 and Method 9.2 determine a basis in a very simple way; it is merely the set of those cells which are encircled at the end of applying the method. Thus, for Problem 9.2 the basis B is

$$B = \{(1,3),\ (2,2),\ (3,1),\ (3,2),\ (3,3)\}.$$

In Problem 9.3, the basis is

$$B = \{(1,1),\ (1,2),\ (2,2),\ (3,2),\ (3,3)\}.$$

In general, each row and each column has at least one basis cell in it. The symbol $\{\ \}$ is from standard set theory notation. Thus, $\{e, f\}$ refers to the set containing items e and f. In conventional linear programming language, the basis set refers to those x_{ij}'s that are the basic variables in a corresponding tableau format. Thus, in Problem 9.2, the basic variables are x_{13}, x_{22}, x_{31}, x_{32}, and x_{33}; the remaining four x_{ij}'s are nonbasic.

Let us return to the b.f.s. of Problem 9.2 as depicted, e.g., by Table 5 of that problem. We would like to determine if it is possible to obtain an improved b.f.s., i.e., a b.f.s. that has a lower shipping cost w. Let us consider a process that is analogous to Stage Two of the simplex algorithm. That is, suppose that we consider adding a particular cell to the basis and also removing a cell from the basis, thereby forming a new basis. For example, suppose that we try to use cell (2,1), which has a cost 14. Let us attempt to ship an amount s by this route (s items from warehouse 2 to market 1). This is indicated in the following table by putting a square box around the 14 and an s above it. In order that the shipments $(x_{ij}$'s$)$ add up to the proper a_i's and b_j's in the rows and columns, we must *decrease* the amount shipped by the basis cells that are in *the same row and column* as cell (2,1). Thus, we subtract s from the amounts shipped by basis cells (2,2) and (3,1); this is indicated in the table by $49 - s$ and $52 - s$.

26	23	⑩ ^{61}
⟦14⟧ ^{s}	⑬ $^{49-s}$	21
⑯ $^{52-s}$	⑰ $^{19+s}$	㉙ ^{19}

**Table for introducing (2,1) into the
shipping pattern.**

Now, to adjust for these changes in row 3 and column 2, we have to add s to the amount shipped by basis cell $(3,2)$. This is indicated by placing $19 + s$ next to cell $(3,2)$. Note that $a_2 = 49$, $a_3 = 90$, $b_1 = 52$, and $b_2 = 68$ regardless of what s is. Clearly, s is required to be nonnegative. To determine an upper bound for s, we note that $49 - s$ and $52 - s$ must both be nonnegative. (The two preceding statements pertain to the $x_{ij} \geq 0$ requirement of the standard transportation problem.) Thus, the possible range for s is $0 \leq s \leq 49$. If s exceeded 49, then $49 - s$ would be negative.

Problem 9.4

For s in the allowable range $0 \leq s \leq 49$, determine the change in shipping cost. Discuss the effect of such a change and a plausible choice for s.

Solution

The change is from the value calculated in Problem 9.2, namely, \$2953. Shipping s units via route $(2,1)$ will incur a cost of $14s$ dollars, since 14 is the cost per unit. Decreasing the shipment by s units along route $(2,2)$ will cause a decrease in cost of $13s$ dollars. Using the same reasoning for routes $(3,1)$ and $(3,2)$, the net change in shipping cost is

$$14s - 13s + 17s - 16s = (14 - 13 + 17 - 16)s = 2s.$$

Thus, the shipping cost will be *increased* if s is positive. Because the goal is to minimize w, s is chosen to be zero. In other words, cell $(2,1)$ should *not be used* because using it would only lead to a solution that is worse than the one already available. By worse, we mean a solution of higher w (recall that we wish to minimize the cost w).

Problem 9.5

Refer to the basis as given in Table 5 of Problem 9.2. Discuss the effects of attempting to introduce cell $(2,3)$ into the shipping pattern. Indicate the best possible shipment s to be associated with cell $(2,3)$. Repeat the analysis for the other cells not in the basis, namely, cells $(1,1)$ and $(1,2)$.

Solution

If we try to introduce cell $(2,3)$, the resulting shipping changes are as shown in Table 1:

26	23	⑩ 61
14	⑬ 49−s	21 s
⑯ 52	⑰ 19+s	㉙ 19−s

(T.1)

**Table for Introducing Cell (2,3) into
the Shipping Pattern.**

Here the net change in shipping cost associated with a shipment of s units via cell (2,3) is

$$21s - 13s + 17s - 29s = (21 - 13 + 17 - 29)s = -4s.$$

Thus, it is plausible to make s as positive as possible. Since the expressions $49 - s$ and $19 - s$ must both be nonnegative, s is confined to the interval $0 \le s \le 19$ (making s larger than 19 causes $19 - s$ to be negative). Thus, the most desirable value of s, subject to this restriction, is $s = 19$.

If we repeat the analysis for cell (1,1), we obtain the following table:

$\boxed{26}$ ^{s}	23	⑩ $^{61\ -\ s}$
14	⑬ ^{49}	21
⑯ $^{52-s}$	⑰ ^{19}	㉙ $^{19+s}$

(T.2)

**Table for Introducing Cell (1,1) into
the Shipping Pattern.**

Since the expressions $52 - s$ and $61 - s$ must both be nonnegative, s has to be in the interval $0 \le s \le 52$. The net change in shipping cost associated with the shipment of s units by cell (1,1) is

$$(26 - 10 + 29 - 16)s = 29s.$$

This means that the shipping cost will be *increased* if s is positive. Since this runs counter to our goal, the best choice for s in this case is zero, i.e., cell (1,1) should not be incorporated into the b.f.s. given in Table 5 of Problem 9.2.

If we attempt to incorporate cell (1,2) into the shipping pattern, the basis cells that become involved are (1,3), (3,2), and (3,3). To compensate for s units shipped by cell (1,2), there has to be a decrease of s units via cells (1,3) and (3,2) and an increase in s units via cell (3,3). Corresponding to all this, there would be a net *increase* in shipping cost of

$$(23 - 10 + 29 - 17)s = 25s.$$

Thus, the best choice for s is also zero in this case; in other words, cell (1,2) should not be used.

From the preceding two problems, it is clear that it would pay to introduce cell (2,3) into the shipping pattern to the greatest extent possible. It would not be wise, however, to introduce any of the other nonbasis cells into the shipping pattern.

Problem 9.6

Introduce cell (2,3) into the shipping pattern of Problem 9.2 to the greatest extent possible. Determine the new basis that results, calculate the new shipping cost, and determine whether the b.f.s. corresponding to the new basis is the optimal solution.

Solution

Referring to the first part of Problem 9.5, it is clear that the best choice for s is $s = 19$. Thus, the amount shipped by cell (3,3) becomes zero, and cell (3,3) is *removed* from the basis. The new table corresponding to the choice $s = 19$ in Table 1 of Problem 9.5 is given as follows:

26	23	⑩ 61	
14	⑬ 30	㉑ 19	(T.1)
⑯ 52	⑰ 38	29	

A New Basis for Problem 9.2.

This new basis consists of cells (1,3), (2,2), (2,3), (3,1), and (3,2). The new b.f.s. corresponding to this basis is as follows:

$x_{13} = 61$, $x_{22} = 30$, $x_{23} = 19$, $x_{31} = 52$, $x_{32} = 38$, and all other $x_{ij} = 0$.

The cost associated with this basis can be computed directly in the following way:

$$w = 10 \cdot 61 + 13 \cdot 30 + \cdots + 17 \cdot 38 = \$2877.$$

It can be computed with less difficulty by using the result from Problem 9.5 that the *change* in shipping cost associated with shipping s units by cell (2,3) is $-4s$. Thus, the new shipping cost is given by $\$2953 - 4 \cdot 19 = \$2877$, since the former cost was $\$2953$ and $s = 19$ here.

The question now is whether or not there is still another basis that is better than the one just found, i.e., better in the sense that it has a lower shipping cost associated with it. If we attempt to introduce any of the cells (1,1), (1,2), (2,1), and (3,3) into the basis, an *increase* in shipping cost will result. For example, for cell (1,2), there will be an increase of the form $21s$. The calculation is done in the same way as in Problem 9.5; thus, $(23 - 10 + 21 - 13)s = 21s$. In other words, for each of the four cells just cited, the change in shipping cost will be of the form $v \cdot s$ in each case, where v is positive. The respective v values, which are called *cycle values,* are given in parentheses in the following table:

26(25)	23(21)	⑩ 61	
14(2)	⑬ 30	㉑ 19	(T.2)
⑯ 52	⑰ 38	29(4)	

**Cycle Values Associated with the
New Basis.**

There is a theorem which indicates that when all such values associated with nonbasis cells are nonnegative, then the existing basis yields the optimal solution. (A formal statement of this theorem will be given shortly.) Thus, the optimal solution to Problem 9.1 is the b.f.s. given right after Table 1, and Min w = \$2877. In other words, the optimal shipping plan is to ship 61 typewriters from Cleveland (W_1) to Chicago (M_3), 30 from Pittsburgh (W_2) to Detroit (M_2), 19 from Pittsburgh to Chicago, 52 from Baltimore (W_3) to New York (M_1), and 38 from Baltimore to Detroit.

In Problem 9.5, when we tried to introduce cell (2,3) into the basis, we found that a *cycle* was formed by it and the cells in the current basis to which we added $\pm s$. A similar statement applies to the cell (1,1). Denoting a cycle by an uppercase C, we have

$$C(2,3) = \{(2,3), (3,3), (3,2), (2,2)\}$$

and

$$C(1,1) = \{(1,1), (1,3), (3,3), (3,1)\},$$

i.e., the cycle $C(2,3)$—the cycle formed by cell (2,3)—is the *set* that contains the cells (2,3), (3,3), (3,2), and (2,2).

It is instructive to reconsider how a cycle is formed. Referring back to Table 1 of Problem 9.5, the goal there was to form a cycle corresponding to the cell (2,3) in order that the shipping plan might be changed. First, we can move in the same column (column 3) until a basis cell is reached. Thus, cell (3,3) is in the cycle. Then we move in the row of cell (3,3) until another basis cell is reached; there are two possibilities here, namely, (3,2) and (3,1). The latter is not fruitful, however, since it is the only basis cell in its column. Thus, (3,2) is in the cycle. Then we move in the column of (3,2) until basis cell (2,2) is reached. Finally, we move in the row of (2.2) until we reach either another basis cell or the entering cell. In this case, we encounter the entering cell (2,3). The cycle $C(2,3)$ is thus completely determined.

Note: We could also first move in row 2, rather than in column 3. In this case, we would put cell (2,2) in the cycle after putting the entering cell (2,3) in. Then cells (3,2) and (3,3) would be put in, to complete the cycle. The same cycle as above is thus obtained.

The reader is warned that it is sometimes not so easy to determine a cycle corresponding to a nonbasis cell. It is hoped that further commentary and more examples will help to resolve any difficulties associated with determining cycles.

We now present the formal definition of a cycle.

Definition 9.1. A cycle C is a subset of cells of the data matrix with the property that each row and each column of the original data matrix contains either zero or two cells of C.

The cycles that we are interested in are most often of a special type. One of the cells is *not* in the current basis, while the remaining cycle cells *are* in the current basis. The cell (p,q) not in the basis is said to be the one that "forms" or "generates" the cycle $C(p,q)$.

Other examples of cycles are shown in Figure 9.1. In the problems covered in this chapter, most of the cycles that arise are not too difficult to find. The following theorems indicate some of the key properties of cycles as they pertain to the work here. The proofs are given in the *Solutions Manual* of [3].

Theorem 9.2. *Given a current basis B and any cell (p,q) that is not in the current basis. There is a unique cycle formed by (p,q) and some of the cells of the current basis.*

Theorem 9.3. *The basis B of a b.f.s., which is a set of $m + n - 1$ cells, contains no cycles among its cells.*

FIGURE 9.1 Examples of cycles in data matrices (tables) of transportation problems. *Note:* the □ denotes the cell not in the current basis while the ○ refers to cells in the current basis.

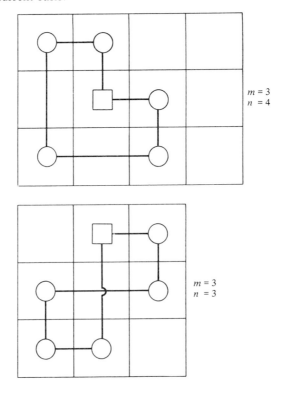

Notes:
a. As an example of Theorem 9.2, the cycle $C(2,3)$ cited just after Problem 9.6 is the *only* cycle that can be formed by cell $(2,3)$ and the cells of the basis of Problem 9.2.
b. For problems in which m and n are large, it is often not an easy matter to determine a cycle. There is a method, known as the *Crossing Out Algorithm* (described in [3]), that will always find the cycle. This fact is also used as the basis for proving Theorem 9.2.

In forming a new basis in Problem 9.6, an outside cell (in Problem 9.6, cell $(2,3)$) was added to the old basis. The new basis was formed after another cell of the cycle $C(2,3)$ was discarded from the old basis. The discarding was done in such a way that the shipment x_{23} could be made as large as possible but still not cause any other x_{ij} to become negative. The reason that $(2,3)$ was introduced is because the "value" associated with its cycle was negative; therefore, introducing $(2,3)$ into the basis would cause the desired effect of reducing the overall shipping cost w.

To make the process of finding a better (i.e., lower cost) b.f.s. from a current b.f.s. more systematic, let us develop some additional terminology and symbolism.

Definition 9.2: Value of a Cycle. Let $C(p,q)$ be the unique cycle formed when cell (p,q) is combined with cells of the basis B. Starting with cell (p,q) and going around the cycle (in either direction), we alternately label cells "getter" and "giver" cells, where (p,q) is a getter cell. The value of the cycle, denoted by v_{pq}, is the sum of the costs associated with the cells in $C(p,q)$, where the costs of the getter cells are given plus signs and the costs of the giver cells are given minus signs.

Problem 9.7

Refer back to Problems 9.4 and 9.5. Determine the values of the cycles formed in those problems.

Solution

The cycle associated with the cell $(2,1)$ consists of $(2,1)$, $(2,2)$, $(3,2)$, and $(3,1)$; they are alternately listed here as getter, giver, getter, giver. Thus, the value of the cycle associated with cell $(2,1)$ is

$$v_{21} = c_{21} - c_{22} + c_{32} - c_{31} = 14 - 13 + 17 - 16 = +2.$$

In a similar way, we see that

$$v_{23} = c_{23} - c_{33} + c_{32} - c_{22} = 21 - 29 + 17 - 13 = -4.$$

In addition,

$$v_{11} = c_{11} - c_{13} + c_{33} - c_{31} = 26 - 10 + 29 - 16 = 29$$

and

$$v_{12} = c_{12} - c_{13} + c_{33} - c_{32} = 23 - 10 + 29 - 17 = 25.$$

Problem 9.8

Determine the values of the cycles that appear in Figure 9.1.

Solution

For the top diagram, cell $(2,2)$ is the one introduced into the basis and hence is a getter cell. Thus,

$$v_{22} = c_{22} - c_{23} + c_{33} - c_{31} + c_{11} - c_{12}.$$

In the bottom diagram, cell $(1,2)$ is introduced into the basis and hence is a getter cell. Thus,

$$v_{12} = c_{12} - c_{13} + c_{23} - c_{21} + c_{31} - c_{32}.$$

The process of determining the value of a cycle is also known as *cell evaluation,* where the cell referred to is the one introduced into the basis.

In Problems 9.4 and 9.5 we found that if a cycle has a positive value, it does not pay to introduce the cell (generating the cycle) into the basis, since the total shipping cost will only increase. On the other hand, if a cycle has a negative value, it does pay to introduce the generating cell into the basis, since that will reduce the total shipping cost. This can be seen from the following:

$$\Delta w = x'_{pq} \cdot v_{pq}, \tag{1}$$

where Δw denotes the change in shipping cost associated with introducing cell (p,q) into the basis. The term x'_{pq} denotes the shipment associated with cell route (p,q) *after* cell (p,q) is introduced into the basis. Since (usually) $x'_{pq} > 0$, it follows that Δw will be negative when v_{pq} is. Equation (1) is just a generalization of the following statement:

$$\text{``Change in shipping cost''} = s \cdot v. \tag{1a}$$

A statement of this type appeared, e.g., in Problems 9.4 and 9.5.

When considering which cells to introduce into an existing basis, those cells associated with negative cycle values will be considered as logical *candidates* to introduce. Some authors (e.g., [1] and [4]) choose the cell associated with the most negative cycle value as the one to enter the basis. Such a procedure is analogous to applying Method 7.2 to Stage Two of the simplex algorithm.

It is useful to indicate the changes in shipments and basis cells that occur when a *candidate* cell (p,q) is introduced into the basis. Here x_{ij} denotes a shipment before (p,q) is introduced, and x'_{ij} denotes a shipment afterward. Clearly, $x_{pq} = 0$, since there is no shipment via route (p,q) before cell (p,q) is introduced into the basis.

Method 9.3: Change in Basis and Shipping Pattern

The only shipments that are affected are those associated with cells in the *cycle* determined by *candidate* cell (p,q).

a. x'_{pq} = minimum x_{ij} of giver cells in cycle.
b. For getter cells, $x'_{ij} = x_{ij} + x'_{pq}$.
  For giver cells, $x'_{ij} = x_{ij} - x'_{pq}$.
c. The cell (p,q) is introduced into the basis, and a minimum giver cell (i.e., a giver cell with a minimum x_{ij}) is removed from the basis.

Problem 9.9

Illustrate the use of Method 9.3 with respect to Problem 9.6, i.e., determine the changes brought about by introducing cell $(2,3)$ into the shipping pattern.

Solution

The preliminary shipping pattern is that displayed by Table 5 of Problem 9.2, which is reproduced below as Table 1 (this shipping pattern is none other than the preliminary b.f.s. as determined by the Minimum Entry Method):

$$
\begin{array}{|ccc|}
\hline
26 & 23 & ⑩\ ^{61} \\
14 & ⑬\ ^{49} & 21 \\
⑯\ ^{52} & ⑰\ ^{19} & ㉙\ ^{19} \\
\hline
\end{array}
\qquad \text{(T.1)}
$$

Shipping Pattern Before Cell (2,3) is Introduced.

The cycle determined by cell $(2,3)$ is $(2,3)$, $(3,3)$, $(3,2)$, and $(2,2)$ in the order getter, giver, getter, giver.

Let us apply Method 9.3 to determine the new b.f.s. From part (a) we have

$$x'_{23} = \text{minimum of } x_{33} \text{ and } x_{22},$$

i.e.,

$$x'_{23} = \text{minimum of 19 and 49} = 19.$$

Part (b) indicates that

$$x'_{32} = 19 + 19 = 38,$$
$$x'_{33} = 19 - 19 = 0,$$

and

$$x'_{22} = 49 - 19 = 30.$$

Finally, part (c) indicates that cell (3,3) is removed from the basis. The new basis and shipping pattern (b.f.s.) are indicated in the following table:

26	23	⑩ 61
14	⑬ 30	㉑ 19
⑯ 52	⑰ 38	29

(T.2)

Shipping Pattern After Cell (2,3) is
Introduced.

The shipping patterns depicted in Table 2 of both Problem 9.6 and this problem are identical.

The following development pertains directly to the Stepping Stone Method of solving transportation problems. This algorithm, labeled as Method 9.4, will be given shortly.

Definition 9.3. A transportation problem is *nondegenerate* if for each b.f.s. we have $x_{ij} > 0$ for all $m + n - 1$ cells of the corresponding basis, B. (The basis B consists of those cells in the b.f.s. under consideration.)

Now, it can be shown (e.g., see [3]) that any degenerate transportation problem can be transformed into an equivalent nondegenerate problem. The following theorem provides the cornerstone of the Stepping Stone Method.

Theorem 9.4. *Suppose that a b.f.s. to a transportation problem is attained such that the values of the cycles (see Definition 9.2) associated with all cells not in the basis are nonnegative. Then the b.f.s. is an optimal solution.*

Problem 9.10

Illustrate Theorem 9.4 with respect to Problem 9.6.

Solution

In Table 2 of Problem 9.6, a b.f.s. (with associated basis) was developed for which all *values of cycles* associated with cells not in the basis were positive. The relevant four cycle values were indicated in parentheses in that table. Thus, Theorem 9.4 indicates that the b.f.s. is also an optimal solution of the problem.

Problem 9.11

Prove Theorem 9.4.

Solution

First of all, note that the following equation applies:

$$\Delta w = x'_{pq} \cdot v_{pq}. \tag{1}$$

Here x'_{pq} denotes the new shipment associated with cell route (p,q) when cell (p,q) is added to the basis. This (i.e., x'_{pq}) is obtained from Method 9.3, part (a). In addition, v_{pq} is the value of the cycle formed by (p,q)—see Definition 9.2. The term Δw denotes the change in total shipping cost when the external cell (p,q) is added to the basis.

Suppose that the hypothesis of the theorem holds. Then if we attempt to introduce any cell (p,q) not in the basis, the value of the cycle formed by (p,q) is nonnegative. Furthermore, from Method 9.3, part (a), x'_{pq} is nonnegative. Thus, the shipping cost w will either increase or stay the same. It can be shown that if two or more external cells are introduced simultaneously, then the effects are additive; thus, such a combination would also either increase or not change the shipping cost. It is now possible to go from the current b.f.s. (and corresponding basis) to any other b.f.s. by adding cells to the current basis and adjusting the amounts shipped accordingly. Therefore, there is no other b.f.s. that has lower shipping cost than the current one. In other words, the current b.f.s. is an optimal solution.

The converse of Theorem 9.4 is also true for the case of a nondegenerate transportation problem. We now present one way of stating the converse.

Theorem 9.4′. *Given a b.f.s. to a nondegenerate transportation problem. Suppose that there is a cell not in the basis, which determines a cycle with negative value. Then introducing such a cell into the basis will yield a new b.f.s. for which the shipping cost is lower. In other words, the current b.f.s. is not an optimal solution.*

Problem 9.12

Prove Theorem 9.4′.

Solution

Refer to Equation (1) of Problem 9.11. Suppose that we consider introducing a cell (p,q) with associated negative value into the basis. From Method 9.3, part (a) and the assumption of nondegeneracy, it follows that x'_{pq} is positive. Thus, from Equation (1) it follows that the change in shipping cost incurred when cell (p,q) is introduced into the basis is negative, i.e., the shipping cost associated with the new b.f.s. is lower.

The ideas and procedures that pertain to Method 9.3 and Theorems 9.4 and 9.4' form the foundation for the Stepping Stone Method of solving transportation problems.

Method 9.4: Stepping Stone Method

1. Set up the initial data table (matrix).
2. Use the Minimum Entry Method (Method 9.1) or some other method (e.g., Method 9.2) to obtain an initial b.f.s., with corresponding basis B.
3. Calculate the values of cycles corresponding to cells not in the current basis.
   (a) If all such values are nonnegative, stop. The current b.f.s. is an optimal solution.
   (b) If there is a cell (p,q) that has a negative value of cycle, then it is a candidate for entering a new basis. Go to step 4.
4. Introduce a candidate cell (p,q) into the basis and form a new basis and shipping pattern according to the rules of Method 9.3. Return to step 3.

An example of the use of Method 9.4 is available if we consider Problems 9.2 and 9.6. In the former, an initial b.f.s. to Problem 9.1 was found by using the Minimum Entry Method. Problem 9.6 corresponded to applying steps 3 and 4 and then 3 again. The cell (2,3) was introduced into the basis because the cycle value corresponding to it was negative. Then, in the new basis, it was discovered that all relevant cycle values were positive. Thus, the new basis corresponded to an optimal b.f.s.

Problem 9.13

Apply the Stepping Stone Method to Problem 9.1, but first use the Northwest Corner Rule to obtain an initial b.f.s.

Solution

The application of the Northwest Corner Rule has already been done in Problem 9.3. The initial b.f.s., as determined there, is presented again in the following table:

$$
\begin{array}{|ccc|}
\hline
\text{\textcircled{26}}^{\,52} & \text{\textcircled{23}}^{\,9} & 10 \\
14 & \text{\textcircled{13}}^{\,49} & 21 \\
16 & \text{\textcircled{17}}^{\,10} & \text{\textcircled{29}}^{\,80} \\
\hline
\end{array}
\qquad (\text{T.1})
$$

The cost associated with this b.f.s. is $w = \$4686$, as calculated in Problem 9.3. At this point, steps 1 and 2 of Method 9.4 have been carried out.

Turning to step 3, we evaluate cycle values corresponding to cells not in the basis of Table 1. For example, cell (1,3) generates the cycle containing (1,3), (3,3), (3,2), and (1,2). The value of this cycle is $10 - 29 + 17 - 23 = -25$. Repeating this type of calculation for the other nonbasis cells leads to Table 1', in which cycle values are indicated in parentheses following cell unit costs (in other respects, Table 1' is identical to Table 1):

$$
\begin{array}{ccc}
\text{\textcircled{26}}\ ^{52} & \text{\textcircled{23}}\ ^{9} & 10(-25) \\[2mm]
14(-2) & \text{\textcircled{13}}\ ^{49} & 21(-4) \\[2mm]
16(-4) & \text{\textcircled{17}}\ ^{10} & \text{\textcircled{29}}\ ^{80}
\end{array}
\qquad \text{(T.1')}
$$

We bring (1,3) into the basis. The smallest giver is (1,2), which can give 9. Thus, we remove (1,2) from the basis and form the remainder of the new shipping pattern by following Method 9.3. In other words, we alternately add and subtract 9 from the shipments around the cycle. We have thus performed step 4 of Method 9.4. The new cost can be computed by observing that $\Delta w = x'_{pq} \cdot v_{pq} = 9 \cdot (-25) = -225$. Thus, the new cost is given by $w_{\text{new}} = w_{\text{old}} + \Delta w = 4686 - 225 = \$4461$.

Returning to step 3, we compute cycle values corresponding to non-basis cells. The new shipments and cycle values (in parentheses) appear in Table 2:

$$
\begin{array}{ccc}
\text{\textcircled{26}}\ ^{52} & 23(25) & \text{\textcircled{10}}\ ^{9} \\[2mm]
14(-27) & \text{\textcircled{13}}\ ^{49} & 21(-4) \\[2mm]
16(-29) & \text{\textcircled{17}}\ ^{19} & \text{\textcircled{29}}\ ^{71}
\end{array}
\qquad \text{(T.2)}
$$

Note that an alternate calculation of the cost associated with Table 2 is $w_{\text{new}} = 26 \cdot 52 + 10 \cdot 9 + \cdots + 29 \cdot 71 = \$4461$, in agreement with the previous calculation.

The b.f.s. corresponding to Table 2 is not optimal, since some of the cycle values are negative. Note that the cycle corresponding to (2,1) is quite interesting. The cells it contains, in alternating getter–giver order, are (2,1), (2,2), (3,2), (3,3), (1,3), and (1,1). In other words, the cycle $C(2,1)$ contains all the cells of the current basis.

The cycle generated by cell (3,1), namely, $C(3,1)$, contains cells (3,1), (3,3), (1,3), and (1,1). Introducing (3,1) into the current basis causes the smallest giver cell—(1,1)—to leave. The new shipments and cycle values associated with the new basis appear in Table 3:

26(29)	23(25)	⑩ 61
14(2)	⑬ 49	21(−4)
⑯ 52	⑰ 19	㉙ 19

(T.3)

The only cycle that has negative value is the one generated by cell (2,3). Thus, it pays to bring cell (2,3) into the basis. The smallest giver in the cycle is (3,3), which can give 19. After forming a new basis and shipping pattern, we arrive at Table 4, in which the new cycle values are listed:

26(25)	23(21)	⑩ 61
14(2)	⑬ 30	㉑ 19
⑯ 52	⑰ 38	29(4)

(T.4)

The b.f.s. corresponding to Table 4 is optimal, since all the relevant cycle values are positive. The identical table was derived in Problem 9.6. In that problem, only one table was needed beyond the initial b.f.s. found by the Minimum Entry Method.

Each time we perform the loop steps 3 and 4 of Method 9.4 we make a change that is analogous to the pivoting operation of the simplex algorithm. This process of forming a new basis will be called *pivoting*.

We provide one further note on terminology: Henceforth, we shall sometimes refer to the supply (a_i) and demand (b_j) conditions collectively as *rim conditions*.

It is useful at this point to review. First we considered a model transportation problem, which was originally stated in Problem 9.1. We found initial basic feasible solutions by the Minimum Entry Method (Method 9.1) in Problem 9.2 and by the Northwest Corner Rule (Method 9.2) in Problem 9.3. Then we applied the Stepping Stone Method (Method 9.4) to find the optimal solution of our model problem (see Problems 9.6 and 9.13). The optimal solution is as follows: $x_{13} = 61$, $x_{22} = 30$, $x_{23} = 19$, $x_{31} = 52$, $x_{32} = 38$, and all other $x_{ij} = 0$; also, Min $w = \$2877$. The same solution was obtained in Problem 8.15 by employing the EASYLP computer package.

The initial model transportation problem—Problem 5.6—was solved by employing Stages Zero, One, and Two of the simplex algorithm. It is instructive to solve that problem using Method 9.4.

Problem 9.14

Use Method 9.4 to solve Problem 5.6.

Solution

The initial data table, together with rim conditions, is presented below:

$$
\begin{array}{|cc|l}
\hline
5 & 8 & a_1 = 15 \\
4 & 10 & a_2 = 5 \\
\hline
\end{array}
\qquad \text{(T.1)}
$$

$$b_1 = 12 \quad b_2 = 8$$

We will now perform step 2 of Method 9.4 by using the Minimum Entry Method. The calculations and the resulting table are indicated as follows:

$$
\begin{array}{c}
\sqrt{2} \qquad \sqrt{3} \\
\begin{array}{|cc|l}
\hline
\textcircled{5}^7 & \textcircled{8}^8 & 1\cancel{5}\ \cancel{8}\ 0 \\
\textcircled{4}^5 & 10 & \cancel{5}\ 0 \\
\hline
\end{array} \\
\sqrt{1} \\
\begin{array}{cc}
1\cancel{2} & \cancel{8} \\
7 & 0 \\
0 & \\
\end{array}
\end{array}
\qquad \text{(T.2)}
$$

The only relevant cycle is that corresponding to cell (2,2); the value of cycle is $10 - 4 + 5 - 8 = +3$. Thus, the b.f.s. pertaining to Table 2 is an optimal solution. The solution values for the x_{ij}'s are $x_{11} = 7$, $x_{12} = 8$, $x_{21} = 5$, and $x_{22} = 0$. These check out with the results of Problem 5.6. The amount of effort involved in finding the optimal solution is much smaller in the current approach.

The next problem indicates what happens when degeneracies occur.

Problem 9.15

An automobile company wishes to transport cars from three supply outlets to three different sales locations. The supply outlets have available 20, 48, and 12 cars, respectively. The demands at the three sales locations are 20, 48, and 12 cars, respectively. The costs per car to ship from supply to sales locations are given in the following table, where the monetary unit is $10, i.e., the 16 in row 1 and column 1 represents $160. The supply outlets are designated W_1, W_2, and W_3, while the sales locations are labeled M_1, M_2, and M_3. The rim conditions are given along the boundaries.

	M_1	M_2	M_3	
W_1	16	24	28	20
W_2	17	18	20	48
W_3	31	19	22	12
	20	48	12	

Initial Data Table for Problem 9.15

Find the optimal solution and minimum total shipping cost.

Solution

First we employ the Minimum Entry Method to find an initial b.f.s. The starting smallest entry is the 16 in cell (1,1); the maximum amount (20) is shipped by this route, thereby satisfying both W_1 and M_1. According to rule 3(c) of Method 9.1, row 1 is then checked ($\sqrt{1}$). We now observe that the 17 in cell (2,1) is the smallest entry in the unchecked table. The amount in supply outlet W_2 is 48, while the amount demanded by M_1 is 0. Thus, 0 is shipped by the route (2,1), after which a check ($\sqrt{2}$) is applied to column 1. The table that results from these two operations is given below:

	$\sqrt{2}$				
$\sqrt{1}$	⑯ ^{20}	24	28	2̸0̸ 0	
	⑰ ^{0}	18	20	4̸8̸ 48	(T.1a)
	31	19	22	12	
	2̸0̸	48	12		
	0̸				
	0				

Continuing with Method 9.1, we next ship 48 via cell (2,2) in the unchecked table, after which row 2 is checked ($\sqrt{3}$). The remaining unchecked table now consists of cells (3,2) and (3,3). We next ship 0 by the cell (3,2) and then check column 2. Finally, 12 is shipped by cell (3,3), after which column 3 is checked. The results of these operations, as well as the initial b.f.s., are indicated in Table 1b, which follows:

$$
\begin{array}{c|ccc|c}
 & \sqrt{2} & \sqrt{4} & \sqrt{5} & \\
\hline
\sqrt{1} & \text{⑯}^{\,20} & 24 & 28 & 0 \\
\sqrt{3} & \text{⑰}^{\,0} & \text{⑱}^{\,48} & 20 & \cancel{48}\ 0 \\
 & 31 & \text{⑲}^{\,0} & \text{㉒}^{\,12} & \cancel{12}\ \cancel{12}\ 0 \\
\hline
 & 0 & \cancel{48} & \cancel{12} & \\
 & & \cancel{0} & 0 & \\
 & & 0 & &
\end{array}
$$

(T.1b)

The cost corresponding to this initial b.f.s. is $16 \cdot 20 + 18 \cdot 48 + 22 \cdot 12 = 1448$. Note that the initial b.f.s. has two basic x_{ij}'s that are zero, namely, x_{21} and x_{32}. Clearly, this b.f.s. is degenerate.

We now evaluate values of cycles generated by nonbasis cells. The results of these calculations are presented in Table 1', which is otherwise the same as Table 1b:

$$
\begin{array}{|ccc|}
\hline
\text{⑯}^{\,20} & 24(7) & 28(9) \\
\text{⑰}^{\,0} & \text{⑱}^{\,48} & 20(-1) \\
31(13) & \text{⑲}^{\,0} & \text{㉒}^{\,12} \\
\hline
\end{array}
$$

(T.1')

The only cycle value that is negative is that of cell (2,3). Thus, it is brought into the basis in accordance with step 4 of the Stepping Stone Method. The resulting table is given below (relevant cycle values are also presented within parentheses):

$$
\begin{array}{|ccc|}
\hline
\text{⑯}^{\,20} & 24(7) & 28(9) \\
\text{⑰}^{\,0} & \text{⑱}^{\,36} & \text{⑳}^{\,12} \\
31(13) & \text{⑲}^{\,12} & 22(1) \\
\hline
\end{array}
$$

(T.2)

Since the cycle values are now all positive, this table corresponds to the optimal solution. The change in w from Table 1' to Table 2 is given by $\Delta w = x'_{23} \cdot v_{23} = 12 \cdot (-1) = -12$. Thus, since $w = 1448$ for Table 1', $w = 1448 - 12 = 1436$ for Table 2; this is the minimum cost. Since each monetary unit is $10, Min w = $14,360. The basic x_{ij}'s for the optimal solution are as follows:

$$x_{11} = 20,\ x_{21} = 0,\ x_{22} = 36,\ x_{23} = 12,\ \text{and}\ x_{32} = 12.$$

Thus, in this degenerate solution, 20 cars should be shipped from W_1 to M_1, 36 from W_2 to M_2, 12 from W_2 to M_3, and 12 from W_3 to M_2.

It occasionally becomes impossible to use certain routes in a transportation problem. Possible reasons for this include natural disasters (flooded regions, areas hit by tornados), weight limits on bridges, height limits on waterways, road construction, and local traffic laws. One way of handling a *prohibited route* is to assign an extremely large, but unspecified, unit transportation cost M to a cell with a prohibited route. After that, the usual Stepping Stone approach is used to find the optimal solution. We illustrate this in the next problem.

Problem 9.16

A car rental agency serving seven cities has a surplus of cars in three cities (labeled W_1, W_2 and W_3) and a need for cars in four cities (labeled M_1, M_2, M_3, and M_4). The excess of cars in W_1, W_2, and W_3 is 20, 20, and 32, respectively, while the needs of cities M_1 through M_4 are 16, 20, 20, and 16, respectively. The table of distances (in miles) between the cities and the rim conditions are as follows:

	M_1	M_2	M_3	M_4	
W_1	17	23	20	M	20
W_2	23	15	23	20	20
W_3	25	M	13	21	32
	16	20	20	16	

The two occurrences of M's in the cells (1,4) and (3,2) indicates that it is not possible to transport cars from W_1 to M_4 or from W_3 to M_4 because of, e.g., road-repair work. Thus, M denotes an extremely large distance. How should the agency transport its cars to satisfy the demands and minimize the total distance driven?

Solution

The quantity to be minimized is total distance; we denote this by w. We employ Method 9.4 in the usual way, first using the Minimum Entry Method to find an initial b.f.s.

Tables 1a and 1b indicate the sequence of operations involved in employing the Minimum Entry Method:

	√3		√1		
	⑰ 16	23	20	M	2̸0 4
√2	23	⑮ 20	23	20	2̸0 0
	25	M	⑬ 20	21	3̸2 12
	1̸6	2̸0	2̸0	16	
	0	0	0		

(T.1a)

	√3	√5	√1	√6	
	⑰ 16	㉓ 0	20	Ⓜ 4	4̸ 4̸ 0
√2	23	⑮ 20	23	20	0
√4	25	M	⑬ 20	㉑ 12	1̸2 0
	0	0̸	0	1̸6	
		0		4̸	
				0	

(T.1b)

The initial b.f.s. is depicted in Table 1b. Note that it is degenerate (x_{12} = 0) and also that it contains a prohibited route, namely, cell (1,4). The number of cells in the basis (6) is, of course, equal to $m + n - 1$ as it should be. The evaluation of cells not in the basis is carried out and the cycle values are given in parentheses in Table 1′.

⑰ 16	㉓ 0	20(28 − M)	Ⓜ 4
23(14)	⑮ 20	23(39 − M)	20(28 − M)
25(−13 + M)	M(2M − 44)	⑬ 20	㉑ 12

(T.1′)

In Table 1′, three cells have negative cycle values, namely, (1,3), (2,3), and (2,4). This is so because of the presence of the large negative number −M. The cycle generated by cell (2,3) is interesting. It contains, in alternating getter–giver order, the cells (2,3), (3,3), (3,4), (1,4), (1,2), and (2,2).

We now introduce cell (2,4) into the initial basis. Since cell (1,4) is the minimum giver cell on the cycle formed by (2,4), it leaves the basis. The resulting basis, together with the new cycle values for nonbasis cells, appears in Table 2:

⑰ ^{16}	㉓ ^{4}	20(0)	$M(M-28)$
23(14)	⑮ ^{16}	23(11)	⑳ ^{4}
25(15)	$M(M-16)$	⑬ ^{20}	㉑ ^{12}

(T.2)

The cycle generated by cell (1,3) is interesting. It contains, in alternating getter–giver order, the following cells: (1,3), (3,3), (3,4), (2,4), (2,2), and (1,2). In Table 2, all cycle values for nonbasis cells are nonnegative. Thus, the b.f.s. pertaining to Table 2 is an optimal solution. The pertinent x_{ij} values and the minimum total distance are as follows:

$$x_{11} = 16, \, x_{12} = 4, \, x_{22} = 16,$$

$$x_{24} = 4, \, x_{33} = 20, \, x_{34} = 12, \text{ and}$$

$$\text{Min } w = 17 \cdot 16 + 23 \cdot 4 + \cdots + 21 \cdot 12 = 1196 \text{ miles.}$$

One of the cycle values in Tableau 2 is a zero; this means that there is another optimal b.f.s. to the problem. The situation is analogous to what was discussed in Section 2 of Chapter 7 for standard linear programming problems.

Problem 9.17

Determine an alternate optimal b.f.s. for Problem 9.16.

Solution

The alternate optimal b.f.s. is determined by introducing cell (1,3) into the basis, since the cycle value for this cell is zero. The minimum giver cells in the cycle formed by cell (1,3)—this cycle was listed in Problem 9.16—are (2,4) and (1,2), each of which can give four. Removing the latter cell from the basis and then adjusting shipments via Method 9.3 leads to the basis given in Table 3:

⑰ ^{16}	23(0)	⑳ ^{4}	$M(M-28)$
23(14)	⑮ ^{20}	23(11)	⑳ ^{0}
25(15)	$M(M-16)$	⑬ ^{16}	㉑ ^{16}

(T.3)

Since all the relevant cycle values in Table 3 are nonnegative, this table corresponds to an optimal b.f.s.

The alternate optimal x_{ij}'s are read off from Table 3 as follows:

$$x_{11} = 16, \ x_{13} = 4, \ x_{22} = 20, \ x_{24} = 0, \ x_{33} = 16, \text{ and } x_{34} = 16.$$

Note that this b.f.s. is degenerate. The value for w corresponding to this b.f.s. should be the same as in Problem 9.16. Again calculating w (as a check), we obtain

$$\text{Min } w = 17 \cdot 16 + 20 \cdot 4 + \cdots + 21 \cdot 16 = 1196 \text{ miles.}$$

Note that the value of the cycle formed by cell (1,2), which just left the basis, is zero.

Problem 9.18

A company has its three machines—I, II, and III—serviced by four maintenance crews—A, B, C, and D. Machine I must be serviced 30 times a year; machine II, 40 times a year; and machine III, 30 times a year. Suppose that crews A, B, C, and D are assigned 20, 20, 25, and 35 of these jobs, respectively. The times (in hours) required of each crew to service each machine are given in the following table (along with the rim conditions):

	A	B	C	D	
I	8	10	14	8	30
II	7	11	12	6	40
III	5	8	15	9	30
	20	20	25	35	

Determine the optimal assignment plan, i.e., the plan for which the total servicing time is minimized. If there are alternate optimal solutions, indicate how to determine them.

Solution

In the first part of the Stepping Stone Method we determine an initial b.f.s. The operations involved. from employing the Minimum Entry Method, are indicated in Table 1 (the numbers adjacent to the checks indicate the sequence of operations):

	√1	√4	√6	√2	
	8	(10) 10	(14) 20	8	3̶0̶ 2̶0̶ 0
√5	7	11	(12) 5	(6) 35	4̶0̶ 5̶ 0
√3	(5) 20	(8) 10	15	9	3̶0̶ 1̶0̶ 0
	2̶0̶	2̶0̶	2̶5̶	3̶5̶	
	0	1̶0̶	2̶0̶	0	
		0	0		

(T.1)

In the next table we indicate values of cycles for the nonbasis cells, as well as a retabulation of the shipments from Table 1. Let us do a sample calculation. The cells in the cycle formed by cell (2,1), in alternating getter–giver order, are as follows: (2,1), (2,3), (1,3), (1,2), (3,2), and (3,1). Thus, $v_{21} = 7 - 12 + 14 - 10 + 8 - 5 = 2$. This value and the remaining cycle values are listed in Table 2:

8(1)	(10) 10	(14) 20	8(0)
7(2)	11(3)	(12) 5	(6) 35
(5) 20	(8) 10	15(3)	9(3)

(T.2)

Since all relevant cycle values are nonnegative, the b.f.s. corresponding to Table 2 is optimal. The optimal x_{ij}'s and Min w are as follows:

$$x_{12} = 10, \ x_{13} = 20, \ x_{23} = 5, \ x_{24} = 35,$$

$$x_{31} = 20, \ x_{32} = 10, \text{ and Min } w = 830 \text{ hours.}$$

The presence of a zero cycle value for cell (1,4) indicates that there is an alternate optimal b.f.s. Introducing cell (1,4) into the basis via Method 9.3 leads to Table 3, in which cell (1,3) has left the basis:

8(1)	(10) 10	14(0)	(8) 20
7(2)	11(3)	(12) 25	(6) 15
(5) 20	(8) 10	15(3)	9(3)

(T.3)

The x_{ij}'s for this alternate optimal b.f.s. are read off as follows:

$$x_{12} = 10, \ x_{14} = 20, \ x_{23} = 25,$$

$$x_{24} = 15, \ x_{31} = 20, \text{ and } x_{32} = 10.$$

Again, Min $w = 830$ hours.

As indicated in Section 2 of Chapter 7, any convex combination of the vectors corresponding to the above optimal solutions yields an optimal solution. For example, if $\bar{A}$ denotes the vector solution from Table 2 and $\bar{B}$ the vector solution from Table 3, then $\bar{C} = \frac{1}{2}\bar{A} + \frac{1}{2}\bar{B}$ will yield an optimal (but nonbasic) solution. The relevant x_{ij}'s are

$$x_{12} = 10, \ x_{13} = 10, \ x_{14} = 10, \ x_{23} = 15,$$

$$x_{24} = 25, \ x_{31} = 20, \text{ and } x_{32} = 10.$$

The key calculations are $x_{13} = \frac{1}{2}(20 + 0) = 10$, $x_{14} = \frac{1}{2}(0 + 20) = 10$, $x_{23} = \frac{1}{2}(5 + 25) = 15$, and $x_{24} = \frac{1}{2}(35 + 15) = 25$.

Let us make a partial interpretation relative to the initial statement of Problem 9.18. The optimal solution from Table 2 indicates that machine I should be serviced 10 times a year by crew B, and 20 times a year by crew C, machine II should be serviced 5 times a year by crew C, etc. The minimum total yearly time for the servicing of all machines by all crews is 830 hours.

2. UNBALANCED TRANSPORTATION PROBLEMS

In the work so far in this chapter, the assumption of balance has prevailed; i.e., we have assumed that

$$a_1 + a_2 + \cdots + a_m = b_1 + b_2 + \cdots + b_n.$$

An unbalanced transportation problem is one for which this equation does not hold, i.e., the sum of the supplies does not equal the sum of the demands.

Before we consider unbalanced problems, we have to develop some additional theoretical tools. Suppose that all unit cost terms in the data table of a balanced problem are changed by some constant amount k, i.e.,

$$c'_{ij} = c_{ij} + k \text{ for all } i \text{ and } j, \tag{1}$$

where c'_{ij} refers to "new" unit costs. Suppose also that a particular shipping plan (i.e., a particular set of x_{ij}'s) is applied in both cases. The collection of x_{ij}'s will be indicated symbolically by the matrix symbol X. In matrix notation, $X = (x_{ij})$. The symbols $w(X)$ and $w'(X)$ denote the old and new total costs, respectively, for the shipping plan X. Furthermore, we define

$$J = a_1 + \cdots + a_m = b_1 + \cdots + b_n. \tag{2}$$

Theorem 9.5. *Given a definite shipping plan $X = (x_{ij})$. Suppose that all the unit cost terms c_{ij} are changed by the constant amount k. That is, $c'_{ij} = c_{ij} + k$ for all i and j. Then $w(X)$ and $w'(X)$ are related by*

$$w'(X) = w(X) + k \cdot J. \tag{3}$$

Problem 9.19

Prove Theorem 9.5 for a balanced transportation problem with two sources ($m = 2$) and three destinations ($n = 3$).

Solution

For a given shipping plan $X = (x_{ij})$, we have

$$w(X) = c_{11}x_{11} + c_{12}x_{12} + \cdots + c_{23}x_{23}, \tag{i}$$

and

$$w'(X) = c'_{11}x_{11} + c'_{12}x_{12} + \cdots + c'_{23}x_{23}. \tag{ii}$$

Substituting $c'_{ij} = c_{ij} + k$ into (ii) yields

$$w'(X) = c_{11}x_{11} + c_{12}x_{12} + \cdots + c_{23}x_{23}$$
$$+ k(x_{11} + x_{12} + x_{13} + x_{21} + x_{22} + x_{23}). \tag{iii}$$

However,

$$J = a_1 + a_2 = (x_{11} + x_{12} + x_{13}) + (x_{21} + x_{22} + x_{23}). \tag{iv}$$

Substituting (iv) and (i) into (iii) yields

$$w'(X) = w(X) + k \cdot J. \tag{v}$$

An important aspect of Theorem 9.5 is that it says $w(X)$ and $w'(X)$ differ by a *constant*, i.e., a term independent of X. From a general property of functions it follows that if w is minimized when $X = X^*$, then w' will also be minimized when $X = X^*$. Here $X^* = (x^*_{ij})$ denotes an optimal solution for X. Moreover, Min w' = Min $w + k \cdot J$. These ideas are generalized in the following theorem.

Theorem 9.6. *Given two functions $w(X)$ and $w'(X)$ of the same variable X, that differ by a constant, i.e.,*

$$w'(X) = w(X) + K.$$

Suppose that w is minimized when $X = X^*$, i.e.,

$$\text{Min } w = w(X^*).$$

Then w' is also minimized when $X = X^*$, i.e.,

$$\text{Min } w' = w'(X^*).$$

Moreover, $\text{Min } w' = \text{Min } w + K.$

Problem 9.20

Suppose that all of the unit cost terms in Problem 9.1 are increased by $15. Determine the new optimal solution and minimum total cost.

Solution

Here we have $c'_{ij} = c_{ij} + 15$ (in other words, $k = 15$). Thus, the new unit costs are $c'_{11} = 41$, $c'_{12} = 38$, . . . , $c'_{33} = 44$. The optimal solution and Min w for Problem 9.1 are presented in Problem 9.6. According to Theorem 9.6, the *same* optimal solution applies to the new problem. That is, $x^*_{13} = 61$, $x^*_{22} = 30$, . . . , and $x^*_{32} = 38$ also constitute the optimal solution for the current problem.

From Theorems 9.5 and 9.6 it follows that Min w' and Min w differ by $k \cdot J = 15J$. From Problem 9.1, $J = 200$. Thus,

$$\text{Min } w' = \text{Min } w + k \cdot J = 2877 + 15 \cdot 200 = \$5877,$$

i.e., the minimum total cost for the current problem is $5877.

The following theorem, which is of great practical value, is an extension of Theorem 9.5. Here the new unit cost c'_{ij} is related to the old unit cost c_{ij} by

$$c'_{ij} = c_{ij} - r_i - k_j \text{ for all } i \text{ and } j. \tag{4}$$

The r_i's are dependent only on the row label i, while the k_j's depend only on the column label j. Furthermore, in the following, the *constant* $K(r,k)$ is given by

$$K(r,k) = (r_1 a_1 + \cdots + r_m a_m) + (k_1 b_1 + \cdots + k_n b_n). \tag{5}$$

Theorem 9.7. *Given a definite shipping plan $X = (x_{ij})$, suppose that all of the unit cost terms c_{ij} are changed according to $c'_{ij} = c_{ij} - r_i - k_j$. Then the old and the new total costs $w(X)$ and $w'(X)$ are related by*

$$w'(X) = w(X) - K(r,k). \tag{6}$$

Problem 9.21

Prove Theorem 9.7 for a balanced transportation problem with two sources and three destinations.

Solution

For a given shipping plan $X = (x_{ij})$, we have

$$w(X) = c_{11}x_{11} + c_{12}x_{12} + \cdots + c_{23}x_{23}, \tag{i}$$

and

$$w'(X) = c'_{11}x_{11} + c'_{12}x_{12} + \cdots + c'_{23}x_{23}. \tag{ii}$$

Substituting Equation (4) from above into (ii) yields

$$\begin{aligned}
w'(X) = {} & c_{11}x_{11} + c_{12}x_{12} + \cdots + c_{23}x_{23} \\
& - r_1(x_{11} + x_{12} + x_{13}) - r_2(x_{21} + x_{22} + x_{23}) \\
& - k_1(x_{11} + x_{21}) - k_2(x_{12} + x_{22}) - k_3(x_{13} + x_{23}). \tag{iii}
\end{aligned}$$

However,

$$a_i = x_{i1} + x_{i2} + x_{i3} \text{ for } i = 1, 2 \tag{iv}$$

and

$$b_j = x_{1j} + x_{2j} \qquad \text{for } j = 1, 2, 3. \tag{v}$$

Substituting (i), (iv), and (v) into (iii) yields

$$\begin{aligned}
w'(X) &= w(X) - r_1 a_1 - r_2 a_2 - k_1 b_1 - k_2 b_2 - k_3 b_3 \\
&= w(X) - K(r,k). \tag{vi}
\end{aligned}$$

The following theorem follows directly from Theorems 9.6 and 9.7.

Theorem 9.8. *Suppose that the unit costs of two balanced transportation problems with the same m and n are related by $c'_{ij} = c_{ij} - r_i - k_j$. Then the respective total costs $w(X)$ and $w'(X)$ are both minimized at the same optimal shipping plan X^*, and, moreover,*

$$\text{Min } w' = \text{Min } w - K(r,k)$$

In other words, Min w' and Min w are related by the same type equation as Equation (6) above. We are now in a position to handle unbalanced transportation problems.

Problem 9.22: Supplies Exceed Demands

Consider the following unbalanced transportation problem for which the sum of the supplies (80) exceeds the sum of the demands (78):

	M_1	M_2	M_3	
W_1	4	5	7	$a_1 = 30$
W_2	8	6	4	$a_2 = 50$
	$b_1 = 18$	$b_2 = 40$	$b_3 = 20$	

Develop an approach for finding the optimal solution and minimum total cost.

Solution

We introduce a fictitious destination M_4 whose demand ($b_4 = 2$) will cause balance between supplies and demands. A question arises as to the values to be assigned to c_{14} and c_{24} in the expanded data table. An answer may be provided by a practical interpretation of the problem. Thus, c_{14} can be interpreted as a unit *storage* (or overstocking) cost for storing items at supply location W_1. A similar interpretation holds for c_{24}. The usual procedure is to assume that these storage costs are equal, i.e., $c_{14} = c_{24} = k$. Thus, we obtain the following balanced data table, which corresponds to the original data table:

	M_1	M_2	M_3	M_4	
W_1	4	5	7	k	30
W_2	8	6	4	k	50
	18	40	20	2	

According to Theorem 9.8, we can subtract the constant k from both entries of column 4 without changing the optimal shipping plan. This can be seen clearly if we let $r_i = 0$ for rows 1 and 2, $k_j = 0$ for columns 1, 2, and 3, and let $k_4 = k$ for column 4. In other words, only the entries of column 4 are altered.

Thus, we obtain the following *equivalent* reduced data table, where $c'_{14} = c'_{24} = k - k = 0$, and the other costs are the same as before:

	M_1	M_2	M_3	M_4	
W_1	4	5	7	0	30
W_2	8	6	4	0	50
	18	40	20	2	

The problem can then be solved without difficulty by employing the Stepping Stone Method. The optimal solution is given in the following table:

④ 18	⑤ 12	7(4)	0(1)
8(3)	⑥ 28	④ 20	⓪ 2

The cycle values for nonbasic cells appear in parentheses. The interpretation of the term $x_{24} = 2$ is that two items are to be kept at supply location W_2. The minimum total shipping cost is calculated as follows:

$$\text{Min } w = 4{\cdot}18 + 5{\cdot}12 + 6{\cdot}28 + 4{\cdot}20 = \$380.$$

Note that this excludes the cost of storage at W_2.

Problem 9.22 indicates the procedure to follow when supplies exceed demands if it is reasonable to assume that the costs of storage at the supply locations are all the *same*.

Problem 9.23: Demands Exceed Supplies

Consider the following unbalanced transportation problem for which the sum of the demands (80) exceeds the sum of the supplies (78):

	M_1	M_2	M_3	
W_1	4	5	7	$a_1 = 28$
W_2	8	6	4	$a_2 = 50$
	$b_1 = 20$	$b_2 = 40$	$b_3 = 20$	

Develop an approach for finding the optimal solution and minimum total cost.

Solution

In this case there is no way to satisfy all the demands. The best that can be done is to fall short of the total demands by two units.

We introduce a fictitious source W_3, which "supplies" two units. However, unit costs must now be assigned to c_{31}, c_{32}, and c_{33}. To do this, we interpret c_{3j} as the unit cost of having destination M_j remain unsatisfied. The usual procedure is to assume that these costs are all equal, i.e., $c_{31} = c_{32} = c_{33} = r$. Thus, we obtain the following balanced data table, which corresponds to the original data table:

	M_1	M_2	M_3	
W_1	4	5	7	28
W_2	8	6	4	50
W_3	r	r	r	2
	20	40	20	

According to Theorem 9.8, we can subtract the constant r from all the entries of row 3 without changing the optimal shipping plan. This can be seen clearly if we let $r_i = 0$ for rows 1 and 2, $r_3 = r$ for row 3, and k_j

= 0 for columns 1, 2, and 3. In other words, only the entries of row 3 are altered.

Thus, we obtain the following equivalent reduced data table, where $c'_{31} = c'_{32} = c'_{33} = r - r = 0$, and the other costs are the same as before:

	M_1	M_2	M_3	
W_1	4	5	7	28
W_2	8	6	4	50
W_3	0	0	0	2
	20	40	20	

For this table, the optimal solution is none other than the initial b.f.s. as determined by the Minimum Entry Method. It is given in the following table:

④ 20	⑤ 8	7(4)
8(3)	⑥ 30	④ 20
0(1)	⓪ 2	0(2)

Since $x_{32} = 2$ in the optimal solution, destination M_2 will receive only 38 items, while the other destinations will be satisfied. The minimum total shipping cost (neglecting the cost of not satisfying M_2) is calculated as follows:

$$\text{Min } w = 4 \cdot 20 + 5 \cdot 8 + 6 \cdot 30 + 4 \cdot 20 = \$380.$$

Problem 9.23 indicates the technique to use when demands exceed supplies if it is reasonable to assume that the costs of leaving destinations unsatisfied are all the *same*.

Good treatments of unbalanced transportation problems may be found in [4] and [6].

3. DUALITY

Our goal in this section is to develop the dual problem of a general transportation problem. Duality theory was covered in Section 5 of Chapter 5 for the case of maximum and minimum linear programming problems in standard form. The "data box" format (Figure 5.2) for representing dual linear programming problems will be useful to us here.

The main constraints in the transportation problem are equation constraints. It would be useful, in general, to analyze duality theory for cases when both equation constraints and/or unrestricted variables occur. (Unrestricted variables, discussed in Section 3 of Chapter 5, are those that are *not* required to be nonnegative.)

An equation constraint occurs in Problem 5.4. Let us restate the problem here (first, we relabel the original variables from x, y, and z to x_1, x_2, and x_3):

$$\text{Maximize } u = x_1 + 3x_2 + 2x_3 \tag{1}$$

$$\text{subject to} \qquad x_2 + 2x_3 \le 21 \tag{2}$$

$$3x_1 + 2x_2 + x_3 = 24 \tag{3}$$

and

$$x_1, x_2, x_3 \ge 0.$$

Problem 9.24

Use the data box (Figure 5.2) format to determine the dual problem of Problem 5.4.

Solution

The problem will be in standard maximum form if we replace Equation (3) by the two following $\le$ inequalities, to which it is equivalent:

$$3x_1 + 2x_2 + x_3 \le 24 \tag{3a}$$

$$-3x_1 - 2x_2 - x_3 \le -24. \tag{3b}$$

The principle employed here is that $x = a$ is equivalent to $x \le a$ *and* $x \ge a$; however, the latter is the same as $-x \le -a$. We now line up (2), (3a), (3b), and (1) horizontally in a data box as follows:

	x_1	x_2	x_3	
y_1	0	1	2	21
y_2'	3	2	1	24
y_3'	-3	-2	-1	-24
	1	3	2	

There is a technical reason for using y_2' and y_3' instead of y_2 and y_3 in the minimum problem: The variable y_2 will be used later. We can read the dual minimum problem off vertically from the data box:

$$\text{Minimize} \quad w = 21y_1 + 24y_2' - 24y_3' \tag{4}$$

$$\text{subject to} \quad 3y_2' - 3y_3' \geq 1 \tag{5}$$

$$y_1 + 2y_2' - 2y_3' \geq 3 \tag{6}$$

$$2y_1 + y_2' - y_3' \geq 2 \tag{7}$$

and

$$y_1, y_2', y_3' \geq 0.$$

In (4) through (7), y_2' and y_3' can be placed together in the form $(y_2' - y_3')$. For example, we could write (6) as

$$y_1 + 2(y_2' - y_3') \geq 3,$$

and similarly for (4), (5), and (7). Suppose that we let $y_2 = y_2' - y_3'$. Then the variable y_2, which is the difference of two nonnegative variables, is *unrestricted*. We restate (4) through (7) as follows:

$$\text{Minimize} \quad w = 21y_1 + 24y_2 \tag{4'}$$

$$\text{subject to} \quad 3y_2 \geq 1 \tag{5'}$$

$$y_1 + 2y_2 \geq 3 \tag{6'}$$

$$2y_1 + y_2 \geq 2, \tag{7'}$$

where $y_1 \geq 0$ and y_2 is unrestricted.

There is a simple way of determining the dual form as given by (4') through (7'), which involves making use of the data box in a modified way.

Method 9.5: Dual Problems Containing Equation Constraints and Unrestricted Variables

1a. If the primal problem is maximizing, first convert all the main constraints of inequality type to $a_{i1}x_1 + \cdots + a_{in}x_n \leq b_i$ form. Then line up the main constraints (including equation constraints) horizontally in the data box.

1b. If row k corresponds to an equation constraint, then the dual variable y_k is unrestricted.

1c. If the variable x_ℓ is unrestricted, then the main constraint of the dual problem corresponding to column ℓ is an equation constraint.

2a. If the primal problem is minimizing, first convert all the main constraints of inequality type to $a_{1j}y_1 + \cdots + a_{mj}y_m \geq c_j$ form.

Then line up the main constraints (including equation constraints) vertically in the data box.

2b. If column ℓ corresponds to an equation constraint, then the dual variable x_ℓ is unrestricted.

2c. If the variable y_k is unrestricted, then the main constraint of the dual problem corresponding to row k is an equation constraint.

As an aid to using Method 9.5 in conjunction with the data box format, we shall use the symbol EU to denote rows and columns that indicate the Equation constraint–Unrestricted variable pairing. We illustrate this in the following problem.

Problem 9.25

Obtain the dual form (4′) through (7′) of Problem 9.24 by making use of Method 9.5 and the EU symbolism.

Solution

The proper form for the primal maximum problem is given by (1), (2), and (3), which immediately precede the statement of Problem 9.24. The appropriate data box is given below; the EU symbol is placed in the second row because main constraint (3) is an equation constraint:

	x_1	x_2	x_3		
y_1	0	1	2	21	
y_2	3	2	1	24	EU
	1	3	2		

The EU symbol in row 2 also indicates that y_2 is an unrestricted variable in the dual problem. Reading off the dual problem from the columns of the data box results in the following:

$$\text{Minimize} \quad w = 21y_1 + 24y_2$$
$$\text{subject to} \quad\quad 3y_2 \geq 1$$
$$y_1 + 2y_2 \geq 3$$
$$2y_1 + y_2 \geq 2,$$

where $y_1 \geq 0$ and y_2 is unrestricted.

These are none other than (4′) through (7′) of Problem 9.24.

It is instructive to do another duality problem illustrating the modified use of the data box.

Problem 9.26

Determine the dual problem of the following linear programming problem:

$$\text{Minimize} \quad w = 4y_1 + 5y_2 \tag{1}$$

$$\text{subject to} \quad 3y_1 + 2y_2 = 7 \tag{2}$$

$$y_1 + 4y_2 \geq 8 \tag{3}$$

$$6y_1 + 3y_2 \geq 5, \tag{4}$$

where $y_1 \geq 0$ and y_2 is unrestricted.

Solution

Since the conditions of step (2a) of Method 9.5 are satisfied, the problem is lined up vertically in the following data box. An EU is placed in column 1 because (2) is an equation constraint, while an EU is also placed in row 2 because y_2 is unrestricted:

	x_1	x_2	x_3		
y_1	3	1	6	4	
y_2	2	4	3	5	EU
	7	8	5		
	EU				

The dual maximum problem is read off from the rows of the data box:

$$\text{Maximize} \quad u = 7x_1 + 8x_2 + 5x_3 \tag{5}$$

$$\text{subject to} \quad 3x_1 + x_2 + 6x_3 \leq 4 \tag{6}$$

$$2x_1 + 4x_2 + 3x_3 = 5, \tag{7}$$

where x_1 is unrestricted and $x_2, x_3 \geq 0$.

Thus, the EU symbols indicate that (7) is an equation constraint and that x_1 is unrestricted.

Problem 9.27

Formulate the dual linear programming problem of the general transportation problem for which $m = n = 2$.

Solution

The general transportation problem with two sources and two destinations is as follows:

$$\text{Minimize} \quad w = c_{11}x_{11} + c_{12}x_{12} + c_{21}x_{21} + c_{22}x_{22} \qquad (1)$$

$$\text{subject to} \quad x_{11} + x_{12} \qquad\qquad\qquad = a_1 \qquad (2)$$

$$x_{21} + x_{22} = a_2 \qquad (3)$$

$$x_{11} \qquad + x_{21} \qquad\quad = b_1 \qquad (4)$$

$$x_{12} \qquad\quad + x_{22} = b_2, \qquad (5)$$

where all $x_{ij} \geq 0$.

The problem is lined up in the columns of the following data box, where the EU's indicate the fact that (2) through (5) are equation constraints:

	u_1	u_2	v_1	v_2	
x_{11}	1	0	1	0	c_{11}
x_{12}	1	0	0	1	c_{12}
x_{21}	0	1	1	0	c_{21}
x_{22}	0	1	0	1	c_{22}
	a_1	a_2	b_1	b_2	
	EU	EU	EU	EU	

The variables for the dual problem are chosen to be u_1, u_2, v_1, and v_2 by convention. The objective function is then labeled z to avoid confusion. The dual maximum problem is read off from the rows of the data box:

$$\text{Maximize} \quad z = a_1u_1 + a_2u_2 + b_1v_1 + b_2v_2 \qquad (1')$$

$$\text{subject to} \quad u_1 \quad + v_1 \qquad\quad \leq c_{11} \qquad (2')$$

$$u_1 \qquad\qquad + v_2 \leq c_{12} \qquad (3')$$

$$u_2 + v_1 \qquad\quad \leq c_{21} \qquad (4')$$

$$u_2 \qquad + v_2 \leq c_{22}, \qquad (5')$$

where u_1, u_2, v_1, and v_2 are all unrestricted.

The development of the dual linear programming problem of a general transportation problem with m sources and n destinations involves the same type of approach as used in Problem 9.27. The general transportation problem is given by (1), (2), (3), and (4) at the beginning of Section 1 of this chapter. We specify the form of the dual linear programming problem in the following theorem.

Theorem 9.9. *The dual linear programming problem of a general transportation problem with m sources and n destinations is given as follows:*

Maximize
$$z = a_1 u_1 + a_2 u_2 + \cdots + a_m u_m$$
$$+ b_1 v_1 + b_2 v_2 + \cdots + b_n v_n \tag{1'}$$

subject to
$$u_i + v_j \leq c_{ij} \text{ for } i = 1, 2, \ldots, m$$
$$\text{and } j = 1, 2, \ldots, n, \tag{2'}$$

where

$$u_1, u_2, \ldots, u_m, v_1, v_2, \ldots, v_n \text{ are all unrestricted.} \tag{3'}$$

There are $m \cdot n$ inequalities indicated in (2').

4. THE ASSIGNMENT PROBLEM

The assignment problem is a special type of a linear programming problem. The *balanced* form of the assignment problem is stated as follows: Given n applicants for n jobs, it is desired to assign each applicant to exactly one job, and have each job filled by exactly one applicant. In addition, the total cost of carrying out these assignments is to be minimized. There is available a data table (cost matrix) in which the unit costs (c_{ij}) are given; each c_{ij} is the cost of assigning applicant i to job j. Let the variable x_{ij} be defined as follows:

$$x_{ij} = \begin{cases} 1 \text{ if applicant } i \text{ is assigned to job } j \\ 0 \text{ otherwise.} \end{cases} \tag{A}$$

The mathematical statement of the assignment problem is now given. The symbol w denotes the total cost:

Minimize $w = c_{11}x_{11} + c_{12}x_{12} + \cdots + c_{nn}x_{nn}$ (1)

subject to $x_{i1} + x_{i2} + \cdots + x_{in} = 1$ for $i = 1, 2, \ldots, n$ (2)

$x_{1j} + x_{2j} + \cdots + x_{nj} = 1$ for $j = 1, 2, \ldots, n,$ (3)

where the x_{ij} are as defined above.

The right-hand side of Equation (1) contains $n \cdot n$ terms. The constraints (2) ensure that each applicant is assigned to exactly one job, while constraints (3) ensure that each job is filled by exactly one applicant. The problem as stated is not in the form of a usual linear programming problem because of the restriction that x_{ij} can only take on the values 0 and 1.

Let us tentatively replace condition (A) by the following set of non-negativity conditions (labeled as (4)):

$$x_{ij} \geq 0 \text{ for } i = 1, 2, \ldots, n \text{ and } j = 1, 2, \ldots, n. \tag{4}$$

Thus, we now have the following balanced *transportation* problem:

Minimize $w = c_{11}x_{11} + c_{12}x_{12} + \cdots + c_{nn}x_{nn}$ (1)

subject to $x_{i1} + x_{i2} + \cdots + x_{in} = 1$ for $i = 1, 2, \ldots, n$ (2)

$x_{1j} + x_{2j} + \cdots + x_{nj} = 1$ for $j = 1, 2, \ldots, n,$ (3)

where $x_{ij} \geq 0$ for $i = 1, 2, \ldots, n$ and $j = 1, 2, \ldots, n.$ (4)

The fact that (1) through (4) above constitute a special case of a balanced transportation problem can be seen by making a comparison with (1) through (4) in Section 1 of this chapter. Each supply (a_i) equals one and each demand (b_j) also equals one.

In [1], it is indicated that all basic feasible solutions of (2) through (4) above will be integer-valued (an integer is a whole number). In addition, the constraints of (2) through (4) restrict the integer values of each x_{ij} to be either 0 or 1. Furthermore, the constraint set determined by (2), (3), and (A) is contained within the constraint set determined by (2), (3), and (4) since $x_{ij} = 0$ or 1 implies that $x_{ij} \geq 0$. Thus, an optimal (basic feasible) solution for (1) through (4) will also be an optimal solution for the assignment problem as originally stated (namely, (1) through (3) coupled with (A)). Thus, henceforth we shall take (1) through (4) as a satisfactory mathematical formulation of the assignment problem.

Now because the assignment problem as given by (1) through (4) is a special case of the balanced transportation problem, we should be able to use the Stepping Stone Method (Method 9.4) to solve assignment problems. In fact, this will be done in a sample problem later in the section. However, due to the highly degenerate nature of assignment problems, Method 9.4 turns out to be rather inefficient and tedious to use.

Let us illustrate the high level of degeneracy of the assignment problem. A b.f.s. of (1) through (4) will contain $2n - 1$ basic variables; equivalently, a basis will contain $2n - 1$ cells.* However, it is clear that every b.f.s. will contain exactly n basic variables equal to 1 (e.g., exactly one for each of the n equations of (2)), and thus $(n - 1)$ basic variables equal to 0.

The algorithm recommended for assignment problems is the Hungarian Method, which will be presented shortly. First, however, we introduce a model problem.

---

* A b.f.s. for a balanced transportation problem has $m + n - 1$ basic variables. Equivalently, there are $m + n - 1$ cells in the basis of a balanced transportation problem. Since $m = n$ in the related balanced assignment problem, it will have $2n - 1$ basic variables.

Problem 9.28

Suppose that there are three applicants for three jobs and that the costs incurred by the applicants to fill the jobs are as given in the following table:

	J_1	J_2	J_3
A_1	\$26	\$23*	\$27
A_2	\$23*	\$22	\$24
A_3	\$24	\$20	\$23*

(T.1)

For example, the cost for applicant A_1 to fill job J_1 is \$26. Each applicant is assigned to only one job and each job is to be filled by one applicant only. Determine the assignment of applicants to jobs such that the total cost is minimized.

Solution

We wish to minimize

$$w = 26x_{11} + 23x_{12} + \cdots + 23x_{33}, \tag{1}$$

where each x_{ij} is either 0 or 1. Because of the nature of the problem, only 3 x_{ij}'s can be 1, while the rest must be 0. Since n is small ($n = 3$), this problem can be easily solved by trial and error. There are $3 \cdot 2 \cdot 1 = 3!$ $= 6$ possible (or feasible) assignments. Applicant A_1 can be assigned to any of the three jobs; A_2 can then be assigned to any of the remaining two jobs; and, thus, A_3 can only be assigned to the one remaining job. It is not difficult to see that the optimal assignment results when A_1 is assigned to J_2, A_2 to J_1, and A_3 to J_3. This is indicated by asterisks in Table 1. The minimum total cost is thus

$$\text{Min } w = c_{12}x_{12}^* + c_{21}x_{21}^* + c_{33}x_{33}^*$$

$$= 23 \cdot 1 + 23 \cdot 1 + 23 \cdot 1 = \$69.$$

To demonstrate that the above is the optimal assignment, consider some of the remaining five feasible assignments and their total costs:

A_1 to J_1, A_2 to J_2, and A_3 to J_3: $w = 26 + 22 + 23 = \$71$

A_1 to J_1, A_2 to J_3, and A_3 to J_2: $w = 26 + 24 + 20 = \$70$

A_1 to J_2, A_2 to J_3, and A_3 to J_1: $w = 23 + 24 + 24 = \$71$.

The solution technique employed in Problem 9.28 would be very tedious for larger n. For example, if $n = 6$, then $6! = 720$ total costs would have to be evaluated.

It would be useful to develop methods that would be more efficient for problems with large values of n.

To simplify the analysis, it is assumed that all of the unit costs (c_{ij}) are nonnegative. If this is not the case, we merely increase all the c_{ij}'s by a constant k so that this will be true. One such suitable k is equal to the absolute value of the most negative prior c_{ij}. It follows from Theorems 9.5 and 9.6 that the optimal assignment plan will be unaffected by such a change in the c_{ij}'s.

Furthermore, we agree to convert all the c_{ij}'s to nonnegative whole numbers (integers) if necessary. If it happens that some of the noninteger c_{ij}'s are positive rational numbers, then conversion to integers is possible through multiplication by a large enough positive integer. (Recall that a rational number is a number that can be expressed as a ratio p/q, where both p and q are integers.) For example, a suitable multiplier would be the product of all the denominators of the noninteger c_{ij}'s. It should be noted that the optimal assignment plan is unchanged when all the c_{ij}'s are multiplied by a positive constant. Another point of interest is as follows: It is plausible to assume that irrational c_{ij}'s do not occur in practical assignment problems.

Thus, the algorithmic machinery will be developed for problems in which all the c_{ij}'s are nonnegative integers. The following observation will turn out to be important. Suppose that in an assignment problem (where all the c_{ij}'s are nonnegative) a feasible assignment exists for which all the *corresponding* c_{ij}'s are equal to zero. (Note that a feasible assignment is one for which exactly one x_{ij} for each row—or i value—equals 1, and exactly one x_{ij} for each column—or j value—equals 1. The remaining $(n^2 - n)$ x_{ij}'s equal zero.) This means that for each of the n x_{ij}'s that equals 1, the corresponding c_{ij} equals 0. Then, clearly, this solution must be optimal. After all, the total cost function w would be zero, which is clearly the minimum possible value for w, since we stipulated that all the c_{ij}'s are nonnegative.

The Hungarian Method is based upon the ideas of the previous paragraph and upon Theorem 9.8. The basic idea is to repeatedly use the transformation

$$c'_{ij} = c_{ij} - r_i - k_j \qquad \text{(B)}$$

in some systematic way to generate new cost matrices in which all of the c_{ij}'s are again nonnegative. Finally, when a cost matrix is reached such that there is a feasible assignment for which the corresponding c_{ij}'s equal zero, then that feasible assignment is the optimal assignment.

Definition 9.4. *A permutation set of zeros* is a set of n zeros in a cost matrix such that there is exactly one zero in each row and exactly one zero in each column.

The cells that a permutation set of zeros occupy may be specified as $(1, j_1), (2, j_2), \ldots, (n, j_n)$, where the arrangement of numbers $j_1, j_2, \ldots, j_n$ constitutes a permutation of the n numbers $1, 2, \ldots, n - 1, n$ taken n (or all) at a time.

Thus, the goal of the Hungarian Method is to systematically apply transformation (B) above until a cost matrix that contains a permutation set of zeros is attained. The permutation set of zeros then determines an optimal assignment in the following way: The n x_{ij}'s *corresponding* to the n zeros are set equal to 1.

In the terminology of the above discussion, the optimal assignment would be given by $x_{1j_1} = 1$, $x_{2j_2} = 1, \ldots, x_{nj_n} = 1$, with all other x_{ij}'s equal to zero. In other words, assign applicant 1 to job j_1, applicant 2 to job j_2, etc.

The Hungarian Method was developed by H. Kuhn and is based upon the work of the two Hungarian mathematicians, D. Koñig and J. Egerváry. An extensive discussion of the underlying theory appears in [1]. Informative discussions and illustrative examples that emphasize the use of the algorithm and applications of the assignment problem can be found in [4].

The algorithm involves drawing a minimum number of straight lines horizontally and vertically to cover all zero entries of the cost matrix at a particular stage. This minimum number of straight lines will be denoted by k. Clearly, $k \leq n$, since all zero entries (and, in fact, all entries) of any cost matrix can be covered by n horizontal lines.

For application of the algorithm, it is assumed that all of the c_{ij}'s of the starting cost matrix are nonnegative integers.

We will now state the Hungarian Method.

Method 9.6: Hungarian Method

1. Subtract the smallest entry in each row from every entry of the row and form what will be called the row-reduced matrix. Then subtract the smallest entry in each column of the row-reduced matrix from every entry of the column. Go to step 2.

2. Draw a minimum number k of straight lines to cover all zero entries of the resulting matrix. Go to 3a if $k = n$ and to 3b if $k < n$.

3a. If k equals n, then a permutation set of zeros can be located from among the zeros. The optimal assignment corresponds to the permutation set of zeros. The minimum total cost can then be calculated from the original cost matrix. The problem is thus solved.

3b. If k is less than n, select the smallest uncovered c_{ij} entry. Subtract this entry from all uncovered entries and add this entry to all covered entries at locations where vertical and horizontal lines intersect. Other covered entries are left unchanged. Return to step 2.

After step 1, the process consists of repeating the cycle of steps 2 and 3b as many times as necessary until step 2 leads into step 3a; at that point, the optimal assignment can be determined.

Problem 9.29

Employ the Hungarian Method to solve Problem 9.28.

Solution

Refer to Table 1 of Problem 9.28. Let us apply step 1 of the algorithm. Thus, we subtract the smallest entries of rows 1, 2, and 3 from all the entries of their respective rows (e.g., we subtract the 23 in row 1 from the entries of that row to get 3, 0, and 4 in the resulting row). We are thus led to the following row-reduced matrix:

$$
\begin{array}{|ccc|}
\hline
3 & 0 & 4 \\
1 & 0 & 2 \\
4 & 0 & 3 \\
\hline
\end{array}
\qquad (T.2a)
$$

We now subtract the smallest entries (1, 0, and 2) from the respective columns of Table 2a. This leads to Table 2b:

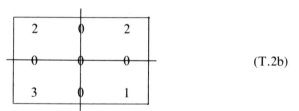

$$(T.2b)$$

The minimum number of lines needed to cover all the zeros is $k = 2$, and the two lines are shown in Table 2b. We have just executed step 2 of the algorithm. Since k is less than n ($n = 3$), we proceed to step 3b. The smallest uncovered c_{ij} is $c_{33} = 1$. We now subtract 1 from all uncovered entries and add 1 to the doubly covered zero in row 2 and column 2. The four other zeros are not altered. The resulting table is as follows:

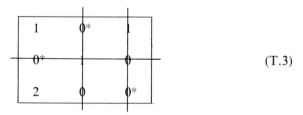

$$(T.3)$$

The minimum number of lines k needed to cover all the zeros is three. One such set of lines is shown in Table 3. The permutation set of zeros contains those zeros in cells (1,2), (2,1), and (3,3). These zeros are indicated by asterisks in Table 3. Thus, the optimal assignment is given by $x_{12} = 1$, $x_{21} = 1$, $x_{33} = 1$, and all other x_{ij}'s $= 0$. In other words, we have reached the same conclusion as in Problem 9.28. Observe that $j_1 = 2$, $j_2 = 1$, and $j_3 = 3$.

Summarizing again, the optimal assignment plan is to assign A_1 to J_2, A_2 to J_1 and A_3 to J_3. This assignment causes the total cost to be minimized at $c_{12} + c_{21} + c_{33} = 23 + 23 + 23 = \$69$.

Notes: a. We indicated that transformation (B) (see the discussion immediately preceding Definition 9.4) was strongly involved in Method 9.6. For example, the effect of going from Table 1 to Table 2b is equivalent to using the transformation with $r_1 = 23$, $r_2 = 22$, $r_3 = 20$, $k_1 = 1$, $k_2 = 0$, and $k_3 = 2$. Furthermore, the effect in going from Table 2b to Table 3 (application of step 3b) is equivalent to using the transformation with $r_1 = 1$, $r_2 = 0$, $r_3 = 1$, $k_1 = 0$, $k_2 = -1$, and $k_3 = 0$. In other words, step 3b is equivalent to letting r_i equal the smallest uncovered c_{ij} for those rows that are uncovered, letting r_i equal zero for covered rows, letting k_j equal zero for uncovered columns, and letting k_j equal the negative of the smallest uncovered c_{ij} for covered columns.

b. From Theorem 9.8 it is clear that the same optimal assignment applies to both Table 1 and Table 3. It is clear from Table 3 that the optimal assignment is given by

$$x_{12} = 1, x_{21} = 1, x_{33} = 1, \text{ and all other } x_{ij} = 0$$

because the w corresponding to this (from Table 3) is

$$w = 0 \cdot 1 + 0 \cdot 1 + 0 \cdot 1 = 0.$$

This zero assignment cost is clearly the minimum possible for Table 3, since all the c_{ij}'s in the table are nonnegative.

It is instructive to solve an assignment problem by using the transportation algorithm (Method 9.4—Stepping Stone Method).

Problem 9.30

Solve Problem 9.28 by employing the Stepping Stone Method (Method 9.4).

Solution

For the first stage of Method 9.4, the Minimum Entry Method is used to find an initial basis and b.f.s. The relevant calculations are given in

Table 1, which follows:

$$
\begin{array}{c|ccc|c}
 & \sqrt{4} & \sqrt{2} & \sqrt{5} & \\
\hline
 & ㉖\,^{0} & 23 & ㉗\,^{1} & 1\ 1\ 0 \\
\sqrt{3} & ㉓\,^{1} & ㉒\,^{0} & 24 & 1\ 1\ 0 \\
\sqrt{1} & 24 & ㉛\,^{1} & 23 & 1\ 0 \\
\hline
 & \begin{matrix}1\\\emptyset\\0\end{matrix} & \begin{matrix}1\\\emptyset\\0\end{matrix} & \begin{matrix}1\\0\end{matrix} &
\end{array}
\qquad\text{(T.1)}
$$

Note that the initial rim conditions (a_i's and b_j's) are all ones. The initial basis contains cells $(1,1)$, $(1,3)$, $(2,1)$, $(2,2)$, and $(3,2)$, of which the first and fourth are degenerate. Cycle values for cells not in the basis are tabulated in parentheses in Table 1a, which follows:

$$
\begin{array}{|ccc|}
\hline
㉖\,^{0} & 23(-2) & ㉗\,^{1} \\
㉓\,^{1} & ㉒\,^{0} & 24(0) \\
24(3) & ㉛\,^{1} & 23(1) \\
\hline
\end{array}
\qquad\text{(T.1a)}
$$

Introducing cell $(1,2)$ into the basis yields the basis of Table 2:

$$
\begin{array}{|ccc|}
\hline
26(2) & ㉓\,^{0} & ㉗\,^{1} \\
㉓\,^{1} & ㉒\,^{0} & 24(-2) \\
24(3) & ㉛\,^{1} & 23(-1) \\
\hline
\end{array}
\qquad\text{(T.2)}
$$

Introducing cell $(2,3)$ into the basis of Table 2 results in Table 3:

$$
\begin{array}{|ccc|}
\hline
26(0) & ㉓\,^{0} & ㉗\,^{1} \\
㉓\,^{1} & 22\ (2) & ㉔\,^{0} \\
24(1) & ㉛\,^{1} & 23(-1) \\
\hline
\end{array}
\qquad\text{(T.3)}
$$

Introducing cell (3,3) into the basis and removing cell (1,3) from the basis of Table 3 leads to the following terminal table, for which all cycle values corresponding to nonbasis cells are nonnegative:

$$
\begin{array}{|ccc|}
\hline
26(1) & \boxed{23}\ ^1 & 27(1) \\[4pt]
\boxed{23}\ ^1 & 22(1) & \boxed{24}\ ^0 \\[4pt]
24(2) & \boxed{20}\ ^0 & \boxed{23}\ ^1 \\
\hline
\end{array}
\qquad (T.4)
$$

The optimal assignment plan is thus $x_{12} = 1$, $x_{21} = 1$, $x_{33} = 1$, and all other x_{ij}'s equal to zero. This is the same result as in Problem 9.29.

Note the large amount of effort involved in solving the above assignment problem by the transportation algorithm (Method 9.4—Stepping Stone Method). The main source of difficulty is the degenerate nature of the problem. For example, in going from the basis of Table 1a to that of Table 2, the nondegenerate x_{ij}'s stayed the same. This type of behavior could lead to an excessive number of pivots if n were larger.

In summary, it is not practical to solve assignment problems by the transportation algorithm for large n.

In transportation problems and in conventional linear programming problems we found situations in which *alternate optimal solutions* could occur. A similar phenomenon can occur in an assignment problem, as the next problem illustrates. In this problem, the initial cost matrix is the same as that of Problem 9.28 except for c_{12} and c_{13}.

Problem 9.31

Find the optimal assignment plan(s) for the assignment problem whose cost matrix is as follows:

$$
\begin{array}{|ccc|}
\hline
26 & 24 & 28 \\[4pt]
23 & 22 & 24 \\[4pt]
24 & 20 & 23 \\
\hline
\end{array}
\qquad (T.1)
$$

Solution

We shall employ the Hungarian Method (Method 9.6). Step 1 leads to the following equivalent cost matrix, for which the zeros can be covered with a minimum of $k = 2$ lines:

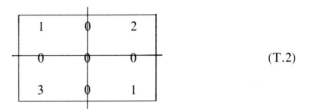

(T.2)

Employing step 3b leads to the following cost matrix, for which $k = n = 3$:

0**	0*	1
0*	1	0**
2	0**	0*

(T.3)

There are two possible optimal assignment plans (corresponding to two possible permutation sets of zeros), and these are given as follows:

First optimal assignment	Cost	Second optimal assignment	Cost
A_1 to J_2	24	A_1 to J_1	26
A_2 to J_1	23	A_2 to J_3	24
A_3 to J_3	23	A_3 to J_2	20
Total Cost	$70	Total Cost	$70

In Table 3, the first optimal assignment is indicated by single asterisks, while the second is indicated by double asterisks. The appropriate costs are obtained from the original cost matrix—Table 1.

Occasionally, the statement of an assignment problem calls for maximizing an objective function. We illustrate how to handle this in the next problem.

Problem 9.32

A firm has five salespeople that it wishes to assign to five territories. The ratings of the salespeople with respect to the different territories are given in the following table:

	T_1	T_2	T_3	T_4	T_5
S_1	12	17	13	14	18
S_2	11	15	13	13	16
S_3	13	16	15	14	17
S_4	12	15	14	12	15
S_5	10	13	11	13	14

(T.1)

For example, salesperson S_1 is rated 12 with respect to his ability to work in territory T_1 (assume that, e.g., the ratings are on a 0 to 20 scale). The firm wishes to assign exactly one salesperson to each territory in such a way as to maximize the sum of the corresponding ratings. Find the optimal assignment plan and the corresponding maximum total rating.

Solution

We can convert this into a minimum problem by multiplying all the ratings by -1. The most negative entry would then be -18 in cell (1,5). Adding $+18$ to all the new entries leads to the following table, in which all entries are nonnegative:

	T_1	T_2	T_3	T_4	T_5
S_1	6	1	5	4	0
S_2	7	3	5	5	2
S_3	5	2	3	4	1
S_4	6	3	4	6	3
S_5	8	5	7	5	4

(T.1′)

Observe that the transformation equation for going from Table 1 to Table 1′ is $c'_{ij} = 18 - c_{ij}$, where c_{ij} refers to the entries of Table 1, while c'_{ij} refers to the new entries of Table 1′. In general, for a similar situation, we would have $c'_{ij} = M - c_{ij}$, where M is the maximum of all the c_{ij}'s of Table 1.

Note: The principle we now use is that the *same collection* of x_{ij}'s which cause a minimizing of $w = \Sigma c'_{ij}x_{ij}$ in Table 1′ will cause a maximizing of $\Sigma c_{ij}x_{ij} = \Sigma(18 - c'_{ij})x_{ij}$ in Table 1. This will also be true if $c'_{ij} = K - c_{ij}$, where K is any constant.

We now employ Method 9.6 on Table 1′. Application of step 1 to Table 1′ leads to Table 2, which follows:

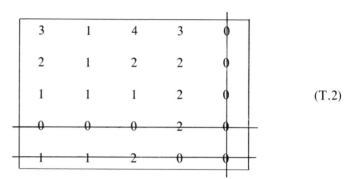

(T.2)

The k value for Table 2 is $k = 3$. Applying step 3b of the Hungarian Method to Table 2 results in Table 3:

2	0*	3	2	0
1	0	1	1	0*
0	0	0*	1	0
0*	0	0	2	1
1	1	2	0*	1

(T.3)

Table 3 contains a permutation set of zeros, which is indicated by asterisks. Thus, an optimal assignment is to assign S_1 to T_2, S_2 to T_5, S_3 to T_3, S_4 to T_1, and S_5 to T_4. (There are 3 alternate optimal assignments.) The maximum total rating can now be calculated by referring back to the original rating table—Table 1. Thus,

Maximum total rating $= 17 + 16 + 15 + 12 + 13 = 73$.

Let us return to the minimum-cost form of the assignment problem. Suppose that the problem is to assign applicants to jobs (as in Problem 9.28), but that certain assignments are not permitted. Such impossible assignments might be the result of, e.g., lack of training, exceedingly high practical costs, geographical limitations, and religious restrictions. If it is not possible to assign the ith applicant to the jth job, we set the cost c_{ij} equal to M, where M is an exceedingly large positive cost.

Problem 9.33

Suppose that the cost matrix for assigning four workers to four jobs is as given in the following table (the three occurrences of M indicate that it is impossible to assign A_1 to J_1, A_2 to J_2, and A_3 to J_4):

M	30	23	25
34	M	16	24
22	19	21	M
21	23	14	20

(T.1)

Determine the optimal assignment and the corresponding minimum total cost.

Solution

We employ the Hungarian Method in the usual way. Application of step 2 leads to Table 2, which follows:

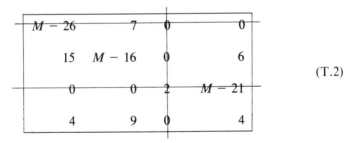

$M - 26$	7	0	0
15	$M - 16$	0	6
0	0	2	$M - 21$
4	9	0	4

(T.2)

The minimum number of lines needed to cover all the zeros is $k = 3$; one such set of lines is indicated in Table 2. Application of step 3b of the Hungarian Method leads to Table 3, which follows:

$M - 26$	7	4	0^*
11	$M - 20$	0^*	2
0	0^*	6	$M - 21$
0^*	5	0	0

(T.3)

A permutation set of zeros is indicated by the asterisks in Table 3. The optimal assignment plan and corresponding minimum total cost are thus given as follows:

Assign A_1 to J_4, A_2 to J_3, A_3 to J_2, and A_4 to J_1.

Min $w = 25 + 16 + 19 + 21 = 81$, where the c_{ij}'s are from Table 1.

5. TRANSPORTATION AND ASSIGNMENT APPLICATIONS

In this section, we shall discuss several additional applications which involve the transportation and assignment models.

Problem 9.34: Unbalanced Assignment Problem

Suppose that there are six people applying for five jobs, and it is desired to fill each job with exactly one person. The costs for filling the jobs with the six people are given in the following table:

	J_1	J_2	J_3	J_4	J_5
P_1	27	23	22	24	27
P_2	28	27	21	26	24
P_3	28	26	24	25	28
P_4	27	25	21	24	24
P_5	25	20	23	26	26
P_6	26	21	21	24	27

(T.1)

Determine the optimal assignment plan, i.e., the plan whereby the total cost of assigning the people is minimized.

Solution

We have a situation that is similar to that of an unbalanced transportation problem where supplies exceed demands (Problem 9.22). Clearly, one of the people cannot be assigned to a job. We introduce a fictitious job J_6. The costs to fill job J_6 represent the costs of not assigning the respective candidates. Let us assume that all such costs are the same, e.g., $c_{i6} = k$ for $i = 1, 2, \ldots, 6$. We can then apply Theorem 9.8 (as in Problem 9.22) and obtain the following table, in which zero cost entries appear in the sixth column:

	J_1	J_2	J_3	J_4	J_5	J_6
P_1	27	23	22	24	27	0
P_2	28	27	21	26	24	0
P_3	28	26	24	25	28	0
P_4	27	25	21	24	24	0
P_5	25	20	23	26	26	0
P_6	26	21	21	24	27	0

(T.1′)

We now apply Method 9.6 to Table 1′. The cost matrix in Table 1′ is already row-reduced because of the zeros in column 6. Thus, application of step 1 consists of subtracting the smallest entry in each column of Table 1′ from all the entries of the respective columns. This leads to Table 2, in which the zeros can be covered by a minimum number of five lines:

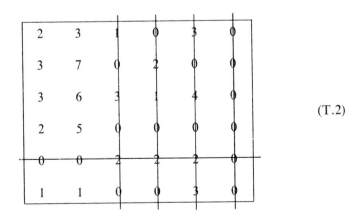

(T.2)

Applying step 3b of Method 9.6 leads to Table 3, which contains a permutation set of zeros:

1	2	1	0*	3	0
2	6	0*	2	0	0
2	5	3	1	4	0*
1	4	0	0	0*	0
0*	0	3	3	3	1
0	0*	0	0	3	0

(T.3)

These zeros, indicated by asterisks, lead to the following optimal assignment plan:

$$P_1 \text{ to } J_4, \ P_2 \text{ to } J_3, \ P_3 \text{ to } J_6, \ P_4 \text{ to } J_5, \ P_5 \text{ to } J_1, \text{ and } P_6 \text{ to } J_2.$$

Note that the assignment of P_3 to J_6 means, in effect, that P_3 will not be assigned. The minimum total cost of assigning the people (excluding the cost of not assigning P_3) can now be determined from the original cost table:

$$\text{Min } w = 24 + 21 + 24 + 25 + 21 = 115.$$

In all of the unbalanced problems done so far, the assumption was made of constant cost terms in the fictitious column or row. Thus, it was then justified to replace the entries in the fictitious column or row by zeros. The following problem illustrates the proper procedure to use if the assumption of constancy is not appropriate.

Problem 9.35

Suppose that in Problem 9.22 the unit storage costs at supply locations W_1 and W_2 are \$5 and \$7, respectively. Using this information, rework the problem.

Solution

In Problem 9.22, which was unbalanced (total supplies $= 80$, while total demands $= 78$), we assumed a constant storage cost at supply locations W_1 and W_2 in dealing with the fictitious destination M_4. Here, however, we use the above storage cost information in developing the following table:

	M_1	M_2	M_3	M_4	
W_1	4	5	7	5	30
W_2	8	6	4	7	50
	18	40	20	2	

(T.1)

Thus, the unit storage cost of $5 at W_1 is equivalent to the cost of "shipping" one unit from W_1 to fictitious destination M_4. The optimal solution corresponding to Table 1 is readily obtained by using the Stepping Stone Method (Method 9.4). It is indicated in the following table:

④^{18}	⑤^{10}	7	⑤^{2}
8	⑥^{30}	④^{20}	7

(T.2)

Note that the current solution differs from that of Problem 9.22. In the present case, two items are stored at storage location W_1, whereas in Problem 9.22, two items were stored at W_2. The minimum total shipping cost here (including the cost of storage at W_1) is computed as follows:

$$\text{Min } w = 4 \cdot 18 + 5 \cdot 10 + 5 \cdot 2 + 6 \cdot 30 + 4 \cdot 20 = \$392.$$

The next problem was already considered in Chapter 8 as Problem 8.8. The approach used there was different from that which follows.

Problem 9.36: Production Scheduling

A certain company manufactures a particular product. The company is planning its production schedule for the months of June, July, and August. The demands for the product are for 70 units in June, 90 units in July, and 120 units in August. The company has manufacturing facilities capable of producing 100 units in each of the three months.

The unit production cost is $15 per unit in June and $16 in both July and August. The inventory storage cost per item produced is zero during the month of production and $3 during any following month. Determine how the company should schedule production in each month in order to minimize total cost.

Solution

The following table summarizes the relevant data. The production capacities are listed in the column on the right, while the demands appear in the bottom row. The time periods June, July, and August are labeled 1, 2, and 3, respectively.

Month	June (1)	July (2)	August (3)	Surplus (4)	Capacity	
June (1)	15	18	21	0	100	
July (2)	M	16	19	0	100	(T.1)
August (3)	M	M	16	0	100	
Demand	70	90	120	20		

The calculations of the unit costs are now illustrated for some typical cases. The cost $c_{22} = 16$ indicates that the cost of production in period 2 (July) is $16 per unit. The cost $c_{23} = 19$ indicates that the unit cost for an item delivered in period 3 is $16 for production in period 2 plus $3 for a one-month storage. The cost $c_{21} = M$ (an exceedingly high positive number) indicates that it is impossible to deliver units that are produced in July in a prior month. Setting $c_{21} = M$ will force x_{21} to equal zero in an optimal solution. Using similar reasoning, all c_{ij}'s for i greater than j are set equal to M in Table 1. Since the sum of the capacities (300) exceeds the sum of the demands (280), a surplus column (column 4) is introduced. This surplus column has the role of a fictitious destination, and all costs in this column are set equal to zero. In addition, the demand associated with this column (20) serves to balance supplies (i.e., production capacities) and demands. The problem is now in the form of a standard balanced transportation problem containing prohibited routes (e.g., see Problem 9.16). Employing the Stepping Stone Method (Method 9.4) leads to the optimal solution, as indicated in the following table:

⑮ 70	18(0)	㉑ 10	⓪ 20
$M(M - 13)$ ⑯ 90		⑲ 10	0(2)
$M(M - 10)$ $M(M - 13)$		⑯ 100	0(5)

(T.2)

The numbers in parentheses are cycle values. In this optimal solution, of the 80 items produced in June, $x_{11} = 70$ are to be delivered in June, while $x_{13} = 10$ are to be stored for two months and then delivered in August. Of the 100 items produced in July, $x_{22} = 90$ are to be delivered in July, while $x_{23} = 10$ are to be delivered in August. Finally, of the 100 items produced in August, all of them ($x_{33} = 100$) are to be delivered in August. Looking at the solution from another point of view, of the 120 items to be delivered in August, 10 are produced in June, 10 in July, and 100 in

August. The minimum total cost (for production and storage) is given as follows:

$$\text{Min } w = 15 \cdot 70 + 21 \cdot 10 + 16 \cdot 90 + 19 \cdot 10 + 16 \cdot 100 = \$4490.$$

Note that there is an alternate optimal solution. This would arise from introducing cell (1,2) into the basis; observe that the cycle value for this cell is zero. The positive x_{ij}'s for this alternate solution are as follows: $x_{11} = 70$, $x_{12} = 10$, $x_{14} = 20$, $x_{22} = 80$, $x_{23} = 20$, and $x_{33} = 100$.

Note: In the first optimal solution, 20 units of the available production capacity for June were not used; this is indicated by the fact that $x_{14} = 20$.

Problem 9.37: The Caterer Problem

In the classical caterer problem, a certain caterer needs to have clean linen napkins for each day over a several-day period. Some of the clean napkins are to be purchased new, while some may be formerly dirty napkins from previous days that have been washed by a laundry service. Suppose that the caterer has no napkins on hand and that he needs 100, 70, and 130 napkins for the next three days. Napkins can be purchased new for $0.60 each. The cost of a fast one-day laundry service is $0.30 per napkin, while the cost of a slow two-day laundry service is $0.20 per napkin. How many napkins should the caterer buy new and how many should he have washed by the different laundry services in order to minimize total costs?

Solution

The following table is a "transportation-problem" interpretation of this particular caterer problem:

	Day 1	Day 2	Day 3	Surplus		
Napkins bought new	.6	.6	.6	0	300	
Dirty napkins from the first day	M	.3	.2	0	100	(T.1)
Dirty napkins from the second day	M	M	.3	0	70	
	100	70	130	170		

In the left-hand margin we have the three "sources" for the napkins, namely, new napkins, dirty napkins from day 1, and dirty napkins from day 2. In the right-hand margin we have the "supplies" from each of these sources. The 300 in the first row is equal to the sum of all the demands for the napkins. (It will turn out in the optimal solution that not all the napkins will be bought new.) The 300 is an upper bound for the number of napkins that could possibly be bought. The second-row supply (100) is the number of dirty napkins from day 1, and the third-row supply (70) is the number of dirty napkins from day 2. These numbers are, of course, the numbers of napkins used on the respective days.

In the columns we list days 1, 2, and 3; these represent "markets." The rim conditions in the bottom margin are the napkin demands for the various days. A surplus column also appears; it represents a fictitious destination whose demand (170) will cause a balance between total supplies and total demands. The costs in this column are all set equal to zero, as is the usual procedure for unbalanced problems.

The first cost in the day 1 column, $c_{11} = 0.6$, denotes the cost per napkin purchased new; these new napkins are used to satisfy the first day's demand. The other costs in column 1 are M's; these exceedingly high costs indicate the impossibility of washing dirty napkins *from* day 1 or 2 to be used on day 1. It will turn out that the x_{ij}'s corresponding to cells with the M's will be zero in any optimal solution.

In the day 2 column, the first cost is the cost per new napkin, and the second entry is the cost per napkin of washing a napkin from day 1 for use on day 2. The third entry, M, indicates that it is impossible to use washed napkins *from* day 2 on day 2. In the day 3 column, the first cost is the cost per new napkin, while the second entry denotes the cost per napkin of washing dirty napkins from day 1 (slow laundry service) for use on day 3. The last cost is the cost per napkin of washing dirty napkins from day 2 for use on day 3.

The solution can be found by using Method 9.4. The initial b.f.s., as determined from the Minimum Entry Method, is displayed in Table 2 (as usual, cycle values for nonbasis cells appear in parentheses):

$(.6)^{100}$	$(.6)^{30}$	$.6(M - .3)$	$(0)^{170}$	
$M(.1)$	$.3(.4 - M)$	$(.2)^{100}$	$0(.7 - M)$	(T.2)
$M(0)$	$(M)^{40}$	$(.3)^{30}$	$0(.6 - M)$	

Note that the prohibited cell (3,2) appears in the basis of Table 2. The b.f.s. of Table 2 is clearly not optimal, since negative cycle values appear. Introducing cell (2,2) into the basis results in Table 3, which follows (the corresponding b.f.s. is optimal):

$$
\begin{array}{|cccc|}
\hline
\textcircled{.6}^{\,100} & \textcircled{.6}^{\,30} & .6(.1) & \textcircled{0}^{\,170} \\[4pt]
M & \textcircled{.3}^{\,40} & \textcircled{.2}^{\,60} & 0(.3) \\[4pt]
M & M & \textcircled{.3}^{\,70} & 0(.2) \\
\hline
\end{array}
\qquad (\text{T.3})
$$

Observe that no prohibited cell appears in the optimal solution. Furthermore, note that cycle values corresponding to prohibited cells will have to be positive; thus, they are not tabulated.

In the optimal solution, 100 napkins are purchased new to satisfy the first day's demand. To satisfy the second day's demand, 30 napkins are purchased new, while 40 napkins from the first day are washed by the fast laundry service. The third day's demand is satisfied by 60 napkins from day 1 (slow two-day laundry service) and 70 napkins from day 2 (fast one-day laundry service). The 170 entry in the first row of the surplus column indicates that 170 napkins of the total demand of 300 napkins are supplied by washed rather than new napkins. In other words, 130 napkins are purchased new, while the remaining demand is satisfied by washed napkins.

The minimum total cost is

$$\text{Min } w = 0.6 \cdot 100 + 0.6 \cdot 30 + \cdots + 0.3 \cdot 70 = \$123.$$

The model of Problem 9.37 can be extended to other applications, as indicated, e.g., in [2] and [3]. In addition to more involved applications involving laundry service, the model is applicable to the servicing of machines by different repair services. The first application of the caterer problem, according to [2, p. 51], was for the repair of airplane engines by either a maintenance ship (fast) or a repair facility on land (slow).

Problem 9.38: Shortest-Path Problem

Consider Figure 9.2, in which the numbers represent distances between the lettered nodes. Suppose that movement is permitted only in the direction of the arrows. (The corresponding arcs are called *directed arcs*.) The starting point is node A and the finishing point is node E. Set up an assignment problem for which the solution will yield the *shortest path* from "start" to "finish."

Solution

This problem can be solved by inspection. However, it is desirable to develop an approach that will be useful for a comparable situation in which many more arcs and nodes occur. The following assignment table

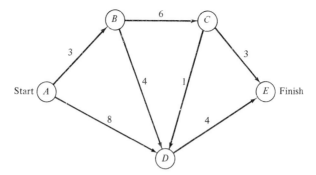

FIGURE 9.2 Path diagram for shortest path problem (Problem 9.38).

gives a representation of the current shortest-path problem:

	B	C	D	E	
Start, A	3^*	M	8	M	1
B	0	6	4^*	M	1
C	M	0^*	1	3	1
D	M	M	0	4^*	1
	1	1	1	1	

(T.1)

The rim conditions are all ones, which is the case for assignment problems. The six occurrences of M denote exceedingly high distances associated with prohibited routes; e.g., in Figure 9.2, it is not possible to
go from node A to node C. Zeros appear in cells which have the same
row and column label. For example, the distance from C to C is zero.
The other cost entries in the table are the directed arc distances that
appear in Figure 9.2. Thus, the distance 8 from A to D appears in the cell
with row label A and column label D.

The optimal solution is attainable either by inspection or by using
Method 9.6 (Hungarian Method). It is indicated by asterisks in Table 1.
The shortest path is thus from A to B, then B to D, and, finally, D to E.
The C to C indication ($x_{32} = 1$) is really not meaningful from a practical
point of view. The shortest-path length is thus $3 + 4 + 4 = 11$ distance
units.

REFERENCES

1. Cooper, L., and Steinberg, D. I. *Methods and Applications of Linear Programming.* Philadelphia, Pennsylvania: W. B. Saunders Company, 1974.

2. Gaver, D. P., and Thompson, G. L. *Programming and Probability Models in Operations Research*. Monterey, California: Brooks/Cole Publishing Company, 1973.
3. Kemeny, J. G., Mirkil, H., Snell, J. L., and Thompson, G. L. *Finite Mathematics with Business Applications*. Second edition. Englewood Cliffs, New Jersey: Prentice-Hall, 1972.
4. Kwak, N. K. *Mathematical Programming with Business Applications*. New York: McGraw-Hill, 1973.
5. Owen, G. *Finite Mathematics*. Philadelphia, Pennsylvania: W. B. Saunders Company, 1970.
6. Strum, J. E. *Introduction to Linear Programming,* San Francisco, California: Holden-Day, 1972.

SUPPLEMENTARY PROBLEMS

The Balanced Transportation Problem

S.P. 9.1:   Use the Minimum Entry Method (Method 9.1) to find an initial basic feasible solution to the following transportation problem:

8	3	40
5	7	60
30	70	

S.P. 9.2:   Repeat S.P. 9.1 for the following problem:

2	4	3	60
5	3	2	80
30	50	60	

S.P. 9.3:   Repeat S.P. 9.1 for the following problem:

10	12	25	45
20	18	5	30
8	6	15	25
28	35	37	

S.P. 9.4:   Use the Northwest Corner Rule (Method 9.2) to find an initial b.f.s. (basic feasible solution) to S.P. 9.1.

S.P. 9.5:   Repeat S.P. 9.4 for S.P. 9.2.

S.P. 9.6:   Repeat S.P. 9.4 for S.P. 9.3.

S.P. 9.7:   By starting with the initial b.f.s. of S.P. 9.1, use the Stepping Stone Method (Method 9.4) to find the optimal solution.

S.P. 9.8:   Repeat S.P. 9.7 for the initial b.f.s. of S.P. 9.2.

S.P. 9.9:   Repeat S.P. 9.7 for the initial b.f.s. of S.P. 9.3.

S.P. 9.10: Repeat S.P. 9.7 for the initial b.f.s. of S.P. 9.4.

S.P. 9.11: Repeat S.P. 9.7 for the initial b.f.s. of S.P. 9.6.

S.P. 9.12: Given the following initial data table. Determine the optimal solution and minimum total cost.

22	17	15	30
20	25	19	50
22	26	24	10
50	30	10	

S.P. 9.13: Given the following initial data table. Determine the optimal solution and minimum total cost. The M's indicate prohibited routes.

M	25	15	30
20	26	13	50
19	24	M	20
55	35	10	

S.P. 9.14: Given the following initial data table. Determine all optimal basic feasible solutions, and the minimum total cost.

30	27	14	60
18	17	25	50
20	21	29	90
52	68	80	

S.P. 9.15: Refer to the initial data table of problem S.P. 9.2. Determine all optimal basic feasible solutions, and Min w for that problem.

Unbalanced Transportation Problems

S.P. 9.16: Given the following initial data table for an unbalanced transportation problem. Find the optimal solution and minimum total cost.

10	15	17	40
18	20	13	60
50	20	25	

S.P. 9.17: Given the following initial data table for an unbalanced transportation problem. Find the optimal solution and minimum total cost.

10	15	17	40
18	20	13	60
50	20	35	

Duality

S.P. 9.18: Determine the dual problem of the following linear programming problem:

Maximize $u = 7x_1 - 4x_2 + 3x_3$

subject to $4x_1 + 2x_2 - 6x_3 \leq 24$

$3x_1 - 6x_2 - 4x_3 \geq 15$

$5x_2 + 3x_3 = 30,$

where $x_1, x_3 \geq 0$ and x_2 is unrestricted.

S.P. 9.19: Determine the dual problem of the following linear programming problem:

Minimize $w = 5y_1 - 4y_2 + 3y_3$

subject to $2y_1 + 7y_3 \geq 8$

$8y_1 + 5y_2 - 4y_3 \leq 15$

$4y_2 + 6y_3 = 30,$

where $y_2, y_3 \geq 0$ and y_1 is unrestricted.

The Assignment Problem

S.P. 9.20: Use the Hungarian Method to solve the assignment problem for which the cost matrix is as follows:

28	25	22
17	26	24
18	22	26

S.P. 9.21: Solve S.P. 9.20 by using the Stepping Stone Method.

S.P. 9.22: Find the optimal assignment plan(s) for the assignment problem for which the cost matrix is as follows:

24	21	23
23	23	24
24	23	26

S.P. 9.23: The ratings of five workers with respect to five jobs are given in the following table:

9	8	7	6	4
8	7	10	6	7
6	9	7	9	5
9	7	6	8	7
6	7	9	7	8

Determine the assignment that will maximize the total rating. (Hint: see Problem 9.32 in text.)

S.P. 9.24: In Problem 9.32 (in text) there are other optimal assignment plans. Determine them.

S.P. 9.25: Find the optimal assignment plan(s) for the assignment problem whose cost matrix is as follows (M indicates exceedingly high cost):

18	24	21	27
M	16	28	26
21	23	14	19
23	17	25	M

Transportation and Assignment Applications

S.P. 9.26: (See also S.P. 8.8.) A manufacturer of a product is planning the production schedule for the months of October, November, and December. The demands for the product are for exactly 80 units in October, 100 in November, and 110 in December. The production facilities are capable of producing at most 100 units in any one month.

The unit production cost is $15 per unit in October, $17 per unit in November, and $18 per unit in December. The storage cost is zero during the production month and $3 per item in any following month. The company has no items in storage on October 1 and wishes to have no items in storage on December 31. Determine the least expensive (minimum cost) way to schedule production and deliveries. *Suggestion*: Use the approach of Problem 9.36.

S.P. 9.27: A caterer has no napkins on hand, and he needs 90, 110, and 100 napkins for each of the next three days. Napkins can be purchased new for $0.80 each. The cost of a fast one-day laundry service is $0.40 per napkin, while the cost of a slow two-day laundry service is $0.30 per napkin. How many napkins should the caterer buy new and how many should he have washed by the different laundry services in order to minimize total costs?

S.P. 9.28: Refer to Figure 9.2. Suppose that the distance from A to B is changed to two and that the arrow from B to D is reversed (the distance from D to B is four), everything else remaining unchanged. Use the Hungarian Method to determine the shortest path from "start" to "finish."

S.P. 9.29: Suppose that there are four people applying for five jobs and that each person is to be assigned to a single job. Furthermore, each job should be filled by one person or not filled at all. The costs for filling the jobs are given in the following table:

	J_1	J_2	J_3	J_4	J_5
P_1	27	23	22	24	27
P_2	28	27	21	26	24
P_3	28	26	24	25	28
P_4	27	25	21	24	24

Determine the optimal assignment plan and the minimum total cost.

Miscellaneous Problems

S.P. 9.30: Refer to the balanced transportation problem as given by (1) to (5) in Section 1 of Chapter 9. Show that (2) and (3) imply (5), i.e., (5) is a necessary condition for (2) and (3).

S.P. 9.31: Again refer to (1) through (5) of Section 1 of Chapter 9. Show that the following statements are equivalent to (2) and (3) provided that (5) holds:

$$x_{i1} + x_{i2} + \cdots + x_{in} \le a_i \text{ for } i = 1, 2, \ldots, m \tag{$\hat{2}$}$$

$$x_{1j} + x_{2j} + \cdots + x_{mj} \ge b_j \text{ for } j = 1, 2, \ldots, n. \tag{$\hat{3}$}$$

ANSWERS TO SUPPLEMENTARY PROBLEMS

S.P. 9.1: $x_{12} = 40$, $x_{21} = 30$, $x_{22} = 30$, and $w = 480$.

S.P. 9.2: $x_{11} = 30$, $x_{12} = 30$, $x_{22} = 20$, $x_{23} = 60$, and $w = 360$.

S.P. 9.3: $x_{11} = 28$, $x_{12} = 10$, $x_{13} = 7$, $x_{23} = 30$, $x_{32} = 25$, and $w = 875$.

S.P. 9.4: $x_{11} = 30$, $x_{12} = 10$, $x_{21} = 60$, and $w = 690$.

S.P. 9.5: $x_{11} = 30$, $x_{12} = 30$, $x_{22} = 20$, $x_{23} = 60$, and $w = 360$.

S.P. 9.6: $x_{11} = 28$, $x_{12} = 17$, $x_{22} = 18$, $x_{23} = 12$, $x_{33} = 25$, and $w = 1243$.

S.P. 9.7: Initial b.f.s. is optimal. Min $w = 480$; $x_{12} = 40$, $x_{21} = 30$, and $x_{22} = 30$.

S.P. 9.8: Initial b.f.s. is optimal. Min $w = 360$; $x_{11} = 30$, $x_{12} = 30$, $x_{22} = 20$, and $x_{23} = 60$.

S.P. 9.9: Min $w = 847$; $x_{11} = 28$, $x_{12} = 17$, $x_{23} = 30$, $x_{32} = 18$, and $x_{33} = 7$.

S.P. 9.10: Min $w = 480$; $x_{12} = 40$, $x_{21} = 30$, and $x_{22} = 30$.

S.P. 9.11: Min $w = 847$; $x_{11} = 28$, $x_{12} = 17$, $x_{23} = 30$, $x_{32} = 18$, and $x_{33} = 7$.

S.P. 9.12: The *degenerate* optimal solution is $x_{12} = 30$, $x_{21} = 40$, $x_{23} = 10$, $x_{31} = 10$, and $x_{32} = 0$. Min $w = 1720$.

S.P. 9.13: Min $w = 2085$; $x_{12} = 30$, $x_{21} = 40$, $x_{23} = 10$, $x_{31} = 15$, and $x_{32} = 5$.

S.P. 9.14: Min $w = 3688$; $x_{13} = 60$, $x_{22} = 50$, $x_{31} = 52$, $x_{32} = 18$, and $x_{33} = 20$. Alternate optimal b.f.s.: $x_{13} = 60$, $x_{22} = 30$, $x_{23} = 20$, $x_{31} = 52$, and $x_{32} = 38$.

S.P. 9.15: Min $w = 360$; $x_{11} = 30$, $x_{12} = 30$, $x_{22} = 20$, and $x_{23} = 60$. Alternate optimal b.f.s.: $x_{11} = 30$, $x_{13} = 30$, $x_{22} = 50$, and $x_{23} = 30$.

S.P. 9.16: Destination M_4 is fictitious. Min $w = 1305$; $x_{11} = 40$, $x_{21} = 10$, $x_{22} = 20$, $x_{23} = 25$, and $x_{24} = 5$. Thus, five items are stored at source W_2.

S.P. 9.17: Source W_3 is fictitious. Min $w = 1335$; $x_{11} = 40$, $x_{21} = 10$, $x_{22} = 15$, $x_{23} = 35$, and $x_{32} = 5$. Thus, destination M_2 will receive five fewer items than demanded.

S.P. 9.18: Minimize $w = 24y_1 - 15y_2 + 30y_3$
subject to $\qquad 4y_1 - 3y_2 \geq 7$
$\qquad\qquad 2y_1 + 6y_2 + 5y_3 = -4$
$\qquad\qquad -6y_1 + 4y_2 + 3y_3 \geq 3$,
where y_1, $y_2 \geq 0$ and y_3 is unrestricted.

S.P. 9.19: Maximize $u = 8x_1 - 15x_2 + 30x_3$
subject to $\qquad 2x_1 - 8x_2 = 5$
$\qquad\qquad -5x_2 + 4x_3 \leq -4$
$\qquad\qquad 7x_1 + 4x_2 + 6x_3 \leq 3$,
where x_1, $x_2 \geq 0$ and x_3 is unrestricted.

S.P. 9.20: Min $w = 61$ and $x_{13} = x_{21} = x_{32} = 1$.

S.P. 9.21: See the answer to S.P. 9.20.

S.P. 9.22: Min $w = 69$ and $x_{13} = x_{21} = x_{32} = 1$. In addition, $x_{12} = x_{23} = x_{31} = 1$.

S.P. 9.23: Maximum total rating equals 44; $x_{11} = x_{23} = x_{32} = x_{44} = x_{55} = 1$. In addition, $x_{12} = x_{23} = x_{34} = x_{41} = x_{55} = 1$.

S.P. 9.24: $x_{15} = x_{22} = x_{33} = x_{41} = x_{54} = 1$; $x_{15} = x_{22} = x_{31} = x_{43} = x_{54} = 1$; $x_{12} = x_{25} = x_{31} = x_{43} = x_{54} = 1$.

S.P. 9.25: Min $w = 75$ and $x_{11} = x_{24} = x_{33} = x_{42} = 1$.

S.P. 9.26: Here x_{ij} denotes the number of items produced in month i and delivered in month j. October, November, and December are labeled 1, 2, and 3, respectively. Surplus destination is labeled 4. The answer: Min $w = \$4910$, $x_{11} = 80$, $x_{12} = 0$, $x_{13} = 10$, $x_{14} = 10$, $x_{22} = 100$, and $x_{33} = 100$. Alternate optimal solution is $x_{11} = 80$, $x_{12} = 10$, $x_{14} = 10$, $x_{22} = 90$, $x_{23} = 10$, and $x_{33} = 100$.

S.P. 9.27: Minimum total cost equals $\$164$, where 90 napkins are purchased new to satisfy the first day's demand. For the second day, 20 are purchased new, while 90 from the first day are washed by the fast laundry service. The third day's demand is satisfied completely by 100 napkins from day 2 washed by the fast laundry service.

S.P. 9.28: Shortest distance is 11—from A to B to C to E. Note that the distance from D to D (zero) appears in the solution.

S.P. 9.29: Minimum total cost equals 93. Assign P_1 to J_2, P_2 to J_3, P_3 to J_4, and P_4 to J_5. Also, J_1 is not filled.

S.P. 9.30: Let Σ_i represent summing from $i = 1$ to $i = m$, and let Σ_j represent summing from $j = 1$ to $j = n$. If we sum the m equations in (2), we obtain

$$\sum_i \sum_j x_{ij} = \sum_i a_i. \tag{2'}$$

If we sum the n equations in (3), we obtain

$$\sum_j \sum_i x_{ij} = \sum_j b_j. \tag{3'}$$

The left-hand-side double sums in (2') and (3') are equal, and thus

$$\sum_i a_i = \sum_j b_j.$$

S.P. 9.31: The sigma (Σ) convention from S.P. 9.30 applies here. If we sum the m inequalities from (2̂) and the n inequalities from (3̂), we obtain

$$\sum_i \sum_j x_{ij} \le \sum_i a_i \tag{2̂'}$$

and

$$\sum_j \sum_i x_{ij} \ge \sum_j b_j. \tag{3̂'}$$

From (5) and the fact that the left-hand sides of (2̂') and (3̂') are equal, it follows that the above two inequalities are, in fact, *equations*. ($B \le J \le A$ and $A = B$ imply $J = A$ and $J = B$). Now refer back to (2̂). Each of the $\le$ statements from $i = 1$ to $i = m$ must, in fact, be an equation. Otherwise, when we sum up the m statements in (2̂), we would obtain

$$\sum_i \sum_j x_{ij} < \sum_i a_i.$$

However, this is a contradiction (we should have $=$ here and not $<$). In a similar way, we can show that each of the $\ge$ statements in (3̂), from $j = 1$ to $j = n$, must, in fact, be an equation.

Index